Springer Series in Reliability Engineering

Series Editor

Hoang Pham, Department of Industrial and Systems Engineering, Rutgers University, Piscataway, NJ, USA

More information about this series at http://www.springer.com/series/6917

Mangey Ram · Hoang Pham
Editors

Advances in Reliability Analysis and its Applications

Springer

Editors
Mangey Ram
Department of Mathematics
Graphic Era Deemed to be University
Dehradun, Uttarakhand, India

Hoang Pham
Department of Industrial
and Systems Engineering
Rutgers University
Piscataway, NJ, USA

ISSN 1614-7839 ISSN 2196-999X (electronic)
Springer Series in Reliability Engineering
ISBN 978-3-030-31377-7 ISBN 978-3-030-31375-3 (eBook)
https://doi.org/10.1007/978-3-030-31375-3

This Springer imprint is published by the registered company Springer Nature Switzerland AG
The registered company address is: Gewerbestrasse 11, 6330 Cham, Switzerland

Preface

Nowadays, in system reliability engineering, advances in reliability analysis is perhaps one of the most multidimensional topics. This quick development has truly changed the environment of system engineering and this global design. Now with the help of simulations and virtual reality technologies, we can start more of the modeling task.

The aspects dealt in chapter "Time Varying Communication Networks: Modelling, Reliability Evaluation and Optimization" are (i) TVCN models for representing features like mobility, links, and topology, (ii) description of the notion of Time-Stamped-Minimal Path Sets (TS-MPS) and Time-stamped-minimal Cut Sets (TS-MCS) for TVCNs as an extension of MPS and MCS, respectively, that are widely used in static networks, (iii) techniques for enumerating TS-MPS and TS-MCS, and evaluating reliability measure(s)—particularly two-terminal reliability, expected hop, and slot counts along with some other related metrics, and (iv) discussion on several recent optimization problems in TVCNs.

In chapter "Methods for Prognosis and Optimization of Energy Plants Efficiency in Starting Step of Life Cycle", appropriate methods are provided for prognosis and optimizing the effectiveness based on the quality of design, production and testing, assembly and trial release, exploitation, and development of procedures for prognosis of the complex systems behavior based on the characteristics of certain constituent elements of the system and the possible impact of human factors and environment itself on the system.

In chapter "Planning Methods for Production Systems Development in the Energy Sector and Energy Efficiency", methods used in planning the development of the electric power system differ with respect to optimization technique (linear programming, nonlinear programming, etc.), type of approximation (linear, nonlinear), and economic valorization (with inflation, without inflation).

In chapter "The Integral Method of Hazard and Risk Assessment for the Production Facilities Operations", problems of creation of integrated index within the development of control methods of HPF industrial safety condition are designated and the problem of such object's management modeling because of precedents, based on classes of states, is solved (there is an event/there is no event).

Chapter "Multi-level Hierarchical Reliability Model of Technical Systems: Theory and Application" describes an assessment methodology for various sustainability indicators of technical systems, such as reliability, availability, fault tolerance, and reliability-associated cost of technical safety-critical systems, based on Multi-Level Hierarchical Reliability Model (MLHRM).

Chapter "Graph Theory Based Reliability Assessment Software Program for Complex Systems" presents the reliability of the theoretical background and graph theory. After that, the developed MATLAB GUI application based on graph theory for the reliability assessment of complex systems has been discussed.

In chapter "Reliability and Vacation: The Critical Issue", a comparative study of different vacation policies on the reliability characteristics of the machining system is presented. For that purpose, the queueing-theoretic approach is employed and the Markovian models are developed for various types of vacation policies, namely, N-policy, single vacation, multiple vacations, Bernoulli vacation, working vacation, vacation interruption, etc.

In chapter "Software Multi Up-Gradation Modeling Based on Different Scenarios", it has been checked out which release performs best for a particular type of real-life scenario using the unified modeling approach. The intent of this chapter is to consider the increasingly ambitious requirements of the customers and the benefits of situating new features in the software.

In Chapter "A Hidden Markov Model for a Day-Ahead Prediction of Half-Hourly Energy Demand in Romanian Electricity Market", Mathematica code, which relies on the maximum likelihood principle in Hidden Markov Model (HMM) environment, has been developed. Also, HMM approach is an efficient way in modeling short-term/day-ahead energy demand prediction, especially during peak period(s), and in accounting for the inherent stochastic nature of demand conditions which has been discussed.

Chapter "A General (Universal) Form of Multivariate Survival Functions in Theoretical and Modeling Aspect of Multicomponent System Reliability Analysis" presents particular bivariate and k-variate new models and also a general method for their construction competitive to the copula methodology. The method follows the invented universal representation of any bivariate and k-variate survival function different from the corresponding copula representation.

Chapter "An Exact Method for Solving a Least-Cost Attack on Networks" focuses on treating a particular problem of intelligent threats. This chapter attempts to identify the optimal attack strategy on a network that completely prevents the flow from reaching its destination.

In chapter "Reliability Analysis of Complex Repairable System in Thermal Power Plant", the performance of the cooling tower of a coal-fired thermal power was analyzed under fuzzy environment. TFM was used to consider the vagueness of the failure and repair time data. The results are useful in framing the optimum maintenance interval for the considered system for improving plant availability.

Chapter "Performance Analysis of Suspension Bridge: A Reliability Approach" investigates the ability to use the Markov process for degradation modeling of suspension bridges by taking some of its important sections, namely, tower foundation, tower, anchor, cable, and deck along with human error. Here, we also identify various factors responsible for the deterioration of the major components of the bridge, which further affects the working of the mainframe structure.

The engineers and the academicians will definitely gain great knowledge with the help of this book entitled "Advances in Reliability Analysis and its Applications". This book also helps them in the analysis of reliability and its applications. The book is meant for those students who have taken reliability engineering as a subject to study. The material is proposed for postgraduate or senior undergraduate level students.

Dehradun, India Mangey Ram
Piscataway, USA Hoang Pham

Acknowledgements The editors acknowledge Springer for this opportunity and professional support. Also, we would like to thank all the chapter authors and reviewers for their availability for this work.

Contents

Time Varying Communication Networks: Modelling, Reliability Evaluation and Optimization

Gaurav Khanna, S. K. Chaturvedi and Sieteng Soh

Abstract In recent times, there has been a tremendous research interests and growth in the direction of time varying communication networks (TVCNs) due to their widespread applications. The examples of such networks include, but not limited to, the networks like mobile ad hoc networks (MANETs), delay tolerant networks (DTNs), vehicular ad hoc networks (VANETs) and opportunistic mobile networks (OMNs). Some formidable challenges posed by such networks are long propagation delay, frequent disruption of communication between any two nodes, high error rates, asymmetric link rates, lack of end-to-end connectivity, routing, etc. Thus, it is vital for TVCN design, modelling and performance evaluation and/or comparison to assess their performance through some quantifiable metrics like packet delivery ratio (PDR), average number of link failures during the routing process, routing requests ratio, average end-to-end (E2E) delay, route lifetime and network reliability. Although a plethora of tools and techniques are available that deal with the design, modelling, analysis and assessment of reliability and other performance metrics of static networks yet the same is not true for the present days' TVCNs. This Chapter describes extension of the reliability assessment techniques and performance metrics used for static networks to the TVCNs. More specifically, the aspects dealt in this chapter are: (i) TVCN models for representing features like mobility, links and topology, (ii) description of the notion of *time-stamped-minimal path sets* (TS-MPS) and *time-stamped-minimal cut sets* (TS-MCS) for TVCNs as an extension of MPS and MCS, respectively that are widely used in static networks, (iii) techniques for enumerating TS-MPS and TS-MCS, and evaluating reliability measure(s)-particularly

G. Khanna · S. K. Chaturvedi (✉)
Subir Chowdhury School of Quality and Reliability, Indian Institute of Technology Kharagpur, Kharagpur, West Bengal, India
e-mail: skcrec@hijli.iitkgp.ernet.in

G. Khanna
e-mail: gauravkhanna@iitkgp.ac.in

G. Khanna · S. Soh
School of Electrical Engineering, Computing and Mathematical Sciences, Curtin University, Perth, Australia
e-mail: s.soh@curtin.edu.au

M. Ram and H. Pham (eds.), *Advances in Reliability Analysis and its Applications*, Springer Series in Reliability Engineering,
https://doi.org/10.1007/978-3-030-31375-3_1

two-terminal reliability, expected hop and slot counts along with some other related metrics, and (iv) discussion on several recent optimization problems in TVCNs.

Keywords Evolving graphs · Network reliability · Sum-of-disjoint products · Time varying communication networks

1 Introduction

Recent advances in communication, computation and sensing capability have resulted in the advent of multi-hop mobile ad hoc networking paradigms viz., Mobile Ad hoc Networks (MANETs) and MANET-born networks like Delay Tolerant Networks (DTNs), Vehicular Ad hoc Networks (VANETs), Opportunistic Mobile Networks (OMNs) and Wireless Mesh Networks (WMNs) [1]. At the basic level, such networks can be formed in a rapid manner by employing the popular devices like smart phones, tabs, iPads, and laptops whose power, versatility and affordability is increasing day-by-day [2]. All these communication networks have a common feature of time-varying or evolving topology. The change in network topology primarily occurs due to a variety of *intrinsic* (predictable and inherent) *interruptions* like *node mobility* and/or *extrinsic* (unpredictable) *interruptions* like *shadowing* and *hardware failures* [3]. Although, modelling of topology changes are formidable tasks for network designers and managers, yet these changes are often considered as an integral part or nature of these systems rather as anomalies [4]. Some other daunting challenges posed by such networks are long propagation delay, frequent disruption of communication between any two nodes, high error rates, asymmetric link rates, lack of end-to-end connectivity, routing, etc. Thus, modelling and performance evaluation of the above stated networks is an extremely challenging task which requires computation of quantifiable metrics like packet delivery ratio (PDR), average number of link failures during the routing process, routing requests ratio, average end-to-end (E2E) delay, route lifetime and network reliability.

Despite the above challenges, multi-hop mobile ad hoc networking paradigms have found a multitude of space and terrestrial applications. A few application areas of such networks include disruption tolerant satellite networks [5], wildlife monitoring (refer ZebraNet [6]), provisioning of internet facility in inaccessible/rural areas of developing countries (see Daknet [7]), updates to online chatting systems (e.g., MSN) or social networking sites (e.g., Facebook, twitter), emails, firmware and software updates, etc. These applications often necessitate that a user can tolerate a slight delay in data delivery or do not need real time service provisioning [8]. There are enormous other potential application areas of networks which vary/evolve with time like neural networks, citation networks, social networks, disease dissemination and many more. Some more worth mentioning works in this direction covering a variety of such networks can be seen in [9–12]. Note that such time evolving/varying networks belong to a variety of domains and are known with many different names in the literature, viz., *temporal graphs*, *temporal networks*, *evolving graphs*, *time-varying graphs* (TVGs), *time-aggregated graphs* (TAGs), *time-stamped graphs*, *dynamic networks*, *dynamic graphs*, *dynamical graphs*, *spatio-temporal networks* and so on [13]. Thus, for the sake of uniformity, here onwards we would refer the multi-hop communication networking paradigms as *time varying communication networks* (TVCNs).

Today we have a plethora of well-defined *state-of-the-art* techniques for the reliability analysis of static networks. Among them, Minimal Path Set (MPS) and Minimal Cut Set (MCS) based methods have played a major role in static networks' reliability modelling and analysis problems. Note that a MPS, in general, is a path between specified set of nodes whereby no nodes are traversed more than once and every link in this set is needed for ensuring a successful communication. In contrast, a MCS is a minimal set of links whose removal would cause a network graph to fail. The path sets (cut sets) may not necessarily be disjoint and independent in the sense that common links might be occurring in many paths (cuts). To this end, various sum-of disjoint-products (SDP) techniques have been employed to make them disjoint and to compute the network's reliability (unreliability) from these disjointed MPS (MCS). In addition, the path sets and cut sets can also be used to assess an upper/lower bound on reliability as well, if the number of paths or cuts is exceedingly large to avoid computational burden.

However, the current *state-of-the-art* for evaluating reliability of TVCNs still appears to be at its infancy stage because the available analytical techniques for network reliability analysis have mainly been developed for static networks and cannot be extended directly on TVCNs without resorting to their substantial modifications. Due to the frequent changes in topology and unpredictability of failures, TVCNs do not possess continuous *end-to-end* connectivity between a set of specified nodes, e.g., Source-Destination (*S-D*) node pair, as is observed in static networks [14]. More specifically, TVCNs adhere to *store-carry-and-forward* mechanism of data transfer by utilizing multi-hop communication via time respecting opportunistic path sets. Consequently, to use widely employed MPS and MCS based methods for reliability analysis in static networks, one has to devise means to enumerate terms similar to MPS (or MCS) for TVCNs. We believe that in many applications of TVCNs like in Low Earth Orbiting (LEO) satellite networks or in reconnaissance applications, network reliability assessment is of great concern. In general, network reliability in TVCN can be defined as the probability of successful transmission of data packets among a set of designated nodes. In other words, their reliability evaluation assesses how reliably data packets sent by source node(s) can be received at the destination node(s).

In this chapter, we will briefly review the models for representing features like mobility, links and topology of TVCNs. Further, we will describe a recent notion of *time-stamped-minimal path sets* (TS-MPS) and *time-stamped-minimal cut sets* (TS-MCS) for TVCNs as an extension of MPS and MCS, respectively of static networks. We will also describe methods to generate TS-MPS and TS-MCS, discuss the results of research studies utilizing TS-MPS or TS-MCS for evaluating the reliability measure(s), particularly *two-terminal reliability* (*2-TR*), expected hop and slot counts along with some other related but useful network performance metrics, and optimization problems in TVCNs.

2 Modelling Techniques

This section presents some existing models for representing various features of TVCNs.

2.1 Overview

Formally, a TVCN could be modelled as a network graph, $G = (V_G,\ E_G)$, where each node and/or link has a presence schedule defined for it. Corresponding to each link $l \in E_G$ (node $v \in V_G$) the link (node) existence schedule indicates the time instances at which a link (node) is present, and possibly other parameters like traversal distance, traversal cost and latency, etc., [15]. With the help of the link and node existence schedules, changes in topology of a TVCN can be studied [16]. A number of models have been proposed in the literature to study/predict the performance characteristic(s) of TVCNs. A comprehensive discussion on them can be found in [17] and the references therein. However, for the sake of completeness, here we briefly summarize some noteworthy observations.

(1) In TVCNs, one of the prime network attribute is *node mobility*. A *mobility model* attempts to capture the motion of mobile nodes with the change in their speeds and directions over time. It is well-known that the real-life mobility patterns of human beings and vehicles can be very complex to model depending on the objectives of the mission [18]. A complex mobility pattern requires more parameters to be included into the mobility model, thereby, making the model very complex. Performance assessment of TVCNs in the presence of node mobility, characterized by some mobility model, generally requires an investigation of the impact of change in *node velocity* and *mobility patterns* on network reliability, protocols and application services. Such an analysis is of prime importance for the better design and implementation of TVCNs. In a very simple classification, mobility models can be categorized as: (i) Synthetic mobility models [19], and (ii) Trace based mobility models [20, 21]. A synthetic mobility model depicts randomly generated movements and creates synthetic traces while a trace-based mobility model is developed by monitoring and extracting the features from the real movement patterns of users carrying mobile nodes, thus epitomizes reality [17]. Generally, a synthetic mobility model requires complex mathematical modelling, but it can be easily applied to an arbitrary number of nodes and over a large scale. Although several synthetic mobility models have been proposed in the literature and many traces have been collected from the real-world experiments to precisely model node movement, yet there is no universal model, which caters all the requirements. Moreover, in the past, mostly the data traces have been collected by deploying the mobile devices in a small region, usually a university campus or a conference room, and require large time overhead (say six months to one year) to collect a good amount of traces to avoid any biased data

from appearing in the data set. Interested readers may refer the recent reviews on mobility models and data-trace repositories in [11, 22] for deeper insights. Besides, it can be concluded from the cited literature that the trace collection is a difficult job at present as deploying mobile phones on a large scale for data collection is impractical and costly. It is evident that both techniques have their own advantages and challenges. For the readers' benefit, a list of the pros and cons, and associated research challenges for both categories have been tabulated in Tables 1 and 2, respectively.

(2) *Random Graph model*, originally introduced by Solomonoff and Rapoport in 1951 [23], can coherently describe many networks of arbitrary sizes. A random graph/network is formed by adding links having some associated probability between randomly selected node pairs. However, in TVCNs such a modelling is not viable as link existence between nodes is not only distance dependent but also depends on the time of the existence of the link. For instance, in a TVCN any node pair having lesser Euclidean distance between them in comparison to their transmission range will have high probability of link availability in comparison to a node pair having a larger distance between them [24].

(3) The *Random Geometric Graph* (RGG) is used to draw a random graph by placing a set of n vertices independently and uniformly on the unit square $[0,\ 1]^2$ in a random fashion, and by connecting two vertices if and only if their Euclidean distance is at most the given radius r [25, 26]. In other words, in RGG, the mobile nodes in the network can communicate with each other if the Euclidean

Table 1 Pros and Cons of trace based and synthetic mobility models

Type	Pros and Cons					
	Represents reality	Deployment cost	Computational overhead	Scalable	Complexity	Availability of models
Trace based models	Yes	High	Low	No	Low	Few
Synthetic models	No	Low	High	Yes	High	Large

Table 2 Research challenges in trace based and synthetic mobility models

Index	Synthetic models	Trace based models
1	Difficult to design a widely acceptable realistic and simple mobility model	Huge variation in collected trace data with respect to day and time
2	Hard to decide a suitable model for a given scenario	Missing data
3	Difficulty in validation via traces	Requires filtering and pre-processing
4	Speed decay problem	Sufficient samples need to be acquired
5	Extremely variable values for the parameters of simulations	

distance between them is less than or equal to the transmission range of the mobile nodes, and the link reliability is theoretically considered as unity. RGGs have found wide application to model making or breaking of links in MANETs and can possibly be utilized in other TVCNs as well.

(4) In reality, even if the mobile nodes are in transmission range of each other, the signal strength deteriorates due to a variety of reasons like noise, fading effect and/or interference etc., thereby, decreasing the probability of a successful communication. The *propagation-based link reliability model* considers the link reliability as a combination of *free-space propagation model* (FS) and *two-ray ground propagation model* (TRG) to incorporate the variation in radio signal due to variation in terrain, frequency of operation, speed of mobile nodes, obstacles and other technical factors. Based on this combination, FS-TRG propagation model for representing the link reliability of a MANET as a function of distance has been proposed in [27], and was later utilized in [28] for evaluation of MANET reliability. However, the application of two-ray model in ad hoc networks seems implausible as here both transmitter and receiver are of comparable heights and the inter separation distance between transmitter and receiver is quite less.

(5) The *Barabási-Albert model* [29] can suitably represent the time growth in many real world networks like social networks, Internet and biological networks, in which random graphs fail [24]! This model depends on two mechanisms, viz., *growth*, as new vertices enter the network at some rate and *preferential attachment*, as newly added nodes form link *preferentially* with nodes having a large degree. Such a scale-free network model, wherein the network size and node degree distribution are independent of each other [24], is not appropriate for modelling TVCNs as it seems irrational to assume that some nodes may have a much higher number of neighbours than other in a coverage area with uniformly distributed nodes.

(6) The *percolation theory* and *probabilistic epidemic algorithms* based models have been proposed in the literature to model *connectivity* [30, 31]. These models can solve two dependent aspects, viz., a good diffusion of information in the network (needed for routing, broadcast and communication) and its connectivity (needed to reach each node) [32, 33]. In [34], a percolation theory based framework was proposed to calculate the network reliability of random network models and real networks with different nodes and/or links lifetime distributions. To the best of our knowledge, the model has not been utilized for the reliability evaluation of TVCNs like VANETs, DTNs, and OMNs.

(7) In *regular lattice graph model,* the radio communications form links between nodes. In this model, the adjacent nodes are connected with a probability p whereas non-adjacent nodes are indirectly connected via intermediate nodes. Besides, the probability of existence of a link varies as a function of the distance between the nodes to capture the decay of radio signal power with the increasing distance between nodes. Thus, this model can be suitable for modelling TVCNs [24].

(8) The *Wireless World Initiative New Radio* (WINNER) model has been utilized in [35] for urban scenarios with low antenna heights to study the energy consumption in Multi-hop Cellular Networks (MCNs). Thus, in future studies, WINNER model can find potential applications in other TVCNs also.

Recently, the notion of *evolving graph* model [15] has garnered considerable attention of the researchers to model and analyse the performance of TVCNs. Therefore, in the following Section we discuss more about the evolving graph model.

2.2 Evolving Graph Model

The evolving graph model can be used to model and analyse a variety of TVCNs having fixed schedule. Few examples of TVCNs having fixed schedule are public buses having fixed tours and schedules, low duty-cycle sensors with sleep/wake up pattern on a regular interval, and LEO satellite systems, where the topology dynamics can be predicted at different time intervals and the information is opportunistically transmitted between a pair of nodes. The basic idea behind evolving graphs is to formalize the *time domain* in graphs. Note that the explicit inclusion of information about the times of interactions in a graph/network makes predictions and mechanistic understanding more accurate. The definition of the evolving graph model used to represent a TVCN can now be formally stated as:

Definition: Let there be a graph $G(V,\ E)$ and an ordered sequence of its subgraphs, $S_G = G_1(V_1,\ E_1), G_2(V_2,\ E_2), \ldots, G_i(V_i,\ E_i), \ldots, G_\tau(V_\tau,\ E_\tau),\ \forall V_i \subseteq V$ and $E_i \subseteq E$, such that $\bigcup_{i=1}^{\tau} G_i(V_i,\ E_i) = G(V,\ E)$ and each $G_i(V_i,\ E_i)$ is the subgraph in place during $[t_{i-1},\ t_i)\forall i \in [1,\ \tau]$. Let t_0 be the starting of the network evolution and $S_T = t_1,\ t_2,\ \ldots,\ t_\tau$ be an ordered sequence of time slots. Then, the system $G = (G(V,\ E),\ S_G,\ S_T)$ is called an evolving graph.

Refer Figs. 1 and 2 to understand the concept of TVCNs modelled as an evolving graph G. Here, Fig. 1a–d could be thought of as four (viz., $G_1,\ G_2,\ G_3$ and G_4) ordered subgraphs, S_G, which represent a sequence of snapshots depicting the time-evolving topologies of a TVCN during four consecutive time instances $t = 1, 2, 3$ and 4 (or in other words four time slots, S_T i.e., t_1 to t_4). Further, in Fig. 1, t_{i-1} and t_i, $\forall i \in [1,\ \tau]$ represent the starting and ending time of a particular time slot. The difference between end time and start time for each time slot gives its length or duration and may not necessarily be equal. In this modelling paradigm, it is assumed that each node is connected to itself in all the snapshots existing within a given mission time window. This is necessary to show the buffering of data packets. For instance, in Fig. 1a only two nodes, i.e., node 2 and node 4, can receive the data packets sent by node 1. Now, in Fig. 1b the data packets retained in the previous time slot by node 2 and node 4 can be transmitted to node 3. The process is repeated in all other snapshots until the packet either expires due to the unavailability of path towards destination or is delivered to its destination node. This reveals that these snapshots have a sort of memory. Figure 2a represents an underlying graph for a

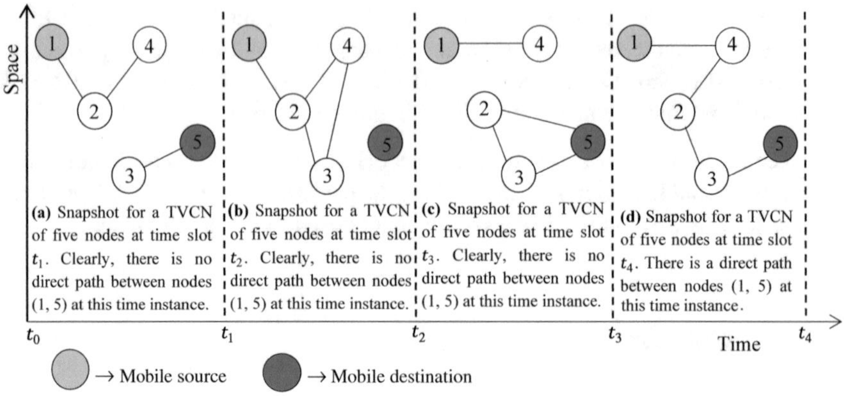

Space
1
4
2
5
3
(a) Snapshot for a TVCN of five nodes at time slot t_1. Clearly, there is no direct path between nodes (1, 5) at this time instance.
(b) Snapshot for a TVCN of five nodes at time slot t_2. Clearly, there is no direct path between nodes (1, 5) at this time instance.
(c) Snapshot for a TVCN of five nodes at time slot t_3. Clearly, there is no direct path between nodes (1, 5) at this time instance.
(d) Snapshot for a TVCN of five nodes at time slot t_4. There is a direct path between nodes (1, 5) at this time instance.
t_0
t_1
t_2
t_3
Time
t_4
→ Mobile source
→ Mobile destination

Fig. 1 Sequence of snapshots of a TVCN of five nodes at four different time slots

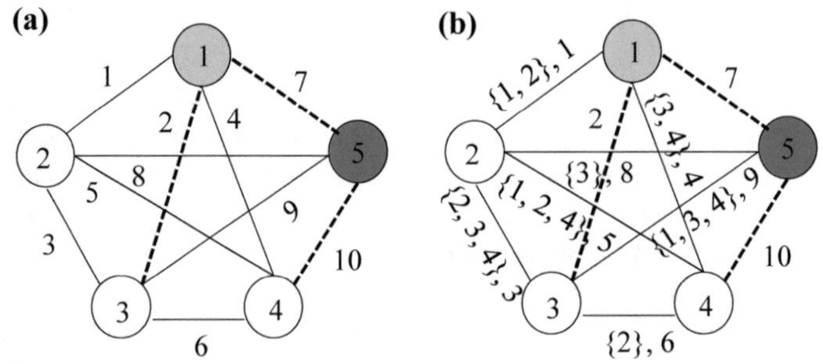

Fig. 2 **a** An underlying graph of a TVCN of five nodes. **b** TAG of the snapshots shown in Fig. 1 at time instances $t = 1, 2, 3$ and 4

TVCN of five nodes shown in Fig. 1. This static graph should be seen as a sort of footprint of TVCN because it flattens the time dimension and indicates only the pairs of nodes that have relations at some time in S_T [12].

Figure 2b shows a *time-aggregated graph* (TAG) representation of the TVCN shown in Fig. 1. A TAG includes all temporal information about the network at $t = 1$, 2, 3 and 4, with link number 2, 7 and 10 not appearing in any time interval (represented as dashed lines). Notice that each label "$\{t_1, t_2, \ldots, t_\tau\}$, l" associated with a link l in a TAG denotes a *time-stamped link* 'l' with its all activity time instances viz., $t_1, t_2, \ldots, t_\tau$. We can denote any such time-stamped link as $l_x^{(y)}$, where x denotes a link which is active at time instance y. Further, a TAG can also be directly transformed into an original temporal reachability information preserving static graph known as *Line Graph* (LG) [36, 37]. This idea of TAG to LG transformation such that a network's original temporal reachability information remains preserved has recently been utilized in [37] for analysing survivability in TVCNs. An example of LG and its application in TS-MPS and TS-MCS enumeration will be discussed later in Sect. 5.

Other and alternative representations of TVCNs could be e.g., as *adjacency tensors*, *space-time graphs*, *affine graphs*, *time line of contacts*, etc., [38]. A review of the other representation styles of TVCNs may also be seen in [39, 40]. Besides, various network visualization tools and packages have been gradually developed for studying the network dynamics and complex processes occurring in large networks coming from different fields of study such as social science, biology, citation networks, WWW, Internet, etc. Pajek, igraph, Time-ordered, Gephi, Kumonote, J-Sim, NS-2, NS-3, Qualnet, Glomosim and Omnet++ are few famous tools/packages used by the researchers studying complex networks like TVCNs [17].

2.3 Path Set and Cut Set Model for TVCNs

As TVCNs do not have an *end-to-end* connectivity like in static networks, they may not have any MPS/MCS; thus, will result in a zero reliability figure. However, it is a false assertion as there may be many opportunistic path sets with respect to time, which may result in successful transit of data packets between designated nodes within a finite window of time. Thus, motivated by the above, the authors in their recent works [8, 36, 39] extended the notion of MPS and MCS of static networks to the *Time-stamped-minimal path set* (TS-MPS) and *Time-stamped-minimal cut set* (TS-MCS), respectively, for TVCNs. The definitions of TS-MPS and TS-MCS have been reproduced below for the sake of completeness.

Definition 1 *Time-stamped-minimal path set* (TS-MPS) is a sub set of a TVCN's links appearing at different instances of time (in non-decreasing order) during an observational period ensuring and establishing a successful communication between a specified set of nodes. This set is said to be minimal in the sense that every element of this set is needed for ensuring a successful communication. In other words, success of every component in at least one TS-MPS ensures communication or network success. In this definition, the assumption of taking instances of time in non-decreasing order makes sure that we do not move into path sets, which have already occurred in the past.

Definition 2 *Time-stamped-minimal cut set* (TS-MCS) is a sub set of a TVCN's links appearing at different instances of time during an observational period such that their removal or non-functioning at those particular instances of time ensures network's dis-connectivity for a specified set of nodes, provided that removal of any proper subset of these links does not disconnects the network graph. In other words, the set is minimal because the system does not fails if any one or more of the *time-stamped links* do not fail. Thus, failure of every component in at least one TS-MCS ensures communication or network failure.

To clearly illustrate the notion of TS-MPS, let us consider the link path set (4–6–9) (refer Fig. 2b) between (*S-D*) node pair (1, 5) which has its links' activity time instances as {3, 4}–{2}–{1, 3, 4}, respectively. So, it can be further expanded

Table 3 Enumerated TS-MPS and TS-MCS between node pair (1, 5) of Fig. 1 or Fig. 2b

TS-MPS	$\{l_1^{(1)} l_8^{(3)}\}, \{l_1^{(2)} l_8^{(3)}\}, \{l_1^{(1)} l_3^{(2)} l_9^{(3)}\}, \{l_1^{(2)} l_3^{(2)} l_9^{(3)}\}, \{l_1^{(1)} l_3^{(3)} l_9^{(3)}\}, \{l_1^{(2)} l_3^{(3)} l_9^{(3)}\},$ $\{l_1^{(1)} l_3^{(2)} l_9^{(4)}\}, \{l_1^{(2)} l_3^{(2)} l_9^{(4)}\}, \{l_1^{(1)} l_3^{(3)} l_9^{(4)}\}, \{l_1^{(2)} l_3^{(3)} l_9^{(4)}\}, \{l_1^{(1)} l_3^{(4)} l_9^{(4)}\}, \{l_1^{(2)} l_3^{(4)} l_9^{(4)}\},$ $\{l_1^{(1)} l_5^{(1)} l_6^{(2)} l_9^{(3)}\}, \{l_1^{(1)} l_5^{(2)} l_6^{(2)} l_9^{(3)}\}, \{l_1^{(2)} l_5^{(2)} l_6^{(2)} l_9^{(3)}\}, \{l_1^{(1)} l_5^{(1)} l_6^{(2)} l_9^{(4)}\}, \{l_1^{(1)} l_5^{(2)} l_6^{(2)} l_9^{(4)}\},$ $\{l_1^{(2)} l_5^{(2)} l_6^{(2)} l_9^{(4)}\}, \{l_4^{(3)} l_5^{(4)} l_3^{(4)} l_9^{(4)}\},$ $\boxed{\{l_4^{(4)} l_5^{(4)} l_3^{(4)} l_9^{(4)}\}}$
TS-MCS	$\{l_1^{(1)} l_1^{(2)} l_3^{(4)}\}, \{l_1^{(1)} l_1^{(2)} l_5^{(4)}\}, \{l_1^{(1)} l_1^{(2)} l_9^{(4)}\}, \{l_8^{(3)} l_9^{(3)} l_9^{(4)}\}, \{l_1^{(1)} l_1^{(2)} l_4^{(3)} l_4^{(4)}\},$ $\{l_3^{(2)} l_3^{(3)} l_3^{(4)} l_6^{(2)} l_8^{(3)}\}, \{l_3^{(2)} l_3^{(3)} l_6^{(2)} l_8^{(3)} l_9^{(4)}\}, \{l_1^{(1)} l_3^{(2)} l_3^{(3)} l_3^{(4)} l_5^{(2)} l_8^{(3)}\},$ $\{l_1^{(1)} l_3^{(2)} l_3^{(3)} l_5^{(2)} l_8^{(3)} l_9^{(4)}\}, \{l_3^{(2)} l_3^{(3)} l_3^{(4)} l_5^{(1)} l_5^{(2)} l_8^{(3)}\}, \{l_3^{(2)} l_3^{(3)} l_5^{(1)} l_5^{(2)} l_8^{(3)} l_9^{(4)}\}$

to generate six possible TS-MPS viz., $l_4^{(3)} l_6^{(2)} l_9^{(1)}$, $l_4^{(3)} l_6^{(2)} l_9^{(3)}$, $l_4^{(3)} l_6^{(2)} l_9^{(4)}$, $l_4^{(4)} l_6^{(2)} l_9^{(1)}$, $l_4^{(4)} l_6^{(2)} l_9^{(3)}$ and $l_4^{(4)} l_6^{(2)} l_9^{(4)}$. Note that in all these six TS-MPS constituting *time-stamped links* are not in non-decreasing order, and as we cannot traverse in the past they all are invalid for establishing connectivity between node pair (1, 5). However, (4–6–9) is a valid MPS for a static network with no time stamps of link activities. A list of all valid TS-MPS between node pair (1, 5) for the example TVCN shown in Fig. 1 has been provided in row 1 of Table 3. Also it is important to note here that recently Fraire et al. [41] highlighted the basic routes in a TVCN using *contact graph* (CG). However, the routes obtained using CG representation fails to explicitly show many potential opportunities of data transfer originating in time, which can easily be shown using TS-MPS. In other words, CG representation obscures TS-MPS, which provides more flexibility in identification of data transfer opportunities by expanding links with respect to time.

Similarly, in case of a TS-MCS the failure of all constituting *time-stamped links* is necessary to snap the (*S*-*D*) node pair connectivity; however, any such TS-MCS may not necessarily belong to a MCS of the underlying static network. For example, $\{l_1^{(1)} l_1^{(2)} l_3^{(4)}\}$ is a TS-MCS for the TVCN, but (1–3) is not a cut set for the underlying static network! Thus, it is worth mentioning here that the inclusion of time in the network brings new constraints into picture, which demands further research efforts for their modelling and performance analysis. Refer row 2 of Table 3 for all valid TS-MCS between node pair (1, 5) for the TVCN shown in Fig. 1. The techniques to enumerate these TS-MPS and TS-MCS will be discussed later in Sect. 5.

3 Difference: TVCN and Static Network

The primary distinctions between TVCN and static network are:

(1) Unlike in static networks, it is possible in TVCNs that some nodes may never ever be able to connect themselves via single (direct) or multiple hops (through relay nodes) yet they are able to communicate with each other via device-to-device communications in *store-carry-and-forward* fashion within a finite time frame.

(2) In each TS-MPS of a TVCN the links have a non-decreasing order of time stamps; but, such a labelling scheme is absent in MPS. It is also important to keep it in mind that in TVCNs the time-stamps are not necessarily consecutive.

(3) A sub path of a shortest MPS is shortest; however, it may not be true for TS-MPS. Thus, existing algorithms for enumerating shortest MPS cannot be directly applied to enumerate TS-MPS [42]. For illustration, in Fig. 2b consider a TS-MPS $l_1^{(1)}l_5^{(1)}l_6^{(2)}$ (though not shortest!) from node 1 to node 3 via node 2 and 4, then its time-stamped sub path $l_1^{(1)}l_5^{(1)}$ from node 1 to node 4 via node 2 is not the shortest time-stamped sub path. It is true as time-stamped link $l_4^{(3)}$ uses only one hop from node 1 to reach node 4 and is shorter; however, $l_4^{(3)}l_6^{(2)}$ is an invalid TS-MPS between node 1 and 3 as the time-stamp of successive link, i.e., $l_6^{(2)}$ is less than that of the previous link, i.e., $l_4^{(3)}$. In other words, TS-MPS $l_4^{(3)}l_6^{(2)}$ is invalid, as its links do not obey the time restriction.

(4) In TVCNs if there exists a path from node v_i to node v_j within a designated time window then it is not necessary that the reverse is also true; however, it is true by default in static networks with bi-directional links. This asymmetry, which arises even though each discrete snapshot of a TVCN is symmetric follows from (2) above, and is a result of the direction induced by the inclusion of time explicitly into the network model. For example, consider a (valid) hypothetical TS-MPS say $l_x^{(a)}l_y^{(b)}l_z^{(c)}$ between a (*S*-*D*) node pair. Recall that the considered TS-MPS will be valid if and only if the time stamp a is less than or equal to the time stamp b which in turn should be less than or equal to the time stamp c. Now, if we reverse the direction of data flow then the TS-MPS shall become invalid. Therefore, if the considered TS-MPS is the only path set between the (*S*-*D*) node pair or if all other TS-MPS also have similar time-ordering of links then there shall be no TS-MPS to transfer data packets in the reverse direction, i.e., from *D* to *S*.

Interested readers are advised to refer [8] for other noteworthy observations from the domain of TVCNs. Further, the present trends show that researchers are focussing on modelling TVCNs along with time by using evolving graph model, as simply aggregating the snapshots without any information about links' activity (refer Fig. 2a) presents a very misleading summary. For example, if the TVCN of Fig. 1 is modelled as an underlying graph shown in Fig. 2a, then node pair (1, 3) will be connected via node 4 using links 4 and 6. However, it is not true as it can be observed from its TAG (refer Fig. 2b) that such a path can never exist in reality.

4 Network Reliability: An Overview

This section presents some important definitions, existing reliability models and techniques for evaluating the network reliability of static networks and TVCNs. It also discusses the role of reliability optimization in the design of TVCNs.

4.1 Important Definitions and Metrics

Network reliability is defined as the probability that a specified set of designated nodes are able to communicate with each other within the intended design features and environments. Depending on the number of nodes in the specified set, the network reliability metric can be termed as *two-terminal*, *k-terminal* or *all-terminal reliability*. The *k-terminal reliability* (k-TR) is more general because $k = 2$ indicates the first metric, while $k = n$ (all nodes of the network) yields *all-terminal reliability*. Network survivability, defined as the ability to survive a certain number of failures [37], is another important metric frequently used to study the fault tolerance of networks. Most works on network reliability and survivability assume that the failures of the units (nodes and/or links) are statistically independent and the failure of a unit does not influence the hazard rate of the surviving units and thereby the subsequent failures [43]. Further, it is commonly assumed that each unit has a probability of either being operational or failed, i.e., has two-states [43]. Besides an operational unit may be assumed to be *perfect*, i.e., theoretically 100% reliable or *imperfect*, i.e., having a reliability of less than 100%. It is also worth mentioning here that link and node reliability usually depends on factors like time, environmental interference, transmission power, signal-to-noise ratio and Euclidean distance between the interacting nodes, while node reliability may also additionally include factors like its capacity, battery power, packet size, etc.

4.2 Reliability Evaluation—Static Networks

Since 1970, evaluation of reliability measures of static networks has drawn a lot of attention from the researchers [43, 44]. Decades of comprehensive research in the domain has resulted in the evolution of many diverse, well-developed and understood graph-theoretic based methods for the reliability modelling and evaluation of static networks. These methods have broadly been categorized as: (i) MPS or MCS based, e.g., *inclusion-exclusion, calculation of bounds, domination theory, sum-of-disjoint products technique* (SDP), *binary decision diagrams* (BDD) and *ordered* BDD (OBDD), and (ii) Non-MPS or non-MCS based, e.g., *state enumeration technique, factoring theorem* and *network transformation* [44]. Both the categories have their well-documented merits and demerits, however, first category is simple, comparatively efficient and economical, and less restrictive than the other.

Numerous methods have been proposed in the literature to enumerate MPS and MCS [45–47]. One can extract from the enumerated MPS, between a pair of nodes, its *foremost*, *shortest*, and the *fastest* MPS to compute *earliest arrival date*, the *minimum number of hops*, and the *minimum delay* (time span), respectively. The enumerated MPS can also be used to compute the expected hop count (EHC) between a (*S-D*) node pair [48]. On the other hand, the MCS, whose number is usually less in comparison to the number of MPS in a complex static network, are directly related to the modes of system failure or weak links in the design. Consequently, MCS are preferred over MPS for reliability evaluation, as they can help in direct identification of the distinct and discrete ways in which a system may fail [49]. Further, the MCS concept has been used and extended to compute the maximum flow [50], and performability metrics like capacity related reliability (CRR) [51] in flow networks.

The techniques utilizing MPS/MCS works in two-steps, viz., enumeration of MPS or MCS between specified set of nodes of the network, and efficiently combining MPS or MCS (logically using the *laws of probability*) thereafter to obtain a compact form of reliability expression or its estimate. The well-known SDP approach is a morphed and compact version of applying probability *law of unionization of events* or *inclusion-exclusion principle* on events. There are two versions of SDP technique viz., *single-variable inversion* (SVI) and *multi-variable inversion* (MVI) [44]. The MVI-SDP is preferred as it inverts a group of variables rather than a single one at a time, and thus results in a more compact reliability expression as compared to the expression rendered by SVI technique.

4.3 Reliability Evaluation—TVCNs

Even though there exists a plethora of methods for assessing the reliability of static networks, very few researchers have attempted to extend the notions from static networks to TVCNs. In the last decade, connectivity, routing protocols, mobility models, energy conservation, security issues and various applications, etc., of TVCNs have widely been studied for analysing their performance. However, their reliability assessment has not been carried out explicitly and possibly because the existing approaches for reliability evaluation of static networks cannot be directly applied to the TVCNs owing to the challenges associated with TVCNs such as frequently changing topology and huge variations in environmental conditions.

The reliability of a TVCN can measure how reliably a packet sent from a source node be delivered to a specified destination node within a specified time under a defined variable topological conditions. In this regard, recently some concerted efforts have been made by the researchers to model and evaluate the reliability of MANETs and DTNs. In [52], authors estimated the reliability of MANETs with the help of Monte Carlo simulation technique. Later, their work was extended in [53] by modifying MANET by making specified set of nodes, (e.g., *S-D*) with perfect nodes as their failure will definitely lead to network failure. In [28], authors utilized Free Space-Two Ray Ground (FS-TRG) propagation model to model link formation, thus, resulting into further modification of the proposed approach of [52]. Authors in [54] evaluated the reliability of capacitated MANETs with the help of Log-Normal Shadowing propagation model. Besides, authors in [48] utilized SDP to compute reliability and *EHC* in wireless communication networks, but did not consider the varying topologies with time. In [55], authors utilized factoring method to evaluate the reliability of MANETs. Rebaiaia and Ait-Kadi [56] proposed polygon-to-chain and series-parallel reduction based approach for evaluating the reliability of MANETs. In [57], Ahmad and Mishra proposed a critical node detection based approach for the reliability evaluation of large scale MANETs. In [58], authors utilized logistic regression based modelling followed by simulation in ns-2.35 simulation software to evaluate the reliability of MANETs. In [59], authors modified the reliability evaluation algorithm for computer networks proposed in [60] to evaluate the reliability of MANETs with imperfect nodes. A stochastic link failure model was used in [61] for the reliability evaluation of wireless multi-hop networks. Recently, authors in [62] evaluated the reliability of MANETs under link and node failure model. Besides, in [63, 64], Meena and Vasanthi proposed a method based on universal generating function for evaluating the reliability of MANETs. Eiza and Ni [65] utilized an evolving graph model to find the most reliable journey in VANETs.

Despite the worth noting efforts for reliability evaluation of MANETs [52, 53], through *end-to-end* connectivity assessment and using Monte Carlo simulation, the researchers fail to consider the fact that there are many other TVCNs like *MANET-born networks* that mostly depend on *store-carry-and-forward* mechanism for data transfer among nodes. This buffering and forwarding of data packets is a feasible approach in environments with sparse connectivity and/or high level of interruptions. Thus, recent works [8, 36, 39], discussed later in Sect. 5, are efforts to efficiently incorporate the *store-carry-and-forward* mechanism, enumerate all TS-MPS/TS-MCS between a given (*S-D*) node pair and evaluate the reliability of TVCNs via SDP approach.

4.4 Reliability Optimization

Reliability optimization for long has been pivotal in network/system design, operation and management. It aims to develop a system adequately meeting the operational requirements, by trading-off between the design parameters and availability of resources, such that it also satisfies the reliability goals [66]. For the system design problems, a single objective optimization problem can be formulated by selecting the most important performance attribute as the objective function. Generally, this attribute could be one amongst the following, i.e., reliability, availability, cost, weight or volume and the rest of the other attributes may be introduced as constraints in the reliability design problem [67]. In [68], author proposed an efficient search procedure to solve integer programming problems arising in reliability design of a system. In

[69] authors proved that maximizing the overall reliability of a complex system subject to certain resource constraints leads to a mixed integer nonlinear programming problem. Literature in the domain shows numerous approximate, exact and heuristic optimization approaches for static networks [70, 71].

4.4.1 TVCN Design and Management

Any modern TVCN can neither be designed for scalability and reliability nor cost-effectiveness as in the case of static networks without proper planning. A well planned and optimized TVCN can often provide better performance with the same infrastructure cost [72]. Thus, network planning is indispensable irrespective of network type. In [73], authors presented a novel optimization models for planning Wireless Mesh Networks (WMNs) with an aim to minimize the network installation cost while providing full coverage to wireless mesh clients. A similar work was carried out in [74] for constructing reliable WMN infrastructure that can resist the failure of a single mesh node and ensures full coverage to all mesh clients. Recently, in [75] authors explored the performance of MCNs with the help of a multi-objective constrained optimization problem and showed that MCNs can improve the quality of service, capacity and energy-efficiency of traditional infrastructure-centric single-hop cellular systems. In [76], authors addressed the problem of energy efficiency in cellular networks and proposed a *store-carry-and-forward* relaying strategy to achieve savings in transmission energy. Similarly, in [77] an optimization problem was formulated for examining the viability of data packet collection from mobile devices, by choosing a subset of relaying (and/or sensing) mobile devices, following *device-to-device* communications. The authors in [78] introduced the problem of feasible *contact plan design* and in [79] they highlighted various challenges involved in *contact plan design* for predictable DTNs, such as satellite constellations, with constrained resources. In space borne TVCNs, a *feasible contact plan design* is crucial for judiciously utilizing the available resources like energy and number of transponders, and thus *contacts*. Usually, a mission and operation control (MOC) centre deterministically computes the expected *contacts* among satellite constellations by using the orbital mechanics and radio models [79]. So, before discussing any further we define the *contact plan* and other related terms in the ensuing paragraph.

Let the sequence of snapshots shown in Fig. 1 denote a *contact topology*. Principally a *contact topology* depicts a set of all potential opportunities (a.k.a. *contacts*), to establish a temporal communication link (or *time-stamped link*) between any two TVCN nodes within a given topology interval. A *feasible contact plan* can be thought of as a subset of contact topology. However, owing to the sparse connectivity and other factors, many times it is assumed that all potential contacts between nodes can belong to the *feasible contact plan.* In other words, a *contact plan* either includes the full *contact topology* or is a subset of it [80]. Therefore, a *feasible contact plan* includes all feasible contacts among other potential ones by satisfying some constraint(s) and meeting the objective(s). Further, it has also been pointed out in [79] that the design of feasible contact plans is still an open area for research. The authors

Table 4 A contact plan for a TVCN shown in Fig. 1

#	From node	To node	Start time	End time	Slot(s)
1/2	**1**	**2**	t_0	t_2	$\{t_1, t_2\}$
3/4	**1**	**4**	t_2	t_4	$\{t_3, t_4\}$
5/6	**2**	**3**	t_1	t_4	$\{t_2, t_3, t_4\}$
7/8	**2**	**4**	t_0	t_2	$\{t_1, t_2\}$
9/10	**2**	**4**	t_3	t_4	$\{t_4\}$
11/12	**2**	**5**	t_2	t_3	$\{t_3\}$
13/14	**3**	**4**	t_1	t_2	$\{t_2\}$
15/16	**3**	**5**	t_0	t_1	$\{t_1\}$
17/18	**3**	**5**	t_2	t_4	$\{t_3, t_4\}$

Bold numbers indicate the communicating node pairs

of [78, 79] later extended their work and proposed different approaches to compute the *feasible contact plans* [5, 80, 81]. Similarly, in [82] authors presented an approach to design a *feasible contact plan* by considering distinctiveness of different missions, energy related issues, and time-varying satellite downlink contacts. They also examined the effect of energy collection rate, battery capacity, and buffer size on the network profit. Table 4 presents an illustrative *contact plan* of a TVCN shown in Fig. 1 by using the full *contact topology* and indicating all contact opportunities, originating over a given time window, for data transfer between different pair of nodes of the considered TVCN. Here, # denotes link number of a bidirectional link in both directions.

In line with the above, authors in [83, 84] investigated topology design issues in predictable DTNs with unreliable links by providing connectivity via space-time path between any pair of nodes with the reliability higher than a required threshold and the lowest cost of the topology. Similarly, Li et al., [85] studied the impact of additional fixed nodes (a.k.a. throwboxes), used to forward data, on TVCN performance. They proposed two optimization problems to choose which throwboxes needs to be activated and also to find the number of required throwboxes. Thus, it is clear that a *feasible contact plan design* problem is of prime importance in TVCN design and reliability optimization.

5 TVCN Reliability Evaluation

Similar to the procedure followed for the reliability evaluation of static networks, the reliability of TVCN can also be assessed by following a two-step process, i.e., (i) Enumerate the TS-MPS/TS-MCS, and (ii) Apply the SDP approach on the enumerated TS-MPS/TS-MCS to compute the network reliability. Note that Step (ii) is straightforward process for which several algorithms are existing in the literature; however, Step (i) needs a careful consideration to capture buffering of data at nodes and the time based link existence scenario of TVCNs. Therefore, at first a network model suitable for such enumerations is selected and thereafter techniques are devised to enumerate TS-MPS and TS-MCS from the model.

5.1 Network Model

Among the several modelling paradigms of TVCNs, the evolving graph model along with its TAG representation (refer Sect. 2.2) can be a choice for modelling any predictable and fixed schedule TVCN and assessing their reliability. Further, as stated earlier that a TAG representation can also be converted to an equivalent LG representation to propose a simple method for enumerating both TS-MPS and TS-MCS effectively.

5.2 TS-MPS Enumeration Techniques

Admittedly, there would be an exponential number of TS-MPS/TS-MCS in a TVCN. Such an exorbitantly large number of TS-MPS/TS-MCS usually depends on the *number of time slots*, *number of nodes* and the *time-stamps for which each link is active*. In this section, we present three novel techniques for enumerating TS-MPS (equivalent to MPS in static network sense) for predictable TVCNs.

5.2.1 Cartesian Product Based Method

This simple enumeration method was developed in [39]. It is based on the Cartesian product of time stamps of each link associated with all the MPS between a given (*S*-*D*) pair of nodes of a network graph. It is based on the notion that any network graph formed during a changing topology scenario would be a subgraph in a complete network graph of n nodes, K_n. Therefore, we can follow a two-step process, i.e., (i) Enumerate the MPS between a given (*S*-*D*) pair of nodes for the complete network graph, K_n; (ii) Enumerate the TS-MPS for a TAG from the MPS obtained in step (i). The flowchart depicting the two-steps of the approach is shown in Fig. 3. Interested readers are requested to refer [39] for intricacies involved in this approach. Recall that the enumerated TS-MPS are made disjointed using SDP technique to obtain the reliability expression or value—discussed in Sect. 5.4 later, and is shown with bold face in Fig. 3.

The Cartesian product based algorithm enumerates all possible TS-MPS between a given (*S*-*D*) pair [39]. Table 3 (provided earlier in Sect. 2.3) presents a list of all 20 TS-MPS generated by following the steps of this algorithm between the node pair

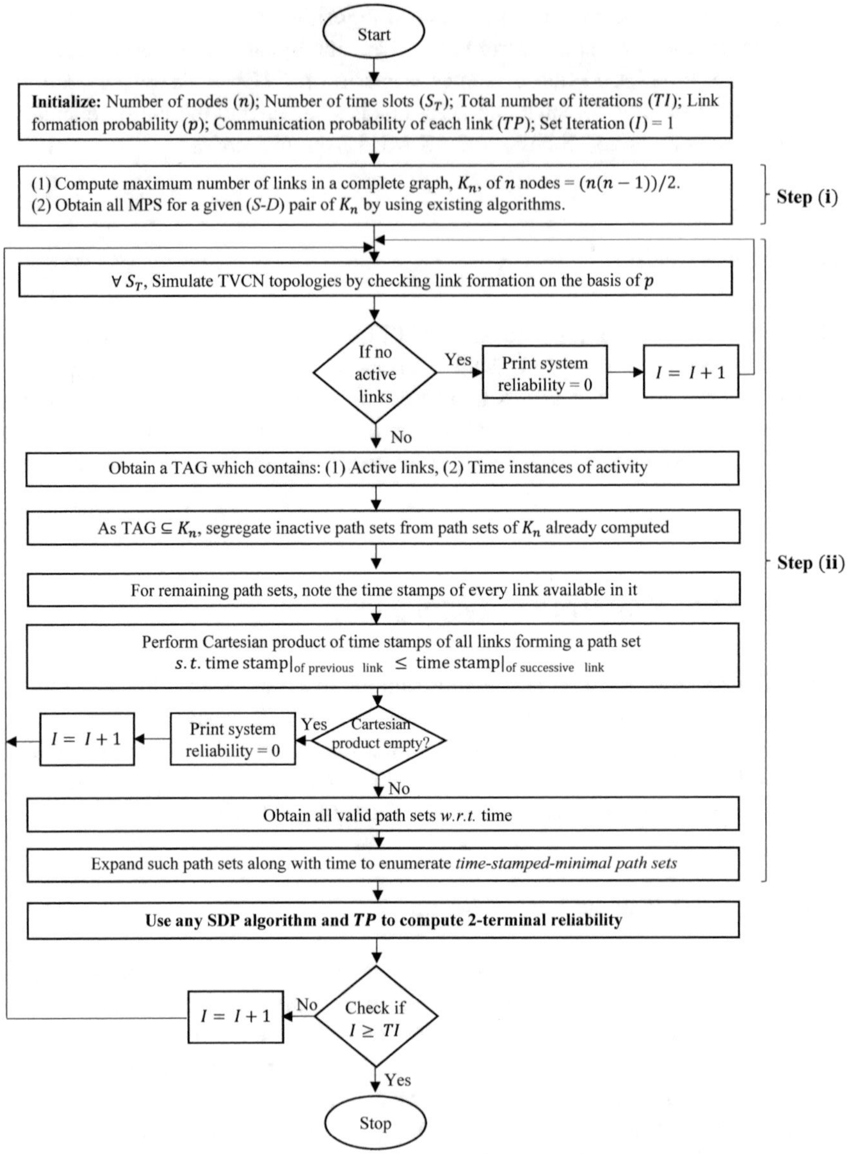

Fig. 3 Flow chart for TS-MPS enumeration and 2-terminal reliability evaluation of TVCNs using Cartesian product based approach

(1, 5) of Fig. 1 or Fig. 2b. Note that the number of TS-MPS between (5, 1) is 13, which shows that in TVCNs it may be possible to have different number of TS-MPS in forward and reverse directions due to presence of the time-dimension. However, the drawback of this approach is that if one needs to enumerate TS-MPS between all node pairs in one go then the entire process has to be repeated for each pair of nodes present in the network—a computational burden with associated costs. The algorithm presented in the next section overcomes this limitation.

5.2.2 Connection Matrix Based Method

This is a matrix based approach [8] capable of enumerating all possible TS-MPS between every pair of nodes of a predictable TVCN. The basic idea behind the development of this technique/algorithm is to exploit all the possible opportunities (in terms of links and/or nodes), which evolve between different pair of nodes with time and lead to the forwarding of data packets towards a specified destination. Recall that the TVCN shown in Fig. 1 has 20 TS-MPS between (*S*-*D*) node pair (1, 5), similarly, it might have a comparable, large or small number of TS-MPS for other node pairs. Hence, for lucid explanation and illustration purpose, we consider a small and sparsely connected network as shown in Fig. 4 and its TAG in Fig. 5. This algorithm requires the connection matrix of the network topology at each instance of time, i.e., it requires the connection details in each snapshot as input.

For the example shown in Fig. 4, let us denote and initialize such connection matrices by $M_t^{(0)}\ \forall t \in [1,\ T]$, where $T = 4$ time units is the entire observation period i.e., $t = 1, 2, \ldots, 4$. Therefore, the connection matrices, viz., $M_1^{(0)}$, $M_2^{(0)}$, $M_3^{(0)}$, and $M_4^{(0)}$ are shown in column 2 of Table 5.

Further, each $M_t^{(0)}$ is processed to obtain $M_t^{(n)} \forall t \in [1,\ T]$, where n is the number of mobile nodes in the network. This involves generation of intermediate matrices $\boldsymbol{M}_t^{(k)} \forall t \in [1,\ T]$ and $\forall k \in [1,\ n]$ whose elements are such that

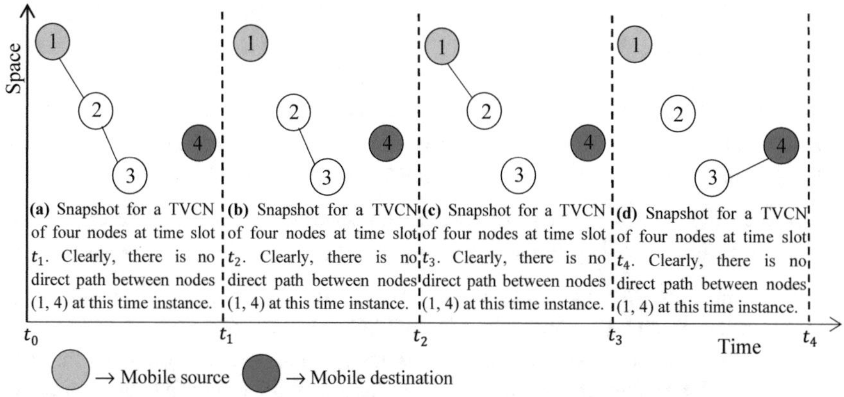

Space
1
2
3
4
(a) Snapshot for a TVCN of four nodes at time slot t_1. Clearly, there is no direct path between nodes (1, 4) at this time instance.
(b) Snapshot for a TVCN of four nodes at time slot t_2. Clearly, there is no direct path between nodes (1, 4) at this time instance.
(c) Snapshot for a TVCN of four nodes at time slot t_3. Clearly, there is no direct path between nodes (1, 4) at this time instance.
(d) Snapshot for a TVCN of four nodes at time slot t_4. Clearly, there is no direct path between nodes (1, 4) at this time instance.
t_0
t_1
t_2
t_3
t_4
Time
→ Mobile source
→ Mobile destination

Fig. 4 Sequence of snapshots of a TVCN of four nodes at four different time slots

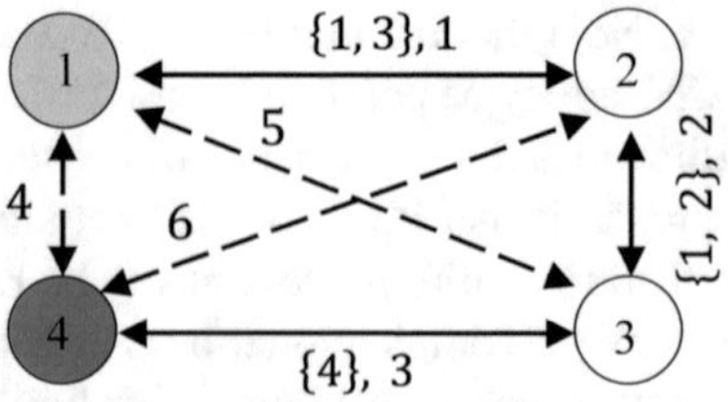

Fig. 5 A TAG of the snapshots shown in Fig. 4 at time instances $t = 1, 2, 3$ and 4

$\boldsymbol{m}_{SD}^{k} = \boldsymbol{m}_{Sk}^{k-1} \,\&\, \boldsymbol{m}_{kD}^{k-1} | \boldsymbol{m}_{SD}^{k-1}$ obtained from the elements in $\boldsymbol{M}_{t}^{(k-1)}$, where S, k and $D \in [1,\ n]$. Here, '**&**' (ampersand) denotes Boolean multiplication and '|' (vertical bar) denotes addition operation, respectively, to obtain such elements. The results at various stages are shown in Table 5. After obtaining $\boldsymbol{M}_{1}^{(n)},\ \boldsymbol{M}_{2}^{(n)},\ \ldots,\ \boldsymbol{M}_{T}^{(n)}$, respectively, we compute TS-MPS by multiplying the matrices and setting the diagonal elements of the resulting matrix to 1 (as diagonal elements with value 1 simply show self-connection/buffering). The obtained TS-MPS between each pair of node can be seen in the matrix below.

$$\text{TS} - \text{MPS} = \begin{bmatrix} 1 & l_1^{(1)} + l_1^{(3)} & l_1^{(1)} l_2^{(1)} + l_1^{(1)} l_2^{(2)} & l_1^{(1)} l_2^{(1)} l_3^{(4)} + l_1^{(1)} l_2^{(2)} l_3^{(4)} \\ l_1^{(1)} + l_1^{(3)} & 1 & l_2^{(1)} + l_2^{(2)} & l_2^{(1)} l_3^{(4)} + l_2^{(2)} l_3^{(4)} \\ l_2^{(1)} l_1^{(1)} + l_2^{(1)} l_1^{(3)} + l_2^{(2)} l_1^{(3)} & l_2^{(1)} + l_2^{(2)} & 1 & l_3^{(4)} \\ 0 & 0 & l_3^{(4)} & 1 \end{bmatrix}$$

For example, the two TS-MPS viz., $l_1^{(1)} l_2^{(1)} l_3^{(4)}$ and $l_1^{(1)} l_2^{(2)} l_3^{(4)}$ between node pair (1, 4) can be observed from location TS-MPS (1, 4) of the TS-MPS matrix. Similarly, we can easily observe that there exists no TS-MPS from node 4 to node 1.

5.2.3 Line Graph Based Method

This method is also accomplished in two-steps: (i) Transform a TAG of TVCN into a Line Graph (LG), and (ii) Enumerate TS-MPS from this LG using any existing enumeration algorithm employed for MPS e.g., [45]. A LG generation begins by creation of two separate nodes for the source and destination node, respectively. This is followed by the creation of a node for each *time-stamped-link* $l_x^{(y)}$ in the TAG. A newly created node, say $l_\alpha^{(t_i)}$, in LG is connected with a directed link to another node $l_\beta^{(t_j)}$ if and only if $t_i \le t_j$. Next, a directed link is added from the node created for the source to $l_\alpha^{(t_i)}$, if $l_\alpha^{(t_i)}$ originates from the source node; similarly a directed link is added from node $l_\beta^{(t_j)}$ to a node created for the destination, if $l_\beta^{(t_j)}$ terminates at the destination. Such a transformation from TAG to LG needs careful consideration to capture the *time based link* (or *time-stamped link*) *existence* scenario. Note that size of LG is dependent on the number of nodes and links in its TAG as well as the number of time slots. Figure 6 shows the LG of a TVCN shown in Fig. 4.

Table 5 Intermediate matrices generation from connection matrices

Time slot	Connection matrix	Intermediate matrices
$t=1$	$M_1^0 = \begin{bmatrix} 1 & l_1^{(1)} & 0 & 0 \\ l_1^{(1)} & 1 & l_2^{(1)} & 0 \\ 0 & l_2^{(1)} & 1 & 0 \\ 0 & 0 & 0 & 1 \end{bmatrix}$	$M_1^{(1)} = \begin{bmatrix} 1 & l_1^{(1)} & 0 & 0 \\ l_1^{(1)} & 1 & l_2^{(1)} & 0 \\ 0 & l_2^{(1)} & 1 & 0 \\ 0 & 0 & 0 & 1 \end{bmatrix}$
		$M_1^{(2)} = \begin{bmatrix} 1 & l_1^{(1)} & l_1^{(1)}l_2^{(1)} & 0 \\ l_1^{(1)} & 1 & l_2^{(1)} & 0 \\ l_1^{(1)}l_2^{(1)} & l_2^{(1)} & 1 & 0 \\ 0 & 0 & 0 & 1 \end{bmatrix}$
		$M_1^{(3)} = \begin{bmatrix} 1 & l_1^{(1)} & l_1^{(1)}l_2^{(1)} & 0 \\ l_1^{(1)} & 1 & l_2^{(1)} & 0 \\ l_1^{(1)}l_2^{(1)} & l_2^{(1)} & 1 & 0 \\ 0 & 0 & 0 & 1 \end{bmatrix}$
		$\boldsymbol{M_1^{(4)}} = \begin{bmatrix} \mathbf{1} & \boldsymbol{l_1^{(1)}} & \boldsymbol{l_1^{(1)}l_2^{(1)}} & \mathbf{0} \\ \boldsymbol{l_1^{(1)}} & \mathbf{1} & \boldsymbol{l_2^{(1)}} & \mathbf{0} \\ \boldsymbol{l_1^{(1)}l_2^{(1)}} & \boldsymbol{l_2^{(1)}} & \mathbf{1} & \mathbf{0} \\ \mathbf{0} & \mathbf{0} & \mathbf{0} & \mathbf{1} \end{bmatrix}$
$t=2$	$M_2^{(0)} = \begin{bmatrix} 1 & 0 & 0 & 0 \\ 0 & 1 & l_2^{(2)} & 0 \\ 0 & l_2^{(2)} & 1 & 0 \\ 0 & 0 & 0 & 1 \end{bmatrix}$	$M_2^{(1)} = \begin{bmatrix} 1 & 0 & 0 & 0 \\ 0 & 1 & l_2^{(2)} & 0 \\ 0 & l_2^{(2)} & 1 & 0 \\ 0 & 0 & 0 & 1 \end{bmatrix}$
		$M_2^{(2)} = \begin{bmatrix} 1 & 0 & 0 & 0 \\ 0 & 1 & l_2^{(2)} & 0 \\ 0 & l_2^{(2)} & 1 & 0 \\ 0 & 0 & 0 & 1 \end{bmatrix}$
		$M_2^{(3)} = \begin{bmatrix} 0 & 0 & 0 & 0 \\ 0 & 1 & l_2^{(2)} & 0 \\ 0 & l_2^{(2)} & 1 & 0 \\ 0 & 0 & 0 & 1 \end{bmatrix}$

(continued)

Table 5 (continued)

Time slot	Connection matrix	Intermediate matrices
		$\boldsymbol{M_2^{(4)}} = \begin{bmatrix} \mathbf{0} & \mathbf{0} & \mathbf{0} & \mathbf{0} \\ \mathbf{0} & \mathbf{1} & \boldsymbol{l_2^{(2)}} & \mathbf{0} \\ \mathbf{0} & \boldsymbol{l_2^{(2)}} & \mathbf{1} & \mathbf{0} \\ \mathbf{0} & \mathbf{0} & \mathbf{0} & \mathbf{1} \end{bmatrix}$
$t = 3$	$M_3^{(0)} = \begin{bmatrix} 1 & l_1^{(3)} & 0 & 0 \\ l_1^{(3)} & 1 & 0 & 0 \\ 0 & 0 & 1 & 0 \\ 0 & 0 & 0 & 1 \end{bmatrix}$	$M_3^{(1)} = \begin{bmatrix} 1 & l_1^{(3)} & 0 & 0 \\ l_1^{(3)} & 1 & 0 & 0 \\ 0 & 0 & 1 & 0 \\ 0 & 0 & 0 & 1 \end{bmatrix}$
		$M_3^{(2)} = \begin{bmatrix} 1 & l_1^{(3)} & 0 & 0 \\ l_1^{(3)} & 1 & 0 & 0 \\ 0 & 0 & 1 & 0 \\ 0 & 0 & 0 & 1 \end{bmatrix}$
		$M_3^{(3)} = \begin{bmatrix} 1 & l_1^{(3)} & 0 & 0 \\ l_1^{(3)} & 1 & 0 & 0 \\ 0 & 0 & 1 & 0 \\ 0 & 0 & 0 & 1 \end{bmatrix}$
		$\boldsymbol{M_3^{(4)}} = \begin{bmatrix} \mathbf{1} & \boldsymbol{l_1^{(3)}} & \mathbf{0} & \mathbf{0} \\ \boldsymbol{l_1^{(3)}} & \mathbf{1} & \mathbf{0} & \mathbf{0} \\ \mathbf{0} & \mathbf{0} & \mathbf{1} & \mathbf{0} \\ \mathbf{0} & \mathbf{0} & \mathbf{0} & \mathbf{1} \end{bmatrix}$
$t = 4$	$M_4^{(0)} = \begin{bmatrix} 1 & 0 & 0 & 0 \\ 0 & 1 & 0 & 0 \\ 0 & 0 & 1 & l_3^{(4)} \\ 0 & 0 & l_3^{(4)} & 1 \end{bmatrix}$	$M_4^{(1)} = \begin{bmatrix} 1 & 0 & 0 & 0 \\ 0 & 1 & 0 & 0 \\ 0 & 0 & 1 & l_3^{(4)} \\ 0 & 0 & l_3^{(4)} & 1 \end{bmatrix}$
		$M_4^{(2)} = \begin{bmatrix} 1 & 0 & 0 & 0 \\ 0 & 1 & 0 & 0 \\ 0 & 0 & 1 & l_3^{(4)} \\ 0 & 0 & l_3^{(4)} & 1 \end{bmatrix}$

(continued)

Table 5 (continued)

Time slot	Connection matrix	Intermediate matrices
		$M_4^{(3)} = \begin{bmatrix} 1 & 0 & 0 & 0 \\ 0 & 1 & 0 & 0 \\ 0 & 0 & 1 & l_3^{(4)} \\ 0 & 0 & l_3^{(4)} & 1 \end{bmatrix}$
		$\mathbf{M_4^{(4)}} = \begin{bmatrix} \mathbf{1} & \mathbf{0} & \mathbf{0} & \mathbf{0} \\ \mathbf{0} & \mathbf{1} & \mathbf{0} & \mathbf{0} \\ \mathbf{0} & \mathbf{0} & \mathbf{1} & \mathbf{l_3^{(4)}} \\ \mathbf{0} & \mathbf{0} & \mathbf{l_3^{(4)}} & \mathbf{1} \end{bmatrix}$

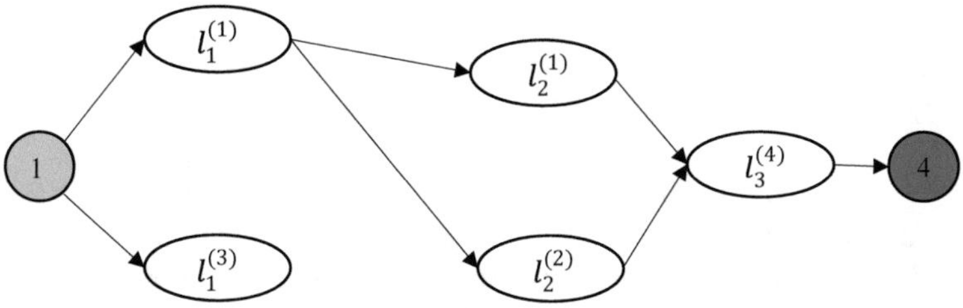

Fig. 6 Line graph of a TVCN shown in Fig. 4

One can easily observe the presence of two TS-MPS between node pair (1, 4) through visual inspection. Note that as source and destination nodes have implicitly been included in the *time-stamped links* viz., $l_1^{(1)}$, $l_1^{(3)}$ and $l_3^{(4)}$, respectively, thus, we require only irredundant constituent nodes in a TS-MPS, i.e., all other nodes except the separate nodes depicting source and destination. Further, it is important to note that we can prune any leaf node, e.g., $l_1^{(3)}$ as it has no outgoing link; therefore, will not lead to the destination [36].

5.3 TS-MCS Enumeration Techniques

The enumeration of TS-MCS is not as straightforward process as it is in the case of TS-MPS enumeration presented in earlier Sect. 5.2.1. This happens because the MCS of the underlying network and the TS-MCS of the TVCN do not bear any apparent relation to the Cartesian product technique; though, the principle of duality does exist between them. The interested readers are encouraged to try it by their own to check the veracity of this claim. In the ensuing paragraph, we discuss two techniques for TS-MCS enumeration.

5.3.1 TS-MPS Based Method

In this method, the authors [39] applied the well-known De Morgan's law in Boolean algebra to examine the principle of duality between TS-MPS and TS-MCS, and to verify the outcome with manually enumerated (using inspection) TS-MCS. It can be seen that the TS-MCS enumerated by inverting the TS-MPS, by applying De Morgan's law, exactly matches with the TS-MCS enumerated using visual inspection; thereby, confirmed the existence of duality between TS-MPS and TS-MCS in TVCNs as well. Table 3 shows the 11 TS-MCS between node pair (1, 5) for the TVCN given in Fig. 1. Besides, one can easily observe from LG (refer Fig. 6) the three TS-MCS viz., $l_1^{(1)}$, $l_3^{(4)}$ and $l_2^{(1)}l_2^{(2)}$ between node pair (1, 4) for the TVCN shown in Fig. 4.

The above method depends on inversion of all enumerated TS-MPS for the TS-MCS enumeration. Thus, it can be said to be a crude approach of TS-MCS generation with huge computational cost.

5.3.2 Line Graph Based Method

In this approach [36], the first step is same as described earlier in Sect. 5.2.3., i.e., transforming TAG to LG. However, in the second step one can use any known and efficient MCS enumerating algorithm like [46, 47] for TS-MCS enumeration. It is important to note here that the performance of this method is highly dependent on the efficiency of the MCS enumerating algorithm utilized in the second step. Further note that, in contrast to the static networks, in TVCNs the number of TS-MCS may not be less in comparison to the number of TS-MPS [36].

5.4 Reliability Related Metrics Evaluation

Once we have TS-MPS or TS-MCS, we can invoke SDP approach to determine the network reliability expression and other reliability metrics. The authors [8, 36, 39] followed the similar approach and utilized MVI-SDP [44] to obtain the reliability expression. The 20 TS-MPS, collated in Table 3, disjointed using MVI-SDP algorithm result in the following reliability (*2-TR*) expression. Here, components with bar on their top represent their respective failure probability. Moreover, each term in the expression represents a probability, e.g., $l_1^{(1)}$ represents the probability that link #1 is active at time slot 1 and $l_8^{(3)}$ represents the probability that link #8 is active at time slot 3. If we assume that the probability of success of each time-stamped link is 0.90, then the expression results in 0.9962 as *2-TR*.

$$
\begin{aligned}
2-TR = {} & l_1^{(1)}l_8^{(3)} + \overline{l_1^{(1)}}l_1^{(2)}l_8^{(3)} + \overline{l_8^{(3)}}l_1^{(1)}l_3^{(2)}l_9^{(3)} + \overline{l_1^{(1)}}\,\overline{l_8^{(3)}}l_1^{(2)}l_3^{(2)}l_9^{(3)} \\
& + \overline{l_3^{(2)}}\,\overline{l_8^{(3)}}l_1^{(1)}l_3^{(3)}l_9^{(3)} + \overline{l_1^{(1)}}\,\overline{l_3^{(2)}}\,\overline{l_8^{(3)}}l_1^{(2)}l_3^{(3)}l_9^{(3)} + \overline{l_8^{(3)}}\,\overline{l_9^{(3)}}l_1^{(1)}l_3^{(2)}l_9^{(4)} \\
& + \overline{l_1^{(1)}}\,\overline{l_8^{(3)}}\,\overline{l_9^{(3)}}l_1^{(2)}l_3^{(2)}l_9^{(4)} + \overline{l_3^{(2)}}\,\overline{l_8^{(3)}}\,\overline{l_9^{(3)}}l_1^{(1)}l_3^{(3)}l_9^{(4)} + \overline{l_1^{(1)}}\,\overline{l_3^{(2)}}\,\overline{l_8^{(3)}}\,\overline{l_9^{(3)}}l_1^{(2)}l_3^{(3)}l_9^{(4)}
\end{aligned}
$$

$$
\begin{aligned}
&+\overline{l_3^{(2)}}\,\overline{l_3^{(3)}}\,\overline{l_8^{(3)}}l_1^{(1)}l_3^{(4)}l_9^{(4)}+\overline{l_1^{(1)}}\,\overline{l_3^{(2)}}\,\overline{l_3^{(3)}}\,\overline{l_8^{(3)}}l_1^{(2)}l_3^{(4)}l_9^{(4)}+\overline{l_3^{(2)}}\,\overline{l_3^{(3)}}\,\overline{l_8^{(3)}}(1-l_3^{(4)}l_9^{(4)})l_1^{(1)}l_5^{(1)}l_6^{(2)}l_9^{(3)}\\
&+\overline{l_3^{(2)}}\,\overline{l_3^{(3)}}\,\overline{l_8^{(3)}}(1-l_3^{(4)}l_9^{(4)})\overline{l_5^{(1)}}l_1^{(1)}l_5^{(2)}l_6^{(2)}l_9^{(3)}+\overline{l_1^{(1)}}\,\overline{l_3^{(2)}}\,\overline{l_3^{(3)}}\,\overline{l_8^{(3)}}(1-l_3^{(4)}l_9^{(4)})l_1^{(2)}l_5^{(2)}l_6^{(2)}l_9^{(3)}\\
&+\overline{l_3^{(2)}}\,\overline{l_3^{(3)}}\,\overline{l_3^{(4)}}\,\overline{l_8^{(3)}}\,\overline{l_9^{(3)}}l_1^{(1)}l_5^{(1)}l_6^{(2)}l_9^{(4)}+\overline{l_3^{(2)}}\,\overline{l_3^{(3)}}\,\overline{l_3^{(4)}}\,\overline{l_8^{(3)}}\,\overline{l_9^{(3)}}\,\overline{l_5^{(1)}}l_1^{(1)}l_5^{(2)}l_6^{(2)}l_9^{(4)}\\
&+\overline{l_1^{(1)}}\,\overline{l_3^{(2)}}\,\overline{l_3^{(3)}}\,\overline{l_3^{(4)}}\,\overline{l_8^{(3)}}\,\overline{l_9^{(3)}}l_1^{(2)}l_5^{(2)}l_6^{(2)}l_9^{(4)}+\overline{l_1^{(1)}}\,\overline{l_1^{(2)}}l_4^{(3)}l_5^{(4)}l_3^{(4)}l_9^{(4)}\\
&+\overline{l_1^{(1)}}\,\overline{l_1^{(2)}}\,\overline{l_4^{(3)}}l_4^{(4)}l_5^{(4)}l_3^{(4)}l_9^{(4)}.
\end{aligned}
$$

Similarly, the 11 TS-MCS, collated in Table 3, result in the below given unreliability (Q) expression. However, here components with bar represent their respective success, instead of failure, probability. Each term like $l_1^{(1)}l_1^{(2)}l_3^{(4)}$ in the given expression represents the probability that link #1 is inactive at both time slots 1 and 2, and link #3 is inactive at slot 4. Besides, on assuming the unreliability of each time-stamped link as 0.10, the expression results in 0.003793636 as Q.

$$
\begin{aligned}
Q=\;&l_1^{(1)}l_1^{(2)}l_3^{(4)}+\overline{l_3^{(4)}}l_1^{(1)}l_1^{(2)}l_5^{(4)}+\overline{l_3^{(4)}}\,\overline{l_5^{(4)}}l_1^{(1)}l_1^{(2)}l_9^{(4)}+(1-l_1^{(1)}l_1^{(2)})l_8^{(3)}l_9^{(3)}l_9^{(4)}\\
&+\overline{l_3^{(4)}}\,\overline{l_5^{(4)}}\,\overline{l_9^{(4)}}l_1^{(1)}l_1^{(2)}l_4^{(3)}l_4^{(4)}+(1-l_1^{(1)}l_1^{(2)})(1-l_9^{(3)}l_9^{(4)})l_3^{(2)}l_3^{(3)}l_3^{(4)}l_6^{(2)}l_8^{(3)}\\
&+(1-l_1^{(1)}l_1^{(2)})\overline{l_3^{(4)}}\,\overline{l_9^{(3)}}l_3^{(2)}l_3^{(3)}l_6^{(2)}l_8^{(3)}l_9^{(4)}+\overline{l_1^{(2)}}\,\overline{l_6^{(2)}}(1-l_9^{(3)}l_9^{(4)})l_1^{(1)}l_3^{(2)}l_3^{(3)}l_3^{(4)}l_5^{(2)}l_8^{(3)}\\
&+\overline{l_1^{(2)}l_3^{(4)}l_6^{(2)}l_9^{(3)}}l_1^{(1)}l_3^{(2)}l_3^{(3)}l_5^{(2)}l_8^{(3)}l_9^{(4)}+\overline{l_1^{(1)}}\,\overline{l_6^{(2)}}(1-l_9^{(3)}l_9^{(4)})l_3^{(2)}l_3^{(3)}l_3^{(4)}l_5^{(1)}l_5^{(2)}l_8^{(3)}\\
&+\overline{l_1^{(1)}}\,\overline{l_3^{(4)}}\,\overline{l_6^{(2)}}\,\overline{l_9^{(3)}}l_3^{(2)}l_3^{(3)}l_5^{(1)}l_5^{(2)}l_8^{(3)}l_9^{(4)}.
\end{aligned}
$$

Note that value of $1-Q$ is equal to the *2-TR* figure. Further, following this approach, the *2-TR* between node pair (5, 1) of the TVCN shown in Fig. 1 turns out to be 0.9779. The difference in reliability figure between (1, 5) and (5, 1) is obvious and occurs due to the appearance of different TS-MPS within a given period of time. Similarly, the *2-TR* between node pair (1, 4) of the TVCN shown in Fig. 4 is 0.8019, while in the opposite direction it is zero, i.e., no two-way communication path exists for data transfer during the same time window. We can infer from the above discussion that though there may be no direct path between a specified (*S-D*) node pair, yet we may achieve an acceptable figure of *2-TR*. Further, in TVCNs *2-TR* value may differ for data transmission between source to destination and from destination to source due to different TS-MPS in both directions.

5.4.1 Other Useful Performance Metrics

The TS-MPS listed in row 1 of Table 3 can be utilized to find the *foremost*, *shortest*, and the *fastest* TS-MPS so as to compute the *earliest arrival date*, the *minimum number of hops*, and the *minimum delay* (time span), respectively. For the sake of completeness and better illustration, we have highlighted the *foremost* and *fastest* TS-MPS in row 1 of Table 3 using bold font and an encircling box, respectively. Among others, the two *shortest* TS-MPS are $\{l_1^{(1)}l_8^{(3)}\}$ and $\{l_1^{(2)}l_8^{(3)}\}$. The algorithms to compute the *foremost*, *shortest*, and the *fastest* TS-MPS have been presented in [8]. Further, one can also compute the *2*-TR value associated with all *fastest*, *shortest* and *foremost* TS-MPS as 0.6561, 0.8910 and 0.9800, respectively, under the assumption of success probability of each link as 0.90.

The enumerated TS-MPS between a (*S*-*D*) node pair can also be used to compute the well-known metric like Expected Hop Count (*EHC*) (to find the *average number of hops*) and new metric like Expected Slot Count (*ESC*) (to find the *average number of time slots*) required to transmit a data packet to the destination node. More specifically, *ESC* can help a designer to assess the *expected delay* in the TVCNs [8]. Using the formulas provided in [8], the *EHC* between node pair (1, 5) of a TVCN shown in Fig. 1 is 2.1130 hops while its *ESC* is 3.0163 slots.

The aforementioned metrics can be very helpful in identifying all possible opportunities of data transmission among different pair of mobile nodes within a given observational period in a variety of TVCNs like DTNs and VANETs. These metrics can also assist network planners to optimize effective routing under different objective functions.

6 Conclusions and Directions for Future Research

In this chapter, we have covered approaches to model and analyse present days' TVCNs from the reliability and performance perspectives. Besides, the chapter also discussed several recent optimization problems in TVCNs. The chapter collated the brief information about different models for representing the TVCNs' features like mobility, links and topology, and brought out some major differences between static networks and TVCNs. This chapter also defined and described the notion of TS-MPS and TS-MCS, respectively, in TVCNs and presented the techniques with suitable illustrations to enumerate them. Further, the chapter explained the use of TS-MPS and TS-MCS in evaluating *2-TR* and various other performance metrics in vogue, viz., *EHC* and *ESC*. We believe that the techniques presented in this chapter for TS-MPS/TS-MCS enumeration and TVCN reliability analysis can be applied to real-world applications like delay tolerant satellite constellations or any vehicular communication network.

In future, the work in the domain of TVCNs can be extended to find bounds on reliability, as its exact evaluation is costly and time consuming. The concept of TS-MPS and TS-MCS can also be extended to study the capacity related reliability (CRR) evaluation in TVCNs where one would need to devise a suitable algorithm to enumerate the set of such TS-MPS/TS-MCS that satisfies a given network flow requirements. This is a challenging task as link capacity is a function of cost and is limited by many other constraints. We are optimistic that various other distinctive directions of study will appear in due course of time in this domain.

Acknowledgements The authors would like to thank the editor and anonymous reviewers for their insightful comments.

References

1. Conti M, Giordano S (2014) Mobile ad hoc networking: milestones, challenges, and new research directions. IEEE Commun Mag 52(1):85–96
2. Khanna G, Chaturvedi SK, Soh S (2019) Reliability evaluation of mobile ad hoc networks by considering link expiration time and border time. Int J Syst Assur Eng Manage
3. Liang Q (2015) Survivability of time-varying networks. Massachusetts Institute of Technology
4. Casteigts A, Flocchini P, Quattrociocchi W, Santoro N (2011) Time-Varying graphs and dynamic networks. In: Frey H, Li X, Ruehrup S (eds) Ad-hoc, mobile, and wireless networks. Springer, Berlin, pp 346–359
5. Fraire JA, Madoery PG, Finochietto JM, Leguizamón G (2017) An evolutionary approach towards contact plan design for disruption-tolerant satellite networks. Appl Soft Comput 52(Supplement C):446–456
6. Juang P, Oki H, Wang Y, Martonosi M, Peh LS, Rubenstein D (2002) Energy-efficient computing for wildlife tracking: design tradeoffs and early experiences with ZebraNet. ACM SIGARCH Comput Archit News 30(5):96
7. Pentland A, Fletcher R, Hasson A (2004) DakNet: rethinking connectivity in developing nations. Computer 37(1):78–83
8. Khanna G, Chaturvedi SK, Soh S (2019) On computing the reliability of opportunistic multihop networks with Mobile relays. Qual Reliab Eng Int 35(4):870–888
9. Holme P, Saramäki J (eds) (2013) Temporal networks. Springer, Berlin
10. Costa L da F et al (2011) Analyzing and modeling real-world phenomena with complex networks: a survey of applications. Adv Phys 60(3):329–412
11. Batabyal S, Bhaumik P (2015) Mobility models, traces and impact of mobility on opportunistic routing algorithms: a survey. IEEE Commun Surv Tutor 17(3):1679–1707
12. Casteigts A, Flocchini P, Quattrociocchi W, Santoro N (2012) Time-varying graphs and dynamic networks. Int J Parallel Emergent Distrib Syst 27(5):387–408
13. Holme P, Saramäki J (2012) Temporal networks. Phys Rep 519(3):97–125
14. Qirtas MM, Faheem Y, Rehmani MH (2017) Throwboxes in delay tolerant networks: a survey of placement strategies, buffering capacity, and mobility models. J Netw Comput Appl 91(Supplement C):89–103
15. Ferreira A (2003) Building a reference combinatorial model for dynamic networks: initial results in evolving graphs
16. Ferreira A (2002) On models and algorithms for dynamic communication networks: the case for evolving graphs. In: In Proceedings of ALGOTEL
17. Khanna G, Chaturvedi SK (2018) A comprehensive survey on multi-hop wireless networks: milestones, changing trends and concomitant challenges. Wirel Pers Commun 101(2):677–722
18. Roy RR (2011) Handbook of mobile ad hoc networks for mobility models. Springer, US, Boston, MA
19. Bai F, Helmy A (2004) A survey of mobility models. Wirel Adhoc Netw Univ South Calif USA 206
20. Aschenbruck N, Munjal A, Camp T (2011) Trace-based mobility modeling for multi-hop wireless networks. Comput Commun 34(6):704–714
21. Munjal A, Camp T, Aschenbruck N (2012) Changing trends in modeling mobility. J Electr Comput Eng 2012:1–16
22. Baudic G, Perennou T, Lochin E (2016) Following the right path: using traces for the study of DTNs. Comput Commun 88:25–33
23. Newman M, Barabasi A-L, Watts DJ (2006) The structure and dynamics of networks. Princeton University Press
24. Hekmat R (2006) Ad-Hoc networks: fundamental properties and network topologies. Springer Science & Business Media
25. Penrose M (2003) Random geometric graphs, vol 5. Oxford University Press, Oxford
26. Díaz J, Mitsche D, Santi P (2011) Theoretical aspects of graph models for MANETs. In: Theoretical aspects of distributed computing in sensor networks Springer, pp 161–190

27. Peiravi A, Kheibari HT (2008) Fast estimation of network reliability using modified Manhattan distance in mobile wireless networks. J Appl Sci 8(23):4303–4311
28. Padmavathy N, Chaturvedi SK (2013) Evaluation of mobile ad hoc network reliability using propagation-based link reliability model. Reliab Eng Syst Saf 115:1–9
29. Caldarelli G (2007) Scale-free networks: complex webs in nature and technology. Oxford University Press, Oxford
30. Silva AP, Hilario MR, Hirata CM, Obraczka K (2015) A percolation-based approach to model DTN congestion control, pp 100–108
31. Chen W (2014) Explosive percolation in random networks. Springer, Berlin
32. Amor SB, Bui M, Lavallée I (2010) Optimizing mobile networks connectivity and routing using percolation theory and epidemic algorithms. In: IICS, pp 63–78
33. Shen C-C, Huang Z, Jaikaeo C (2006) Directional broadcast for mobile ad hoc networks with percolation theory. Mob Comput IEEE Trans 5(4):317–332
34. Li D, Zhang Q, Zio E, Havlin S, Kang R (2015) Network reliability analysis based on percolation theory. Reliab Eng Syst Saf 142:556–562
35. Coll-Perales B, Gozalvez J, Lazaro O, Sepulcre M (2015) Opportunistic multihopping for energy efficiency: opportunistic multihop cellular networking for energy-efficient provision of mobile delay-tolerant services. IEEE Veh Technol Mag 10(2):93–101
36. Khanna G, Chaturvedi SK, Soh S (2019) Two-Terminal reliability analysis for time-evolving and predictable delay-tolerant networks. Recent Adv Electr Electron Eng 12:1
37. Liang Q, Modiano E (2017) Survivability in time-varying networks. IEEE Trans Mob Comput 16(9):2668–2681
38. Holme P (2015) Modern temporal network theory: a colloquium. Eur Phys J B 88(9):1–30
39. Chaturvedi SK, Khanna G, Soh S (2018) Reliability evaluation of time evolving Delay Tolerant Networks based on Sum-of-Disjoint products. Reliab Eng Syst Saf 171:136–151
40. Gottumukkala RN, Venna SR, Raghavan V (2015) Visual analytics of time evolving large-scale graphs. IEEE Intell Inf Bull 16(1):10–16
41. Fraire JA, Madoery PG, Charif A, Finochietto JM (2018) On route table computation strategies in delay-tolerant satellite networks. Ad Hoc Netw 80:31–40
42. Huang S, Cheng J, Wu H (2014) Temporal graph traversals: definitions, algorithms, and applications. ArXiv Prepr. ArXiv14011919@@
43. Misra KB (1992) Reliability analysis and prediction: a methodology oriented treatment. Elsevier, Amsterdam, New York
44. Chaturvedi SK (2016) Network reliability: measures and evaluation. John, Hoboken, New Jersey , Salem, Massachusetts, Scrivener Publishing
45. Chaturvedi SK, Misra KB (2002) An efficient multi-variable inversion algorithm for reliability evaluation of complex systems using path sets. Int J Reliab Qual Saf Eng 09(03):237–259
46. Mishra R, Chaturvedi SK (2009) A cutsets-based unified framework to evaluate network reliability measures. IEEE Trans Reliab 58(4):658–666
47. Ahmad SH (1988) Simple enumeration of minimal cutsets of acyclic directed graph. IEEE Trans Reliab 37(5):484–487
48. Soh S, Lau W, Rai S, Brooks RR (2007) On computing reliability and expected hop count of wireless communication networks. Int J Perform Eng 3(2):267–279
49. Billinton R, Allan RN (1992) Reliability evaluation of engineering systems concepts and techniques. Springer, Boston, MA, USA, Springer, Imprint
50. Clark J, Holton DA (2005) A first look at graph theory. World Scientific, Singapore
51. Soh S, Rai S (2005) An efficient cutset approach for evaluating communication-network reliability with heterogeneous link-capacities. IEEE Trans Reliab 54(1):133–144
52. Cook JL, Ramirez-Marquez JE (2008) Mobility and reliability modeling for a mobile ad hoc network. IIE Trans 41(1):23–31
53. Chaturvedi SK, Padmavathy N (2013) The influence of scenario metrics on network reliability of mobile ad hoc network. Int J Perform Eng 9(1)
54. Padmavathy N, Chaturvedi SK (2015) Reliability evaluation of capacitated mobile ad hoc network using log-normal shadowing propagation model. Int J Reliab Saf 9(1):70–89

55. Migov DA, Shakhov V (2014) Reliability of ad hoc networks with imperfect nodes. Presented at the International workshop on multiple access communications, pp 49–58
56. Rebaiaia M-L, Ait-Kadi D (2015) Reliability evaluation of imperfect K-terminal stochastic networks using polygon-to chain and series-parallel reductions. In: Proceedings of the 11th ACM symposium on QoS and security for wireless and mobile networks, pp 115–122
57. Ahmad M, Mishra DK (2012) A reliability calculations model for large-scale MANETs. Int J Comput Appl 59(9)
58. Singh MM, Baruah M, Mandal JK (2014) Reliability computation of mobile Ad-Hoc network using logistic regression. In: 2014 eleventh international conference on wireless and optical communications networks (WOCN), pp 1–5
59. Kharbash S, Wang W (2007) Computing two-terminal reliability in mobile ad hoc networks. In: Wireless communications and networking conference, 2007. WCNC 2007. IEEE, pp 2831–2836
60. Rai S, Kumar A, Prasad EV (1986) Computing terminal reliability of computer network. Reliab Eng 16(2):109–119
61. Egeland G, Engelstad P (2009) The availability and reliability of wireless multi-hop networks with stochastic link failures. IEEE J Sel Areas Commun 27(7):1132–1146
62. Panda DK, Dash RK (2017) Reliability evaluation and analysis of mobile Ad Hoc networks. Int J Electr Comput Eng IJECE 7(1)
63. Meena KS, Vasanthi T (2016) Reliability analysis of mobile Ad Hoc networks using universal generating function: reliability analysis of MANET using UGF. Qual Reliab Eng Int 32(1):111–122
64. Meena KS, Vasanthi T (2016) Reliability design for a MANET with cluster-head gateway routing protocol. Commun Stat. Theory Methods 45(13):3904–3918
65. Eiza MH, Ni Q (2013) An evolving graph-based reliable routing scheme for VANETs. IEEE Trans Veh Technol 62(4):1493–1504
66. Misra KB (1975) On optimal reliability design: a review. IFAC Proc 8(1):27–36
67. Misra K (1991) Multicriteria redundancy optimization using an efficient search procedure. Int J Syst Sci 22(11):2171–2183
68. Misra K (1991) Search procedure to solve integer programming problems arising in reliability design of a system. Int J Syst Sci 22(11):2153–2169
69. Li D, Sun X, McKinnon K (2005) An exact solution method for reliability optimization in complex systems. Ann Oper Res 133(1–4):129–148
70. Bertsekas DP (1998) Network optimization: continuous and discrete methods. Athena Scientific, Belmont, Mass
71. Misra KB (1991) An algorithm to solve integer programming problems: an efficient tool for reliability design. Microelectron Reliab 31(2–3):285–294
72. Benyamina D, Hafid A, Gendreau M (2008) Wireless mesh network planning: a multi-objective optimization approach. In: 2008 5th international conference on broadband communications, networks and systems, London, United Kingdom, pp 602–609
73. Amaldi E, Capone A, Cesana M, Filippini I, Malucelli F (2008) Optimization models and methods for planning wireless mesh networks. Comput Netw 52(11):2159–2171
74. Benyamina D, Hafid A, Gendreau M, Maureira JC (2011) On the design of reliable wireless mesh network infrastructure with QoS constraints. Comput Netw 55(8):1631–1647
75. Coll-Perales B, Gozalvez J, Friderikos V (2016) Energy-efficient opportunistic forwarding in multi-hop cellular networks using device-to-device communications. Trans Emerg Telecommun Technol 27(2):249–265
76. Kolios P, Friderikos V, Papadaki K (2014) Energy-efficient relaying via store-carry and forward within the cell. IEEE Trans Mob Comput 13(1):202–215
77. Wang Y, Li H, Li T (2016) Participant selection for data collection through device-to-device communications in mobile sensing. Pers Ubiquit Comput
78. Fraire J, Finochietto JM (2015) Routing-aware fair contact plan design for predictable delay tolerant networks. Ad Hoc Netw 25:303–313

79. Fraire JA, Finochietto JM (2015) Design challenges in contact plans for disruption-tolerant satellite networks. IEEE Commun Mag 53(5):163–169
80. Fraire JA, Madoery PG, Finochietto JM (2014) On the design and analysis of fair contact plans in predictable delay-tolerant networks. IEEE Sens J 14(11):3874–3882
81. Fraire JA, Madoery PG, Finochietto JM (2016) Traffic-aware contact plan design for disruption-tolerant space sensor networks. Ad Hoc Netw 47:41–52
82. Zhou D, Sheng M, Wang X, Xu C, Liu R, Li J (2017) Mission aware contact plan design in resource-limited small satellite networks. IEEE Trans Commun 65(6):2451–2466
83. Li F, Chen S, Huang M, Yin Z, Zhang C, Wang Y (2015) Reliable topology design in time-evolving delay-tolerant networks with unreliable links. IEEE Trans Mob Comput 14(6):1301–1314
84. Chen H, Shi K (2015) Topology control for predictable delay-tolerant networks based on probability. Ad Hoc Netw 24:147–159
85. Li F, Yin Z, Tang S, Cheng Y, Wang Y (2016) Optimization problems in throwbox-assisted delay tolerant networks: which throwboxes to activate? how many active ones i need? IEEE Trans Comput 65(5):1663–1670

Methods for Prognosis and Optimization of Energy Plants Efficiency in Starting Step of Life Cycle

Z. N. Milovanović, Lj. R. Papić, V. Z. Janičić Milovanović, S. Z. Milovanović, S. R. Dumonjić-Milovanović and D. Lj. Branković

Abstract The probability that an energy system will successfully enter into operation and perform the required function of the criteria within the allowed tolerances for a given period of time and given environmental conditions (working temperature, pressure, humidity, permissible vibrations, noise and shock, changes in operating parameters of labor, etc.) represents its effectiveness. The effectiveness indicator is characterized by a unit (unit parameter) or several effectiveness properties (complex parameter), such as: *reliability* (the ability of system to maintain continuous working ability within the limits of allowed deviations during the calendar period of time, quantified through indicators: probability of operation without cancellation, medium time *in work,* intensity of cancellation and cancellation rate), *maintenance convenience* (ability to prevent and detect cancellation and damage, to restore working ability and correctness through technical service and technical repairs, quantified

Z. N. Milovanović (✉)
Department of Hydro and Thermal Engineering, Faculty of Mechanical Engineering Banja Luka, University of Banja Luka, Stepe Stepanovića71, Banja Luka, Bosnia and Herzegovina
e-mail: zdravko.milovanovic@mf.unibl.org

Lj. R. Papić
DQM Research Center, Poštanski fah 132, Prijevor, Čačak, Serbia
e-mail: dqmcenter@mts.rs

V. Z. Janičić Milovanović
Routing Ltd, Prvog Krajiškog Korpusa 16, Banja Luka, Bosnia and Herzegovina
e-mail: valentina.mil@live.com

S. Z. Milovanović
Department of Construction Project Organisation, Technology and Management, Faculty of Architecture, Civil Engineering and Geodesy, University of Banja Luka, Stepe Stepanovića 77/3, Banja Luka, Bosnia and Herzegovina
e-mail: snjezana.milovanovic@aggf.unibl.org

S. R. Dumonjić-Milovanović
Partner Engineering Ltd, Kralja Nikole 25, Banja Luka, Bosnia and Herzegovina
e-mail: svetlanadm@ymail.com

D. Lj. Branković
SHP Celex, Veljka Mlađenovića bb, P.O. Box 142, Banja Luka, Bosnia and Herzegovina
e-mail: dejan.brankovic@shpgroup.eu

M. Ram and H. Pham (eds.), *Advances in Reliability Analysis and its Applications*, Springer Series in Reliability Engineering,
https://doi.org/10.1007/978-3-030-31375-3_2

through: probability of renewal for the given calendar period of time, medium recovery time and the intensity of renewal), *durability* (the system's ability to maintain its working ability from the very beginning of its application or exploitation until the transition to limit states in which certain stop is possible in the realization of certain activities for technical service maintenance and repairs, defined through indicators: medium resource, gamma-percentage resource, medium expiration date, gamma-percentage life), *stability* (the system's ability to continuously maintain hot reserves, storage and/or transport). Optimal management of complex technical systems must be based on the assessment and complex optimization of the reliability indicators, depending on how they are provided and the hierarchical level of detailing, as well as the current phase of the life cycle. For these reasons, the optimization process includes the basic structural, parametric and constructive solutions related to the technical system itself through the change of its most important characteristics: efficiency (mostly energy), maneuverability, reliability and economic effectiveness in general. The set of goals of optimization is concluded in the overall choice of reliability indicators and possible ways to secure them, and given the already established rules regarding the higher hierarchical level of the system. Creating effectiveness is closely related to the concrete energy system, its behavior at a certain time and the environment in which it works (the power system as a higher hierarchical system). For this realization, it is necessary to provide the appropriate backgrounds, first of all the database, then to define based on the function of the goal and the available database of the interconnection and the relationship between the individual elements of the system. Within this chapter, appropriate methods will be provided for prognosis and optimizing the effectiveness based on the quality of design, production and testing, assembly and trial release, exploitation, development of procedures for prognosis of the complex systems behavior based on the characteristics of certain constituent elements of the system and the possible impact of human factors and environment itself on the system. Starting from the initial stage of development, design and conquering the production of certain types of thermal energy equipment in the goal of fulfilling all requirements without limitations, if it is based on its purpose, a multivariate assembly is set up before the designer, with the need to optimize according to certain already adopted algorithms. The goal is to create such a facility that has a satisfactory structure in terms of reliability indicators, with minimal costs of maintenance during the projected working life. Defining the plant that best meets the set requirements related to reliability and the process of exploitation and maintenance itself must be the result of the implemented optimization process (in this case, on the basis of the selected minimum investment criterion). The sum consumption of all devices that ensure the normal operation of the thermal power plant is called own consumption. General consumption consists of all other devices that do not have a direct impact on the technological process at the plant. Preserving the continuity in supplying of own power consumption is essential for safe operation under normal operating conditions, in the case of short-term transitions, as well as in starting and normal stopping, it is particularly important in case of stopping in case of disorder or failure of the system. With the growing unit block strength, there is also the growth of unit power of electric motors with their own consumption, and therefore requirements related to the power supply. The basic problem is to achieve safe power supply in a variety of drive operation situations, with fewer short-circuit currents and the voltage drop when starting large synchronous motors. The solution is achieved by the proper choice of the transformer of own consumption and voltage level.

Keywords Life cycle · Energy system · Effectiveness · Effectiveness indicators · Optimal management · Optimization methods · Estimates prognosis

1 Introduction

A technical system means a machine or a plant, i.e. an integrated set of parts that represent a whole with a unique working function. Thus, the whole means the mutual conditionality and connection of the constituent elements as a whole, and on the basis of the set function of the goal and the relation between the individual elements and their characteristics. The system is considered incapable of operation and exploitation if the state of the system is such that the value of any of the given parameters that characterize the ability to perform the corresponding function of the target does not correspond to the values defined by the normative-technical documentation, [1].

Beside estimating, data for determining reliability can be obtained forcibly by budget and verification or by natural means (non-forcibly), through user experience, own production and other experiences and through the data of appropriate service organizations engaged in maintenance work. If the observed object is complex (for example, the system of a thermal power plant), then the problem of determining reliability is solved if the reliability of the components or at least their "most critical" parts, their interconnection (structure) and operating conditions (constraints and environmental conditions) are known, [2].

It should be noted that the reliability verification or hypothesis testing in practice is carried out in all stages of development, design, construction and exploitation of the facility, and is mainly related to several basic limiting factors—money and time, i.e., environmental conditions and other technical constraints. The reliability verification itself is followed by the corresponding mathematical apparatus, with a certain level of confidence in the tested parameters. The unsuitable level of reliability during the exploitation of the complex technical system, the existence of irrational investments on the basis of work by removing the consequences rather than the causes clearly indicate the necessity of aligning the existing methods for achieving optimal reliability and their adjustment to the system, with the prior definition and elaboration of the corresponding algorithm.

The basic goal of the system users is to keep the system as long as possible in the state of working ability. In order to achieve this, it is desirable for the system to "get help" through the execution of certain maintenance tasks. Important decisions about the responsibilities, obligations, content and timing of the implementation of certain maintenance tasks define the methodology or maintenance philosophy.

From the aspect of philosophy or methodological approach to maintenance, there are two schools that have attracted the most attention lately: Reliability Centered Maintenance (RCM) and Total Productive Maintenance (TPM), [3]. Maintenance methodology according to reliability also includes failure analysis in the decision-making process on maintenance.

In most of the various industries such as nuclear power plants, thermal power plants and petrochemical industries, now companies are required to assess risk probability, including all significant risks across the entire surface. These systems will have thousands of protection devices, such as fire alarms, gas detectors, critical switches, pressure relief valves, overload protection, protective switches and related equipment designed to prevent malfunctions on main devices. Most of these devices do not fail in safe conditions, and regular checks are required to verify that the overall protection system remains *in operation*. As these systems become extremely complex with increasing the number of protective devices and alarm systems, it is very difficult to estimate the risk of increasing the probability of failure, [4].

2 Review of Previous Research

In 1983, Kirkpatrick, Gellat and Vecchi introduced one of the simplest and most comprehensive techniques of heuristic optimization, which proved to be one of the most efficient—simulated annealing method—Simulated Annealing, [5]. This algorithm imitates the crystallization process during cooling or annealing. When the material is warm, the particles have high kinetic energy and move more or less accidentally in relation to their or the positions of the other particles. By cooling the material, more particles are directed towards the directions that minimize the energy balance.

The simulation algorithm of annealing works the same when it searches for optimal values of variable optimization: It repeatedly suggests random modifications of the current solution, but progressively retains only those that improve the current solutions. Dueck and Scheuer [6] proposed a deterministic rule of acceptability, making the algorithm even simpler: Accept any random modification until the resulting deterioration exceeds a certain threshold; this threshold is reduced through iterations. This algorithm is known as the acceptance threshold—Threshold Accepting, [6].

One of the first algorithms related to optimization problems is the evolutionary strategy of author Rechenberg [7]. Here, the population of the P vectors of the initial solutions is generated. In each of the following iterative steps, each individual is treated as a parent who produces one offspring by adding random modifications to the parent solution. From now on, duplicated populations, only the P best solutions are selected that will constitute the parent population in the next generation.

Methods based on evolution have been substantially confirmed by the advantages of genetic algorithms. Holland attributed the probability of reproduction to individual chromosomes, which show their relatively good status within the population, [8]. In terms of the "survival of the strongest" principle, high capacity increases chances (multiple) reproduction, and low capacity will ultimately lead to extinction.

Ants system and optimization by the Ant Colony method are related to the research of Doriga et al. [9], who transformed an example of the behavior of ants "searching" for food with an appropriate colony (two or more ants) to a heuristic optimization called ants colony, so that the population of artificial ants searches a graph, where nodes correspond to locations, and arches represent the amount of pheromones, i.e. the attraction of choosing the path that links these locations, [9].

Moscato [10] proposed a method that combines the advantages of both concepts using a population of variables that individually perform local searches similar to annealing simulation method. In addition to searching the environment independently of the previous solutions, these methods compete and co-operate: Competitions are held in tournaments where one solution is challenged by another and if it wins, it imposes a solution; cooperation can be achieved by combining solutions with crossing operations as known, e.g. from a genetic algorithm, [10].

In optimization tasks, there is often a situation where the quality of the solution cannot be expressed by only one criterion. Different criteria are usually uncoordinated, i.e. improving one criterion causes deterioration of at least one of the other criteria. Such a situation is formulated as a multi-criteria optimization, also called multi-functional, multi-aiming or vector optimization. The task of multi-criteria optimization is to determine a tolerable solution that is "best" not only for one, but for several functions of objectives. The first formulation of the problem of multi-criteria optimization was given by Pareto [11]. A significant role in multi-criteria optimization has a "decision maker". The method most commonly used in the initial analysis of the selection of the most favorable locations and the assessment of the cost-effectiveness of the investment projects in the energy industry (the previous study and the study of economic justification) is the analytical hierarchical method (AHP—Analytic Hierarchy Process, Saaty), [12]. The main disadvantage in applying this is the use of scale from 1 to 9 [12], which is not suitable for decision-making in situations of uncertainty and other estimates that cannot be expressed by numbers. On the other hand, the defined criteria are often subjective and qualitative in nature, which negatively affects the decision-maker in terms of and expressing their own preference with numerical values and subsequent comparison of estimates [13]. The Fuzzy version of AHP methods (FAHP) is a systematic approach to selecting alternatives and solving problems using the concept of the theory of Fuzzy assemblage, and is used to estimate risk and uncertainty (FAHP, Zadeh, 1965 and AHP method, which is implemented using triangular Fuzzy numbers, Chang), [14]. There are several methods of phasing AHP methods in the literature. Thus, VanLaarhoven and Pedrcyz [15] suggests the first study in which they used three-stranded Fuzzy numbers, and which introduces the principles of Fuzzy logics in AHP method. Buckley [16] initiates that estimation of decision makers can be expressed by using trapezoidal Fuzzy numbers. Boender et al. [17] presented the modification to the phase of the multi-criteria method proposed by Van Laarhoven and Pedrcyz [15]. Chang [14] introduced an expanded analytical method as a new approach to solving FAHP using triangular Fuzzy numbers. Chan and Kumar [13] are developing a Fuzzy-expanded AHP selection model. Feng et al. [18] have proposed a comprehensive method based on the Fuzzy-theory of decision-making for solving the supply chain management problem. Haq and Kannan [19] aimed to show how the phase AHP model helps to solve the problem of selecting suppliers in practice. While, Chamodrakas et al. [20] use the Fuzzy AHP for the selection of suppliers in the electronic sector, while Onut et al. [21] for the selection of suppliers in the telecommunications sector. The application of the AHP/ANP method in the selection of sustainable energy solutions is related to the evaluation of social, economic and environmental protection criteria in the case study [22], evaluation of internal and external economic indicators in the car industry of Thailand [23], identification and analysis of barriers to higher energy efficiency of the plant [24], sustainable development indicators in the steel industry [25], the development of sustainable solutions in construction in South Africa [26], the application of the Fuzzy AHP method for the promotion of clean production and optimal organizational structure in companies in Taiwan [27], the criteria evaluation of environmental protection as well as the technological and economic criteria are used for typical development projects of power plants [28], the use of technical economic and safety assessment criteria with sensitivity analysis in the process industry [29], the selection of the criteria for the application of the Fuzzy Delphi method using the ANP method for determining the weight factors in the semiconductor industry [30] selection of factors for the use of AHP/ANP methods for the treatment of waste in the food industry [31] and others. Using MCDM—Multy-criteria Decision Making applications in the planning and development of energy projects are initially connected to the work of Hobbes and Meirer [32] (applicability of multi criteria methods in energy planning with experimental comparison of results), Huanga et al. [33] (analysis of methods used for energy modeling and environmental modeling), Landhelma et al. [34] (use of MCDM in environmental planning and related management), and later for works for the use of renewable energy resources [35–37]. For the development of the AHP method, the following research should be mentioned [38–42].

Often, when solving the management problem in many technical systems, part of the theory of graphs is used, which analyzes the topic of the realization of extremes on graphs, that is, analyzes and determines the *critical path*, using numerical methods and computers to solve them. The first methods of network planning are related to 1956 and organisations *Du Pont de Nemours and Co* and *the Univac Division* (within the *Remington Rand Corporation*), which dealt with the system of general remounting planning and the maintenance of chemical plants. A few years later, within the POLARIS project of *Booz Allen and Hamilton Company* and *the Missile Systems Division* of the *Lockheed Aircraft Company*, they developed the *Project Evaluation and Review Technique*—PERT, [43, 44]. At the same time in France, the MPM method was developed, which adapts to the application of the computer and gets a new name PRECEDENCE or PDM method, or *"network diagram with previous activities"*, [45, 46].

Creating a performance measurement system is, therefore, a complex task; what can be considered as an optimal performance measurement system, it will vary from case to case [47]. Thus, it is important to understand how performance measurement systems should be developed and integrate into organizational management models. These attitudes are consistent with the basic paradigm of the multi-criteria decision-making process, which resulted in numerous studies of the possibility of applying multi-criteria decision-making in the process of calculating and evaluating organizational performance [48, 49].

3 Theoretical Reviews

Maintenance throughout the life cycle of the technical systems combines a series of accompanying activities, starting with the idea and definition of the concept, the evaluation of their economy, realization, exploitation, and up to the write-off of the system from use. The preparing of the maintenance system, through maintenance-based design, is conditioned by the development of the productive forces of the society and it is aimed at extending the lifespan, while achieving optimal connections between technical and technological and economic characteristics, [1].

The process of maintaining the means of operation as one of the most important parts of the total production process has the task of preventing and eliminating system failures, primarily through the rationalization and optimization of their use, and the increase in productivity and cost-effectiveness in the process of production or exploitation.

The existence of a large number of different optimization methods puts the question of choice to the constructor which is the most suitable for solving a specific problem. It can be said that in the optimization of machine systems there are few tasks to which the methods and software can be directly applied. On the other hand, to develop a new method for each specific problem is impossible and meaningless. This means that for most optimization tasks, existing methods are used with fewer or greater changes and adjustments to the specific task, [2].

When choosing the optimization method, the following rule must be respected: the task set should satisfy the mathematical and formal constraints of the method used, and the method used should provide reliable arrival to the correct solutions for as short a time as possible and with as little effort and cost as possible. The most important factors that affect the efficiency of the optimization method are: the ability to solve all the optimization tasks from the set for which the method is intended, the proximity of the solution obtained to the actual optimum and the size of the engaged means of the used method for its application, how long the optimization process is taking place, how much computer memory is used, etc.

The choice of the optimization method is influenced by a number of factors, such as: time spent and other costs required for program development, calendar time required for program development, cost of execution time to solve the desired optimization problem, expected reliability of the program in finding the desired solution, flexibility of the program (can it be used in various ways to solve a general problem), the generalization of the program, can it be used to solve other problems and the easiness with which the program and its outputs can be used, [3, 4].

3.1 Life Cycle of Energy Power Plants

The life cycle begins when the idea of a new technical system is born and ends when it is withdrawn from use. The main processes that help the system through the life cycle phases are: marketing (specifications), design, production, use, and finally—withdrawal from use. Life cycle analysis is a systematic and analytical approach to identifying the resources needed to support the processes of design, production, use and withdrawal from use, [3, 5]. Therefore, life cycle analysis is a life cycle engineering tool whose main objectives are: the impact on design from life cycle aspect, the identification and quantification of total resources related to life cycle processes, and analytically managed life-cycle activities. In other words, life cycle engineering should enable the decision-making process to achieve the best compromise between the investment and the provision of the necessary resources for design, production, use and withdrawal from use. In addition, this approach enables: an early and continuous impact on the design of the system from the aspect of the life cycle, reducing the cost of the system life cycle by limiting the main cost generators throughout the life cycle, and identifying the resources that accompany all the processes (phases) of the life cycle of the system.

Generally speaking, traditional (sequential) engineering is mainly focused on the performance of the system as the main goal rather than the development of a general integrated approach. The latest knowledge and experiences gained in the last decades indicate that the proper performance of the target function, i.e. the required degree of competitiveness of the system cannot be ensured by investing effort mainly after their production and maturity in the use phase, which is most commonly done. Much more important is that engineers are sensitive to see the consequences of potential mistakes that may arise during early stages of system design and development. This means that engineers should be able to take responsibility for the life cycle engineering (competitive, simultaneous engineering), which is most often neglected earlier, [3].

The unavoidable part of the life cycle is also the revitalization, reconstruction and modernization of technical systems, that is, the process of extending the working life of these facilities with modernization and reconstruction, with further improvement of technical, economic and ecological acceptability. This process is very complex in its structure and is often compared to the ranking of the implementation of the new technical facility.

The very process of planning and implementing the revitalization process and the exploitation of the plant itself within the framework of the system under consideration is realized with the aim of achieving a high level of operational safety, which implies defining and detecting possible sources of unreliability. Measures for removing and mitigating their effects must also be defined, and the criterion used is the most frequently used economic criterion. Such a systematic and comprehensive process at a technical facility or plant is an unavoidable and logical process in the lifetime of the building.

The connection of the reengineering process in maintenance of technical systems with the aim of achieving appropriate advantages and improving the reliability of the system is given through the following characteristic elements: analysis of maintenance costs and readiness/availability of the system (as one of the more important characteristics of effectiveness), determining general aspects related to motives and the justification of revitalization, as well as the scope and definition of the most optimal term for the realization of this process. Particularly worth highlighting is the influence of the characteristics, concerning reliability and availability of the system object, on the application of the principles of reengineering through the process of system maintenance, i.e. the systemic approach to the revitalization of some of its capacities.

Planning, development, construction and exploitation, with the maintenance of facilities and systems in technology, brings with itself a large number of phenomena that can cause damage and endanger the health and life of people directly engaged in the facility as well as in the wider environments, [2]. In short, there is a high degree of risk of occurrence of adverse events and their consequences. In the case of more complex technical systems that have great interdependence between their subsystems and elements, the cancellation of any of these can mean an automatic shutdown of the entire system, or a work with reduced power (or, as a more common case, work on a technical minimum), which may result in increasing the costs of the system itself, thermal and other overloads, as well as greater damage in system failures. For these reasons it is necessary that this complex system is reliable in operation. Security of technical systems can be considered from two aspects. The first and most important aspect is the protection of the operator (human) from injuries during the operation of the system. The second aspect is the protection of the system against damage caused by the action of external causes. The advantage in the study is the safety of the operator. Thus, these two aspects are not unconditionally complementary, and the increase in operator security can be achieved at the expense of system security, [3].

For any technical system, even if the target function is within the limits of the permitted deviation, damage can be caused if it is misused. The main causes (components) of the risk of the operator are: the seizing of parts of the body such as arms in the process of operation of the system, the carelessness during the operation of the rotational parts of the system (especially weakly attached units), contact with sharp and abrasive surfaces, the influence of operator static on mobile objects or vice versa, as well as disposing of waste material (especially in production) in the form of colder, shavings, sparks or melted metal.

Sources of system risk are diverse and numerous, and at the design stage, the consequences of critical types of failures must be minimized by predicting protective devices during the operation of the technical system. Risks of technical systems include: impacts, vibrations, corrosion, environment, fire, and mismanagement (overload or work below the technical minimum level), [4].

Many methods of reliability analysis, such as: the analysis of causes and consequences of failure, analysis of the failure tree, the analysis of the importance in terms of the reliability of the whole of the system, can be successfully applied to determine the security features of the system, such as: primary and secondary events, peak event, the probability of the peak event, the minimum set of cross sections, the degree of criticality of the type of failure and the system integrity.

The causes of unwanted events are stochastic phenomena, because they are dependent on a number of specific but also random factors, whose effects cannot be fully observed. Preventive measures are in some way planned activities of suppression and possible reactions to this group of factors. The ability to operate technical systems without failure in stationary and non-stationary operating modes, economic and technical convenience for overhaul of elements, as well as the system as a whole, limitations that monitor the exploitation of the system (environment or superior system, environmental protection, financial resources, etc.) the ability to use appropriate analogue-based solutions with similar plants, norms for control and diagnostics—these are all features that do not have a detailed budget and experimentally argued database that relates to availability and reliability, [5].

On the other hand, with the increasing complexity of technical systems, the problem of their optimal functionality is emerging as an accompanying problem, especially if it is known that such systems can often cause major economic losses or jeopardize the security of the wider macro-region and the people who serve them.

Research aimed at raising the level of reliability and managing the reliability throughout the lifetime of the object is aimed at defining a system of protection measures and optimizing them from the aspect of simultaneous provision of cost-effectiveness of exploitation and implementation of complex regulations related to environmental protection and security of both micro and macro regions.

Defining the basic characteristics of the reliability of technical installations in theoretical and practical considerations is the starting point for giving a forecast or assessing the correctness of the system as a whole, the rest of its lifetime or the lifetime of its most critical elements, and the definition of preventive corrective measures and assessment of the convenience, i.e., determining the probability that the observed part or system as a whole is brought from the state of the failure to the state of operation in the shortest period of time.

3.2 Effectiveness of the Energy System as a Complex Technical System

Due to the stochastic nature of the problem and the independence of a large number of influencing factors, ensuring the effectiveness of the energy system is an extremely complex task, which requires defined and precisely designed procedures that aim to achieve the ability to get into operation (*Readiness* or *Availability*—$G(t)$), then work within the limits of tolerance (*Reliability*—$R(t)$), as well as certain adjustments to environmental conditions and process and work time disorders (*Functional Flexibility*—FF), [1, 2]. A basic standard for assessing system performance and assessing the effectiveness of determining the technical resource of a system is the *function of*

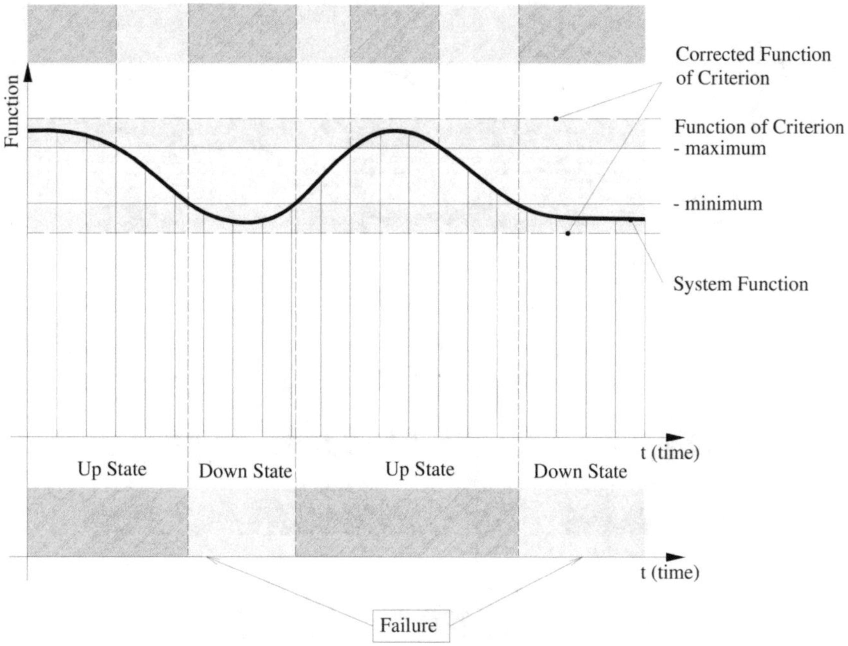

Fig. 1 The function of the criteria, [1]

criteria, which includes parameters of the function (work activity) and acceptability criteria, Fig. 1.

During the course of operation, the technical system achieves intermittent *operating* and *failing* states, which enables the expression of the effectiveness of the system as a probability that the system will successfully perform the required function of the criteria after activation and make certain adjustments to the conditions in the estimated time of operation, Fig. 2.

If the effectiveness of the technical systems is considered through the phases for achieving the planned function of the criterion (Fig. 3), the overall effectiveness of the technical systems is divided into the probability of achieving the criterion function of the technical system, reaching the output demanding quantities into the range of allowed deviations according to the defined function of the criterion, the reliability of holding the function in the area defined by the function of criteria in the projected time period, as well as the ability to adapt the technical system to changes in working and environmental conditions.

The study of the effectiveness of energy systems is based on the investigation of the causes of failure, the definition of the form of their distribution, and the method of predicting the condition in system *failure*. *Availability of the system* presents the probability that the system will start the function within the defined area of the function of the criterion, while the probability of maintaining the required system function in the boundaries of the criterion function within a certain time period is

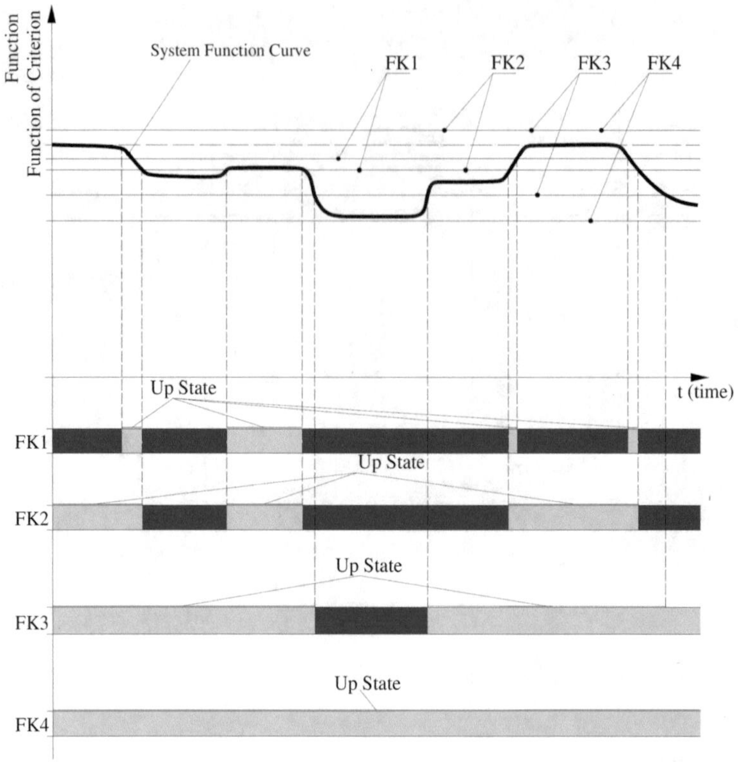

Fig. 2 The influence of the function of the criteria on the working state of the technical system, [1]

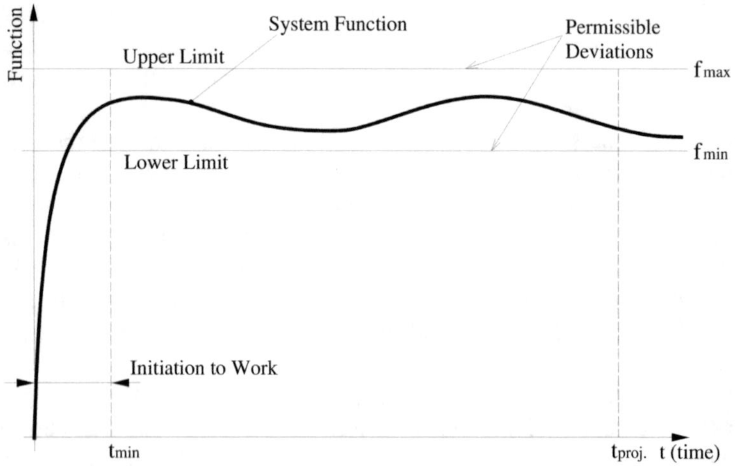

Fig. 3 Function of the technical system, [2]

the *reliability* of the energy systems. *Flexibility* is the ability of technical systems to adapt to changing load, capacity, power and external conditions, [3].

The quality of the system, as the level of realization of the criterion function at a given moment of time, represents a static size and is evaluated after the development of technical systems, while the *effectiveness* of the system, as probability of fulfillment of the function of the criteria for a certain period of time, can be safely evaluated during the period of the system exploitation and represents the dynamic size, [1].

Effectiveness in the process of designing energy technical systems is considered only as a basis for the project request according to the project task. The effectiveness analysis aims at defining possible relationships for presenting the impact of the achieved effectiveness on the costs of development and production of the system, the costs of procurement and commissioning (for some systems of trial drive), as well as the costs of exploitation of the system and its maintenance, or its final disposal after the expiration and an extended period of use of the system, [1].

Costs of the effectiveness of energy technical systems include the costs of obtaining the availability and costs of achieving the required reliability, Fig. 4. In addition, the visible part of the cost of the life cycle of the system consists of procurement costs regardless of whether it develops (research, development, project, testing and production) or buys as a rounded whole and mounts according to the given installation instruction, [2].

On the other hand, the invisible part of the costs of the life cycle of energy systems consists of the costs of distribution and manipulation (transport, handling, manipulation), education costs of repairers and operators, maintenance and repair costs (preventive and corrective maintenance), technical documentation costs (technical assistance, manuals, books, catalogs, prospectuses, scientific and expert papers, etc.),

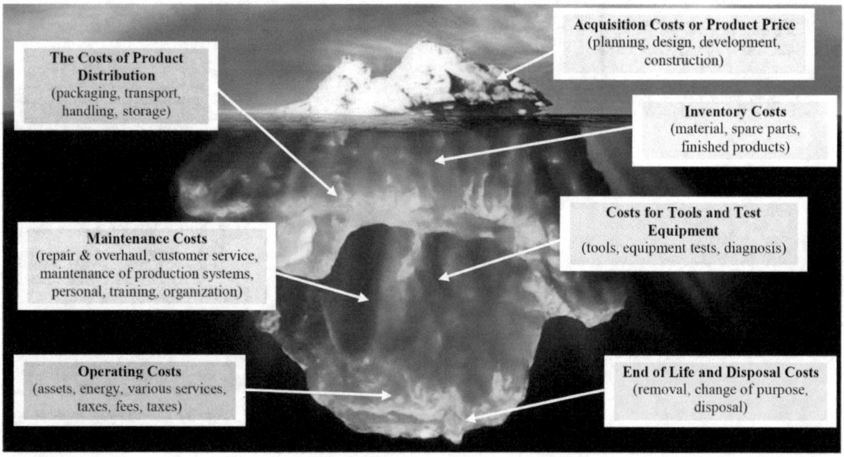

Fig. 4 Costs of the life cycle of technical systems

supply costs (spare parts, supplies and repro material, packaging), costs of revitalization, reconstruction and modernization (extended life of technical systems), as well as expense of disposal (decommissioning, disposal and/or recycling of technical systems), Fig. 5.

Increasing the required level of efficiency of the system decreases the exploitation costs, while the costs of development and production grow. Similar observations apply to the analysis of *maintenance* benefits, [1, 2]. Defining of optimal efficiency of energy technical systems is based on knowledge of the functional dependence

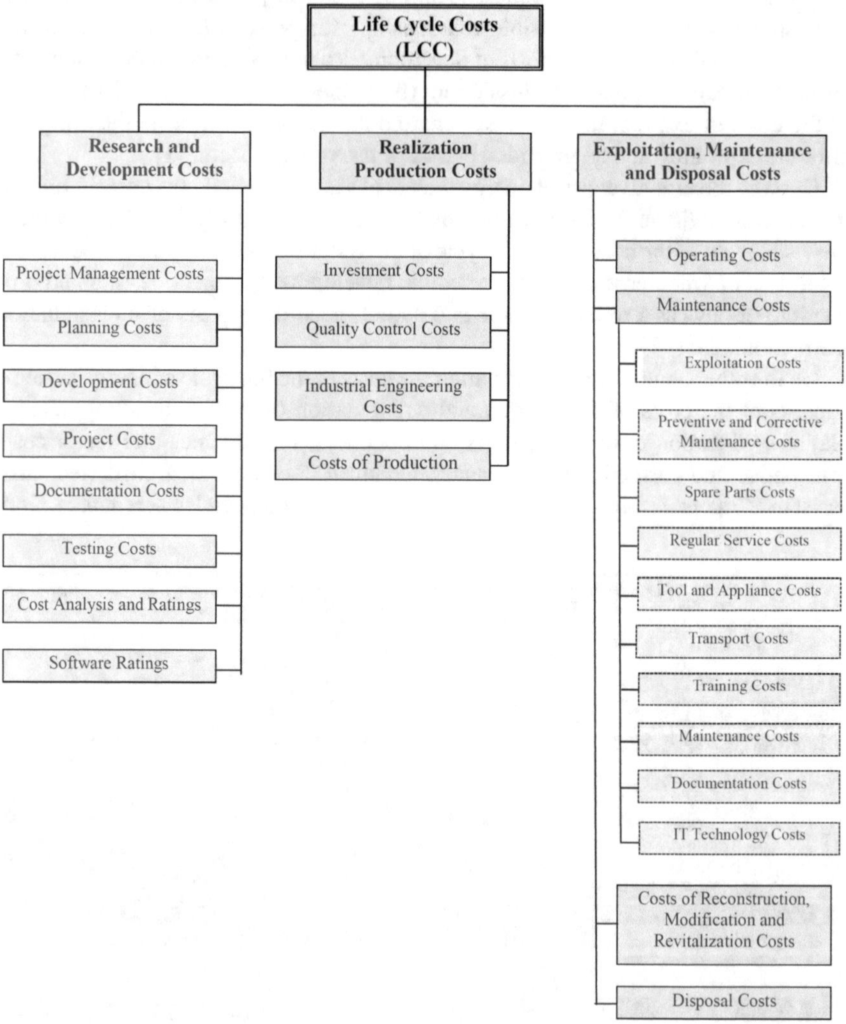

Fig. 5 Life cycle costs of technical systems by activities and phases

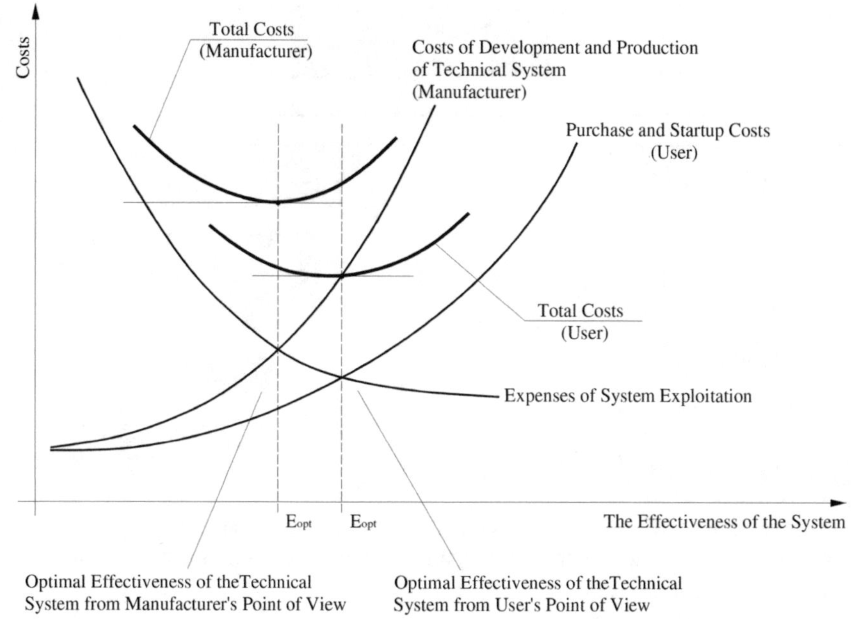

Fig. 6 Diagram of the optimal size of the effectiveness of technical systems, [2]

between the system's effectiveness and the costs of exploitation, the costs of acquisition and commissioning, the development costs and the entire production of the system as a whole. As a consequence of the impact of the market, the optimal efficiency from the user and manufacturer's point of view are different (Fig. 6), whereby the optimal efficiency from the aspect of the user is higher than the optimal efficiency from the aspect of the producer, which is conditioned by the influence of competition and market competition in the liberal market (not monopolistic market position), [1].

The *readiness* or *availability* of energy technical systems is the probability that the system will start the function and have for its result—work within the boundaries defined by the function of the criteria. The achievement of the final requirements is accompanied by a transient occurrence in individual system components (non-stationary operating modes when commissioning and stopping, increased friction due to reduced lubrication, increased loads due to acceleration, increased vibration, etc.).

The classification of time and maintenance for the calculation of readiness or availability is given in Fig. 7.

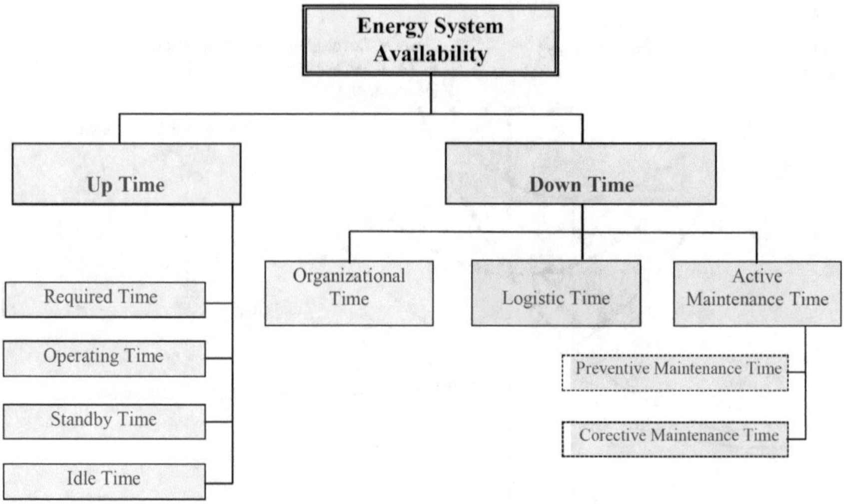

(a) classification of time for the calculation of readiness (availability) for the state of the technical systems correct (in operation) and incorrect (in the cancellation)

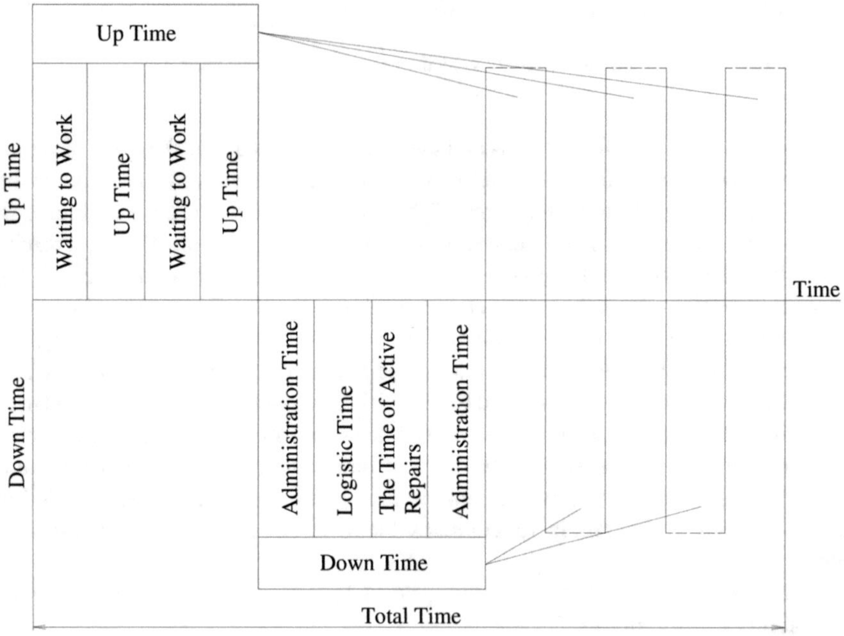

(b) classification of working time and maintenance for the calculation of readiness for the state of technical systems correct (in operation) and incorrect (in the cancellation)

Fig. 7 Diagram of the optimal size of the effectiveness of technical systems

The following dependency is valid, [2]:

$$G(t) = \frac{T_{cor}}{T_{tot}} = \frac{T_{cor}}{T_{cor} + T_{can}} = \frac{\sum_{i=1}^{n} t_{i.cor}}{\sum_{i=1}^{n} t_{i.tes} + \sum_{j=1}^{m} t_{j.can}}, \tag{1}$$

where are: T_{tot}—total calendar time (system life); T_{cor}—total time in proper condition; T_{can}—total time in cancellation; $t_{i.cor}$—i time in the correct state;$t_{j.can}$—j time in cancellation; *n*, *m*—the total number of time segments in the correct state and the time of cancellation in the total time (respectively).

Depending on the time segments of the *correct state*, *in operation* and *in cancellation* are observed, as well as from the depth of the analysis of the time taken, there is a difference between *operational availability, achieved availability*, and *inherent availability*, [2]. It stands:

$$G_{oper}(t) = \frac{MTBM + MRT}{MTBM + MRT + MDT};$$
$$G_{ostv}(t) = \frac{MTBF}{MTBF + \bar{M}};$$
$$G_{unu}(t) = \frac{MTBF}{MTBF + M_{ct}} = \frac{MTBF}{MTBF + MTTR}, \tag{2}$$

where: $MTBM$—Mean Time Between Maintenance; MDT—Mean Down Time; MRT—Mean Redens Time; $\overline{M}$—Medium preventive and corrective time of active maintenance, i.e. Mean Maintenance Active Repair; M_{ct}—Mean Active Corrective Maintenance Time;$MTBF$—Mean Time Between Failures, or between corrective maintenance; $MTTR$—Mean Time To Repair or Medium Corrective Time.

The effectiveness of technical systems based on the time picture of the state of the system is represented by the equation:

$$R(t) + F(t) = 1, \tag{3}$$

whereby the $R(t)$ is marked probability that the system will be during the observed time in state of *in operation* (reliability), and with $F(t)$ the probability that the system will be during the observed time in state of *cancellation* (unreliability), [2].

Differentiating the Eq. (3) by timing and labeling, in the sense of like in Fig. 7a, $dF(t)/dt = f(t)$ and $dR(t)/dt = \rho(t)$ for continuous (ceaseless) changes, i.e. changes in which variables can take any value from the interval (0.0 ÷ 1.0), the Eq. (3) gets the form:

$$\rho(t) + f(t) = 1, \tag{4}$$

where are with $\rho(t)$ or $f(t)$ given the functions of the probability of the occurence state of the system *in failure*, or the function of the system reliability occurence.

In the case of discrete (discontinuous) changes, the functions $\rho(t)$ and $f(t)$ have the form:

$$f(t) = \frac{N_{fail}}{N_0 \cdot \Delta t}; \quad \rho(t) = \frac{N_{oper}}{N_0 \cdot \Delta t}, \tag{5}$$

where: N_{fail}—the number of occurence *in failure* or the number of components that come into the state *in failure* during the observed period of time; N_0—the total number of components that were *in operation* (correct at the beginning of the observed period of time); Δt—the time period in which the reliability of the technical system is analyzed; N_{oper}—the number of components of the technical system that remain correct (state *in operation*) at the end of the observed period of time. Starting from the Eq. (3), the given confidence relation for a certain period of operation and an appropriate unreliability is illustrated in Fig. 7b.

During the operation of the technical system, the reliability of faultless work is continuously reduced, and the cumulative function of unreliability is constantly increasing, Fig. 8.

The cumulative function of the remaining reliability of technical systems can be expressed as:

$$R(t) = 1 - F(t) = 1 - \int_0^{t_1} f(t) \cdot dt, \tag{6}$$

i.e., expressed through a cumulative function of unreliability, the remaining cumulative function of reliability in time $t = t_1$ is equal to the acquired value of the cumulative function of uncertainty in the passed operating time of the technical system $\Delta t = t_1 - t_0$.

At the initial moment of the observed period, the reliability has a value equal to the unit, in order to fall during the time and at the end when the $t \to \infty$ reliability has a value equal to zero, that is:

$$R(t) = \begin{Bmatrix} za\ t = 0 & R(t) = 1 \\ za\ t = \infty & R(t) = 0 \end{Bmatrix}, \tag{7}$$

or for unreliability (Fig. 9):

$$F(t) = \begin{Bmatrix} za\ t = 0 & F(t) = 0 \\ za\ t = \infty & F(t) = 1 \end{Bmatrix}. \tag{8}$$

Starting from the definition of reliability of technical systems, as the probability of maintaining a certain system function in a certain area defined by the boundaries of the criterion function for a certain period of time, applying it on the analysis of technical systems consisting of a large number of components (where it is enough for the proper operation of the system, but it is not the necessary condition for the correct

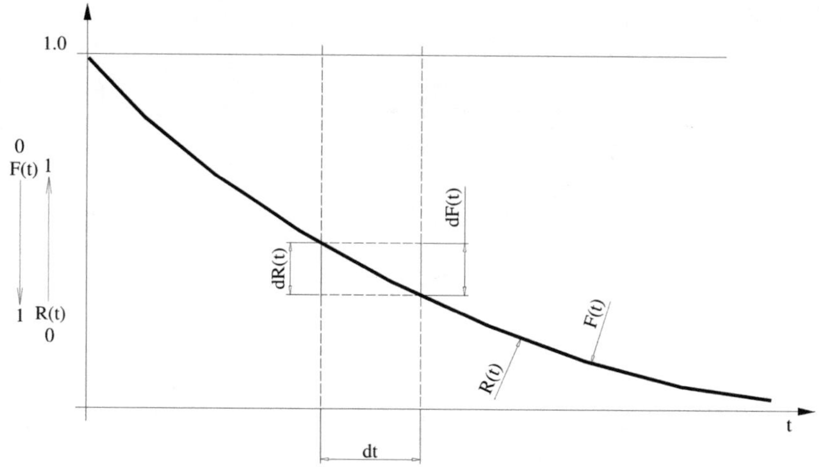

(a) increase in unreliability and decrease in reliability of technical systems as a function of time

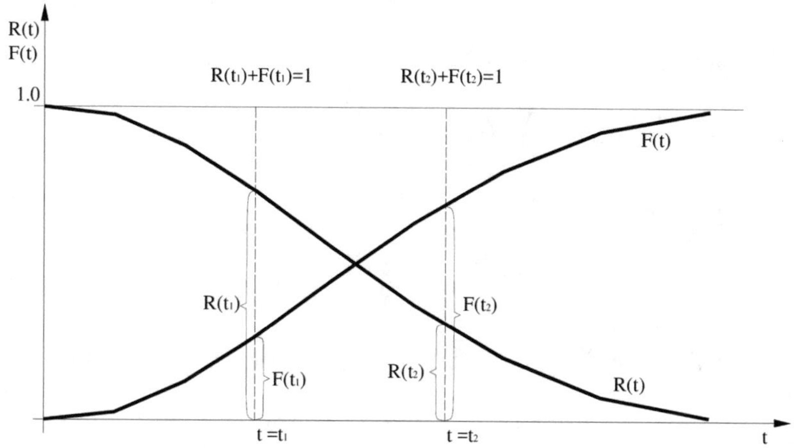

(b) reliability and unreliability as a function of working time of technical systems

Fig. 8 Reliability and unreliability of technical systems, [2]

operation of all components within the limits defined by their criteria functions), in the analysis of complex technical systems the reliability of the system is defined as the probability that a certain number of components N_{isp} of the total number of components of the technical system N_0 will remain valid during a certain period of operation (number of components which were canceled after the end of the observed period is marked with N_{otk}).

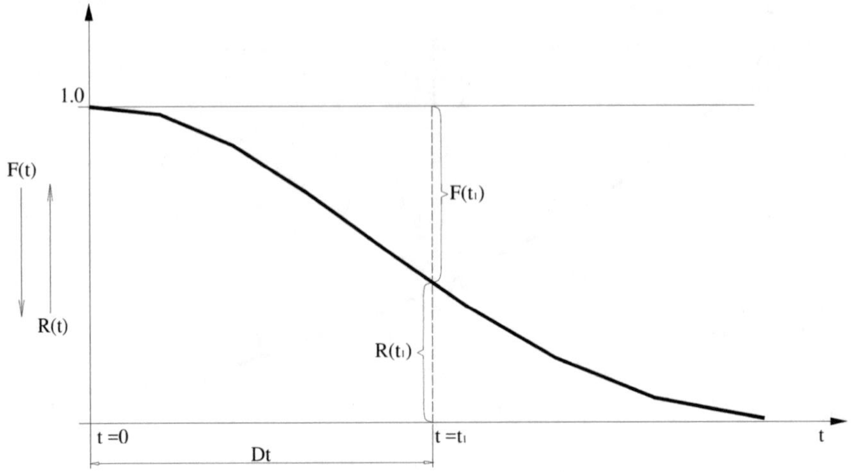

Fig. 9 Display of accumulated unreliability and residual reliability of technical systems, [1]

The following expression is valid:

$$R(t) = \frac{N_{oper}}{N_0} = \frac{N_0 - N_{fail}}{N_0} = 1 - \frac{N_{fail}}{N_0}, \tag{9}$$

i.e., after differentiation by time:

$$\frac{dR(t)}{dt} = -\frac{1}{N_0} \cdot \frac{dN_{fail}}{dt}. \tag{10}$$

If the previous equation is split with N_0/N_{oper}, using the expression (9), we obtain:

$$\frac{1}{R(t)} \cdot \frac{dR(t)}{dt} = -\frac{1}{N_{oper}} \cdot \frac{dN_{fail}}{dt} = -\lambda, \tag{11}$$

where the intensity is marked with $\lambda = \frac{1}{N_{oper}} \cdot \frac{dN_{fail}}{dt}$ (rate, index or rate of occurrence) of the failure, Fig. 10.

Practically, this is the number of failures according to the total population of the technical system in the unit of time.

Similarly is defined the time elapsed between two successive failures in relation to the size of the technical systems, and the same is defined as the reciprocal value of λ, i.e., it is:

$$m = \frac{1}{\lambda}. \tag{12}$$

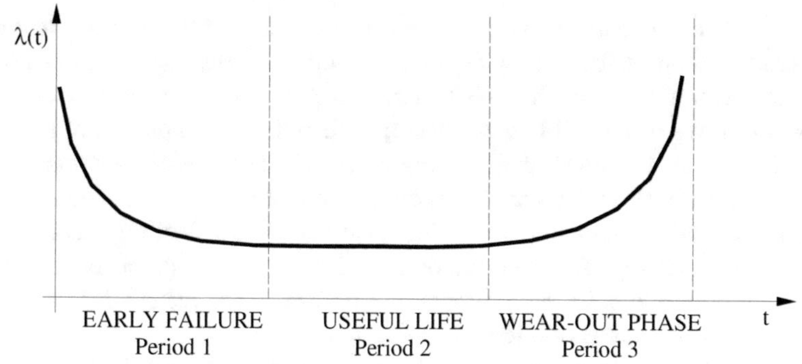

Fig. 10 Characteristic curve of failure intensity of technical systems (so-called shape of "bath"), [50]

If the intensity of the failure is observed as a constant size (the period of random failure), then, by integrating the expression (11) in time, the reliability of the technical systems can be expressed by the exponential function:

$$R(t) = e^{-\lambda \cdot t} = e^{-\frac{t}{m}}. \tag{13}$$

This enables the display of a standardized reliability curve for the mean time between failures, $R = f(m)$, Fig. 11. It is not difficult to show that this relation is also valid:

$$f(t) = \lambda \cdot R(t), \tag{14}$$

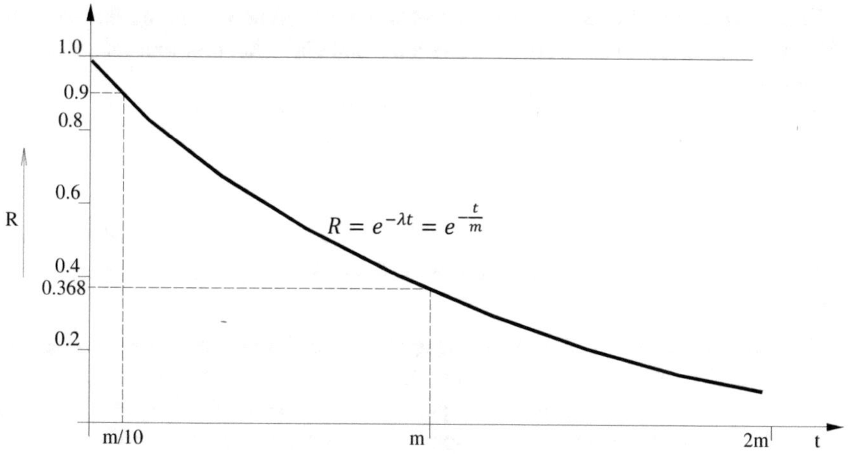

Fig. 11 Reliability curve in case of exponential distribution of random variable, [50]

i.e. the conclusion that at any time the $t = t_i$ index or intensity of the failure is equal to the relation of the density of the failure $f(t_i)$ and reliability $R(t_i)$ at the same time.

Determination of reliability characteristics consists of three interconnected phases: estimates of reliability characteristics, then the determination of the law of distribution, as well as the estimation of parameters and laws of distribution (hypothesis testing). Each of the above phases has several sub-phases, which result in the determination of numerous values of the most common reliability characteristics: mean time of operation T_0, mean standard deviation σ, function of reliability $R(t)$ or function of unreliability $F(t)$, function of occurrence of the condition in failure $f(t)$, as well as functions of intensity of failure $\lambda(t)$, [50].

The very assessment of certain reliability characteristics is practically a matter of making certain conclusions about the behavior of technical systems with unlimited size of the basic set (systems, subsystems, their components and parts) through certain algorithms, based on the behavior of a limited sample size. The final aim of these estimates is to determine which of the known theoretically developed and applicable laws of distribution best corresponds to the experimental data obtained. Applicable distribution laws are most commonly: uniform, normal, Vejbul's, gamma, exponential, hi-squared, beta, Student's, as well as Fischer's distribution. On the basis of previous experience, for technical systems with sudden failure, an exponential distribution has been shown suitable, while the application of normal distribution has proved the most suitable for gradual failures due to aging. Vejbul's distribution has given quite good results in the fields of mechanical and electrical systems (electrical appliances and appliances), [50].

Starting from a certain hypothesis, and based on the algorithm for the selection of the distribution law, the distribution law is defined and the final evaluation of parameters and distribution law is done (mostly often using the Komogorov-Smirnov test).

The reliability of the technical systems for the occurrence of failure due to wear, without changing the worn parts, follows a normal Gaussian distribution according to the equation:

$$f(T) = \frac{1}{\sigma\sqrt{2\pi}} \cdot e^{\frac{(T-M)^2}{2\sigma^2}}, \tag{15}$$

where: T—total accumulated operating time of the technical system; M—average usage lifetime; σ—standard deviation; $\sigma = \frac{\sum(T-M)^2}{N_{fail.1}}$; $N_{fail.1}$—number of cancellations for selected $\sum(T-M)^2$.

The failure occurrence $f(t)$ well illustrates the difference between exponential distribution (for random failure) and normal distribution (for failure due to wear). From Table 1 it can be seen that the exponential distribution for random failures is shown by the failure occurrence function $f(t)$, independent of the working age (T) and dependent on the selected time (t) and is valid until the failure occurs due to wear.

Table 1 Cardiovascular parameters

Distribution	Equation	Distribution wave
Density		
Exponential	$f(t) = \frac{1}{m} \cdot e^{\frac{t}{m}}$; m—the average time between the failure; t—working time	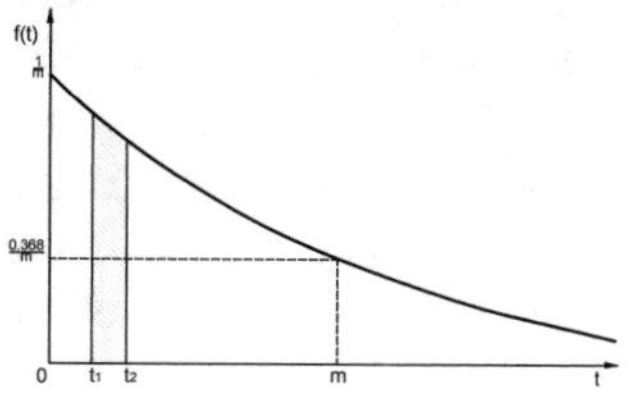
Normal	$f(T) = \frac{1}{\sigma\sqrt{2\pi}} \cdot e^{\frac{(T-M)^2}{2\sigma^2}}$; T—working age (age of the component or system)	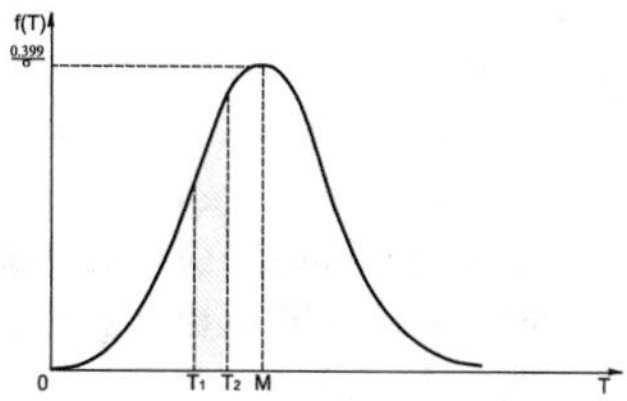
Unreliability		
Exponential	$F(t) = 1 - e^{-\lambda \cdot t}$	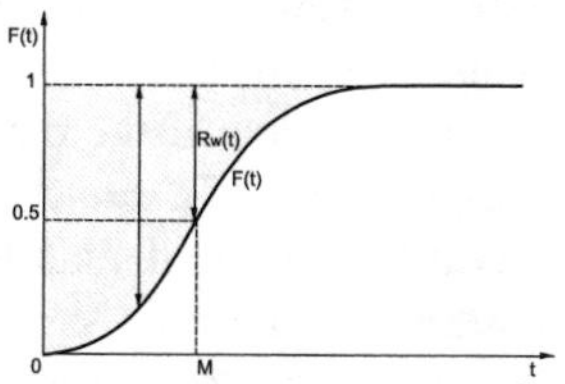
Normal	$F(t) = 1 - \frac{1}{\sigma\sqrt{2\pi}} \int_T^{\infty} e^{\frac{(T-M)^2}{2\sigma^2}} dt$ T—working age (age of the component or system)	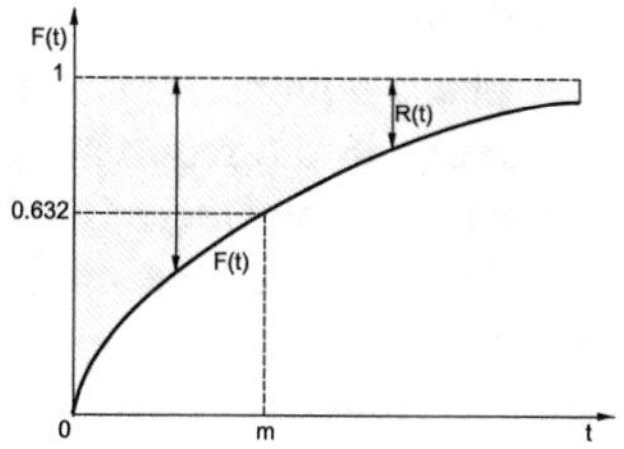

In normal distribution, the failure occurrence of a component sample or a technical system as a whole appears at an average lifetime (M). It should be pointed out that most of the component failure in the exponential distribution occurs up to the average duration between the failures (m), so that the area below the fault occurrence curve in interval $(t_2 - t_1)$, or $(T_2 - T_1)$ gives the uncertainty of the component or the technical system as a whole or the cumulative function of the probability of failure, [2].

Functional adaptability of technical systems is the weakest investigated feature of technical systems, characterized as the ability of technical systems to maintain conditions *in operation* and to change operating conditions or change input parameters as a consequence of the operation of another superior technical system (for example, the operation of an energy facility within the electrical energy system), where working conditions are environmental conditions (temperature, pressure, humidity, dust, vibrations, magnetic and electromagnetic fields, dynamic influences, radiation, etc.). Sufficient reserves of material resources and a constructive solution to technical systems it is possible to keep the system in operation state as well as required economy with acceptable degradation of resource inventories.

On the other hand, the functional adaptability of the system to the change of input, created as a consequence of the work of other systems (complex serial-parallel and parallel-serial configurations of technical systems, quasi-serial and quasi-parallel configuration of technical systems, application of passive and passive-parallel configurations, etc.), is achieved with the functional dependence of the observed technical system within the system from the operational and wider environment. Determination of annual requirements for certain forms of reservation within a higher hierarchical level in the incidence of complete or partial failure of the plant, as well as the amount of costs associated with unplanned maintenance and unplanned remounts, are exclusively incidental. Any change in the level of reliability will directly affect the change in the necessary investments. Additional factors, which can be achieved in certain forms of reservation within a higher hierarchical level, are interlinked with basic effects and investments.

The deterioration of the maneuvering characteristic, in the incidents of partial failure, in the form of a reduction in the rate of increase or decrease in load, results in deviations from the default load graph and its variability. Therefore, within the higher hierarchical level, the existence of additional high maneuvering capacities is also required, which can compensate for the shortcomings mentioned above.

On the other hand, if we view one complex technical system (such as a thermal power plant) within a higher hierarchical system (such as the electricity system), any increase in the required amount of power to cover its own consumption will additionally trigger the possibility of a greater number of failures and a gradual deterioration in the consumption of the plant not only electricity (heat) but also used fuel. There are certain software packages designed for quick calculation of the thermal scheme in the absence of some of its elements (i.e., high-pressure heaters) or changes in certain parameters and characteristics of their parts or components. Examples are the applications of Thermo Flow, Inc., USA: STEAM PRO, STEAM MASTER, RE-MASTER, RECIPRO, THERMO FLEX, QT-PRO^2, [1]. The number and duration of such events are significantly related to the failure rate parameter and the appearance of additional costs for their removal. For these reasons, simplified solutions are often used with the determined mean intervals of the reliability of the type of equipment, as well as the technical and economic indicators of the plant as a whole or parts.

Starting from the initial stage of development, design and gaining of production of certain types of thermal energy equipment in order to meet all the requirements without limitations, if it is based on its purpose, a more varied choice is made to the designer, with the need to optimize according to certain already adopted algorithms. The aim is to produce such a plant with a satisfactory structure in terms of reliability indicators, with minimal maintenance costs during the projected lifetime. Defining the plant that best meets set requirements related to reliability and the process of exploitation and maintenance itself must be the result of the implemented optimization process (in this case, based on the chosen criterion of the minimal investments).

4 Selection of Parameters for the Evaluation of Energy Efficiency of Energy Systems

The realized characteristics in the exploitation of the power plant block are grouped according to the affiliation for the evaluation and analysis of the exploitation effects on the block of a power plant in three groups, [1, 4, 50]:

- Time characteristics achieved in exploitation;
- Energy characteristics achieved in exploitation;
- Techno-economic characteristics realized in exploitation.

4.1 Time Characteristics Achieved in the Exploitation of the Energy System

The time characteristics defined in the exploitation of the energy system for the observed period of time (at the level of the year, the semester, the month or some other period being analyzed) are defined by the coefficient of exploitation of the block K_e and the energy system stop (failure) coefficient (K_e^1). The exploitation coefficient is one of the indicators of the time used for the power plant block and can be defined in two ways. The first method most commonly used in literature is the definition of the block exploitation coefficient over the quotient of the exploitation time or the work of the block on the network (T_e) and the calendar time (T_k) for the observed period (monthly, annual, etc.). For new energy blocks, the annual exploitation factor is 0.85–0.90, while in the old blocks it moves to 0.85, [1]. In this way, the calculation of the coefficient of exploitation is used in comparative analysis, as the remounting time is within the calendar time. The second way of determining the exploitation coefficient on an annual basis is through the ratio between the exploitation time (T_e) and the difference between the calendar time and the time determined for the annual overhaul ($T_k - T_r$). In most of the plants, the duration of the remounts is not standardized, so the data are not applicable with the comparative analyzes of energy efficiency. This calculation method is used more at local level and in the preparation of annual plans for each concrete power plant. The exploitation time in the observed period (annual, semiannual, monthly or some other set of time for analysis) is composed of all the time of the uninterrupted operation of the block without failure—delay ($T_{e.i.}$), where $i = 1, \ldots, n$, and n is the number of the interval of continuous block operation for the observed period.

Failures on installations and vital equipment of the power plant as a result of failure are the result of technological process disturbances and non-stationary modes of operation of the block. Failures on installations and vital equipment that do not allow exploitation personnel to maintain process parameters according to technological guidance and technical regulations of exploitation and that endanger the safe operation of the block, cause delays—failures. Some failures require an immediate blockage of the power plant. These are failures that endanger the safety of plant and personnel in exploitation and maintenance. Therefore, there are protections and blockages on the plant block for instantaneous disconnections with the occurrence of such cases, [1].

Failures or delays of plants and equipment of plant block can be grouped into three basic groups, namely: accidental or unplanned—currently safe exclusion of the block, planned—can extend work for some time to endangering of the technical regulations of exploitation and safety of the plant and environment, as well as planned overhaul delays of the block for annual overhaul. Unplanned failures occur as a result of block disorders caused by abrasion of material, aging and loss of functional properties, thermal overload, improper exploitation and maintenance error, failure to comply with technical instructions and regulations, inadequate remounts, worsening, deteriorated quality of basic fuel, etc. As the plant blocks are a complex unit and composed of a large number of dependent technological units and complex installations, these are also possibilities for unplanned larger failures, and especially for blocks of older performance, [1]. Planned failures include annual remounts and planned actions that occur during exploitation with the aim of increasing energy efficiency.

The analysis of the efficiency and quality of preventive measures of exploitation and maintenance is done through the power failure coefficient. The failure coefficient according to the grouping of the cancellation is made up of: the non-planar failure coefficient (K_{npf}), the plan failure coefficient (K_{pf}), the annual overhaul coefficient (K_{ao}) and the coefficient of suppression (K_{sp}).

Failures can also occur due to the inability of the transmission network to take over the generated electricity due to the occurrence of unplanned surpluses of production in the electricity system in relation to consumers. This phenomenon is most frequently caused by work of hydropower plants of the energy system (favorable hydrology and much cheaper generation of electricity compared to production in the TE). These cases are defined by the coefficient of suppression. The failure coefficient is calculated using the formula:

$$K_{fail} = K_{npf} + K_{pf} + K_{ao} + K_{sp}. \tag{16}$$

4.2 *Energetic Indicators Achieved in the Exploitation of Energy Systems*

The energy characteristics of the energy system give an assessment of the production capacity of the unit and consumed energy for its own needs. The energy characteristics of the block are given in the form of certain parameters: block power utilization coefficient, block capacity utilization coefficient, coefficient of own consumption of power plant block and coefficient of useful effect of power plant.

The block strength utilization coefficient is defined as the ratio of the strength achieved and the nominal (budget) power. For example, the coefficient of exploitation of the block power of the thermal power plant is most often in the range from 0.97 to 1.00, [1].

Power plant production capacity can be analyzed using the block capacity utilization coefficient. This coefficient is determined on the basis of the generated electricity and is defined as the ratio between the realized quantity of produced electricity and the maximum theoretically possible quantities of electricity produced at the 100% utilized capacity of the power plant, [1].

Coefficient of own block consumption of a power plant defines its own needs of block plant with electricity.

Efficiency coefficient of the power plant defines the relation between the produced electric power energy and the total energy input of the primary fuel. In the thermal power plant, when the technological process of electricity production is carried out, in all parts of the transformation of one type of energy into another, there is an irreversible energy loss. These losses are combined and defined by the efficiency coefficient of the thermal power plant—η_{TPP}. The efficiency factor of a power plant by electricity produced (gross) can also be defined as a product of several efficiency coefficients of certain parts of the technological process, as follows:

$$\eta_{TPP}^{B} = \eta_{st.gt}^{B} \cdot \eta_{tg}^{B} \cdot \eta_{pl}, \tag{17}$$

where: $\eta_{st.g}^{B}$—the efficiency coefficient of the boiler plant (gross); η_{tg}^{B}—absolute electric coefficient of turbo generator efficiency (gross); η_{pl}—efficiency coefficient of pipelines (steam piping).

The efficiency factor of a power plant can also be determined by direct method through generated electricity and invested primary energy from fuel. It can be gross and net, and it is defined:

- *gross* (η_{TPP}^{B})—as the ratio of total electricity produced at the power plant generator terminals to the total amount of invested primary energy consumed from the fuel, and
- *net* (η_{TPP}^{P})—as the ratio of electricity produced to the transmission network of the electricity system (production on the threshold) to the total input of primary energy from the fuel.

4.3 Technical and Economic Indicators Achieved in the Exploitation of the Energy System

Technical economic indicators include: the quantities of electricity produced by the generator and delivered to the network of the power system (generator—the threshold of the power plant), the amount of heat input from the energy product for the electricity produced, the exploitation period of the active block operation, and taking into account the hot reserve if it is suppressed from networks. The thermal power plant is defined by the specific consumption of primary fuel and fuel oil, gross and net, specific consumption of heat from primary fuel, specific consumption of decarbonized and demineralized water, as well as the number of labor force per kWh produced electricity (this includes exploitation and maintenance personnel). Exploitation time is defined in time characteristics, as well as the time of pushing from the network. The losses of heat and electricity for their own needs are shown through the corresponding coefficients in the energy characteristics. Part of the weather and energy characteristics could also be included in the technical and economic indicators, because it is impossible to perform a precise division.

As an energy for the production of electricity on a thermal power plant, fuels are used, of which the most common are coal (brown and lignite), oil and gas. Thermal power plants use coal as the primary fuel, and as fuel oil and fuel to support fire in boilers in non-stationary regimes oil or gas are used. Fuel in the boiler plant combusts and its chemical energy during the combustion process is transformed into heat energy that is passed on to the working fluid with the formation of combustion products (flue gases, flying ash and slag). The working fluid in the process of electricity production is demineralized water, whose phase transformation, with the received heat energy from the fuel in the boiler plant, passes into the overheated steam, and in the turbine plant performs work, which further produces electricity in the generator. Demineralized (demi) water is the final product of raw water processing in the chemical preparation of water, and as the intermediate is decarbonized (water), whose consumption is, in significant quantities, in the closed thermodynamic cycle of the block, in the production of electricity of the thermal power plant. Electricity, as the end product further, through the busbar, block of transformers, transmission and distribution network, is delivered to the consumer.

For the monitoring, analysis and optimal management of the production cycle of the thermal power plant, the specific consumption achieved must be maintained within the limits prescribed by the normative tests proven. Specific consumption can be viewed from the aspect of generating electricity on the clamp generator (gross value) and from the aspect of the transmission of electricity to the transmission network (net value). Specific consumption in the production cycle of the thermal power plant can be given as the specific consumption of the basic fuel of the block (gross and net), the specific consumption of the heat of the block (gross and net), the specific consumption of fuel oil (oil fuel or gas), the specific consumption of decarbonized (water) and consumption of demineralized (demi) water. Based on specific consumption, the estimation of the economy of operation of the block of the thermal power plant is carried out, [3].

The specific consumption of the primary fuel of the block (gross and net) is defined as the ratio of the total amount of fuel consumed for the generated electricity and generated electricity on the terminal block generator. According to the needs, the daily, monthly and annual specific fuel consumption of the block can be calculated. It is more used in thermal power plants that have local mines and where there is no quality, technical and chemical analyzes of coal for coal quality assessment.

The specific heat consumption of the block (gross and net) gives the dependence of the consumed heat energy on the achieved parameters in the exploitation of the block and the achieved operating conditions and operating modes. It gives an assessment of the management of exploitation for the observed time from the point of view of economics and utilization of heat energy, and in the transformation of heat energy into electricity, [1].

In general, there is a functional dependence between the heat input of the fuel and the generated power of the block of the thermal power plant, which combines the energy characteristics of these two sizes. This functional dependence is expressed in the form $Q = f(E)$ or $E = f(Q)$, where E is produced electricity and Q is heat energy consumed, [3]. This dependence is uniquely determined by the technical-derived characteristics and is defined by the modes of exploitation of the block of the thermal power plant. However, the deviation from the unidentified dependence of the produced electricity and the consumed heat energy is influenced by a large number of exploitation factors (fresh steam parameters, parameters of inter heated steam, realized flow diagram within the thermodynamic cycle, cold state of the turbine—low pressure turbine outputs with a condenser that depends on the amount of cooling water, the temperature of the cooling water, the cleanliness of the condenser and the penetration of air). The specific heat consumption of the block due to the force at the generator terminals (gross), i.e. the power at the threshold of the block (net), can be expressed in the heat of the working medium, which is accepted in the boiler, as well as the heat of the spent fuel determined by the lower thermal power of the fuel.

The first allows us to, based on the comparison with the corresponding specific heat consumption in the turbine cycle, determine the effect of heat loss of the working medium from the boiler to the turbine, as well as the influence of our own heat consumption. Specific heat consumption (gross and net) serves as a basis for comparison with normative specific heat consumption of fuel. It includes the entire transformation of energy in the block, ranging from fuel to electricity produced. It is directly connected with the specific consumption of heat energy from the primary fuel. Today, this size is increasingly being used for comparative analysis and evaluation of the efficiency of power plants. For its calculation, it is important that the lower thermal power of the basic (primary) fuel is determined with a good quality. Thermal power plants that do not have their own mines get primary fuel by buying in the market and paying for heat units. This coefficient is in direct connection with specific fuel consumption. The specific consumption of fuel for movement and burning can be displayed over the mass (q_{SG}) and through the heat energy consumed from the fuel for movement and burning (q_{SG}). It is defined as the ratio of the total fuel used for burning and movement through the mass or through heat energy and generated electricity. It can be displayed in relation to the produced electricity (gross) and the delivered electricity to the transmission system (net).

This specific consumption is calculated for a certain period of time (daily, monthly and annual). The specific consumption of both deka and demi waters can be demonstrated through volumetric consumption (v_{DC}, v_{DM}) and mass (m^{I}_{DK} and m^{i}_{DM}), [3]. It is defined as the ratio of total consumed deka or demi water consumed and produced electricity (gross) or delivered electricity to the transmission system (net) for the observed period of time (daily, monthly, annual). Gross labor force participation in kWh produced electricity includes personal exploitation and maintenance of a thermal power plant.

5 Main Consumption of Thermal Power Plants

The consumption sum of all devices that ensure the normal operation of the thermal power plant is called self or own consumption. General consumption consists of all other devices that do not have a direct impact on the technological process at the plant. Preserving the continuity in supplying self-consumption with electricity is essential for safe operation under normal operating conditions, in the case of short-term transitions, as well as when starting and stopping, it is especially important in case of stopping in the event of faults and failures of the system. With the growing unit block strength, there is also the growth of unit power of electric motors of their own consumption, and therefore requirements related to the power supply. The main problem is achieving secure power supply in a variety of drive situations, with less short-circuit currents and a drop in voltage when starting large asynchronous motors. The solution is achieved by the proper choice of the transformer of own consumption and voltage level. The coal-fired thermal power plants have by far the most complex system of their own consumption (except own-power plants for nuclear power plants), and it is made up of the following plants, devices and mechanisms of their own consumption with electric power plants, [1]:

- System for the delivery and storage of basic fuel—coal (dumps, unloading cranes, conveyors, etc.);
- Fire and fire support system in the boiler (overloading oil pumps, oil pumps, etc.);
- Coal mining and coal dusting systems (coal mills, dispensers, feeders, etc.);
- Boiler plant (electrostatic filters, fresh air fans, smoke gas fans, re circulated hot air fans, cold gas fans, etc.);
- The system for the slag removal from the boiler (extinguisher, slag crusher, conveyors, etc.);
- Boiler ash extraction system (ventilation fans for fluidization, air compressors for ash transport, etc.);
- Boiler system power supply (electronic pumps, buster pumps, etc.);

- Turbo generator plants (I and II stage condensers, ejector pumps, circulation pumps, network pumps, hydrogen purification pumps, oil pumps, etc.);
- Thermal water treatment plant (plant pumps for heating, return condensate pumps, etc.);
- Auxiliary devices and facilities of the main drive facility (drainage and fire pumps, lifts, ladders, steam and water valves, battery chargers, backup alarm devices, etc.);
- Installation of auxiliary facilities of a thermal power plant (chemical water preparation, workshops, oil stations, compressor plants, etc.).

The electrical equipment of own consumption of thermal power plants includes transformers of their own and general consumption of all voltage levels, medium and low voltage switchgear, electric motors, metal armor and cables for connecting individual parts of their own consumption, control valves, DC power plant, power supply units, and electric lighting. Depending on their function, individual consumers of their own consumption can be of special interest (essential) or auxiliary. The first group of consumers is those whose short-term stopping causes a reduction in electricity production or leads to the stopping of the basic aggregates of the thermal power plant, and in special cases it can cause damage to basic and auxiliary equipment. This group consists of electronic, condensing pumps I and II, circulating (cooling) pumps, fresh air fans, re-circulated hot air fans, cold gas fans, smoke gas fans, feeders and coal dispensers, oil pump, electric valve actuators, etc., [1].

Auxiliary mechanisms (consumers) are those whose short-term stopping causes the reduction of electricity production, and it consists of mechanisms for coal delivery, slag and ash dispatch, etc. Considering the possibility of total voltage loss in a thermal power plant in case of failures, some of the essential mechanisms of own consumption sometimes are equipped with a steam drive by taking off steam from the turbine. These are the most commonly power pumps of block steam boilers.

For the consumers operation of their own consumption, primarily asynchronous motors with short-circuited rotor are used, which, in comparison with other engines, are more reliable, economical, cheaper and simpler. They do not require any special launching devices. The above mentioned advantages completely compensate for some of their defects, such as high values of stroke currents, difficult speed control conditions, etc.), [1].

Beside asynchronous motors, motors of the same direction are also used in power plants for spare oil pumps and some also important consumers powered by dc voltage. If we start from the fact that the total number of engines for the blocks with power of 200 and 300 MW is over 300 motors of different rated power and voltage levels, then it is not difficult to conclude that it is very important to ensure their reliable operation and drive readiness at all times, [1]. As thermal power plants employ motors with nominal power below 1 kW and up to a few MW, average voltage (usually 6 or 10 kV) is used for the supply of large engines (nominal power above 180 kW). For the supply of low voltage engines, 0.4 kV is mostly used, due to the use of standard versions of the engine and also due to the simplified fulfillment of the requirements of protection against indirect contact, [3].

Table 2 gives an overview of the required power units of electric motors for

Table 2 Display of necessary unit power of electric motors (EM) for TPP blocks of different nominal power, [3]

Name of drive mechanism	Block power (MW)			
	413	520	666	827
	Drive unit power of drive EM (kW)			
Power pumps	7000	turbo	5000	10,000
Pumps of cooling water, circular pumps	1250	2500	2000	1500
Condensate pumps	600	1000	1500	1280
Auxiliary low pressure pumps	300	600	1000	1500
Suction fans	–	7000	6000	5000
Fans under pressure	5500	8000	3500	6000
Compressors for general needs	100	400	600	600

TPP blocks of different nominal power. It is noted that the powers of the electric motor does not grow linearly with the forces of the blocks, which can be interpreted by construction solutions and special features, which is especially true for steam generators.

As thermal power plants employ motors with rated power below 1 kW and up to a few MW, for the supply of large engines (rated power above 180 kW) the average voltage (usually 6 or 10 kV) is used. For the supply of low voltage engines, 0.4 kV, both due to the use of standard versions of the engine and because of the simplified fulfillment of the requirements of protection against indirect contact, [3].

6 Methods for the Prognosis and Optimization of Energy Systems Efficiency

The analysis of complex technical systems and their plants (such as energy-process facilities) from the aspect of expected reliability and preventive engineering should provide the following:

- Assessment of the reliability and the load reserve both of the elements and of the technical system as a whole, depending on the technological process and exploitation itself,
- Analysis of the technical solution, with the detection of the so-called, "Bottlenecks" related to reliability, determining the mode of operation and the location of the technical system within a higher hierarchical level,
- Special emphasis is placed on the process of development and design, where there are great opportunities to provide an optimal level of reliability through optimally increasing the reliability of all elements in the structural scheme of the technical system, selecting the preventive overhaul plan with minimal cost savings,

- Display of the reliability indicators of the complex of the technical system in function of the technological scheme and its breakdown, with minimum reduced costs,
- Display and ranking the reliability indicators of the most critical circuits i.e. elements depending on their parameters and characteristics, with minimum reduced costs,
- Creating unique initial data for further research and stochastic analysis and modeling,
- Defining, through the algorithm, the basic way of determining or confirming the level of reliability of a complex technical system,
- Accelerating testing for reliability assessment by increasing the effectiveness of the proposed model (a plan of shortened tests for assessing the reliability of both the elements and the system as a whole) adapted to the complex technical system,
- Defining the necessary activities to improve and/or optimize the reliability of a complex technical system, as well as
- Development of a general modified mathematical model for achieving optimal reliability, development of reengineering processes and defining the level of reliability and maintenance flows with the basic contours of the expert system.

Formation of a database and grouping of scientific-professional methods for reliability assessment, with the critical analysis of the most frequently applied and their adaptation to the specifics of a complex technical system, served to form several modified methods, which as a result have a time dependence of the reliability of operation and the likelihood of failure or failure of the technical system. In doing so, the separation and ranking of the most influential elements within the complex technical system according to their importance in terms of raising the level of reliability has been done.

Optimal management of complex technical systems is mainly based on quantitative assessment and complex optimization of reliability depending on the way it is secured on different inter-stages and plant levels as a complex technical system, [5]. It should be noted that the optimization process represents only one link for long-term optimal management with a higher hierarchical system, which is realized at lower hierarchical levels. The task of optimizing reliability for new plants is reduced to a common choice of reliability indicators and defining the ways of their security.

6.1 Methods for Giving Prognosis Estimates of the Effectiveness of Energy Systems

Research to achieve optimal reliability in the most complex technical systems, such as, for example, a system of thermal power plants, can be classified into three inter-related task groups: *physical-technological, organizational and economic-financial,* [3]. Questions from the first group of tasks related to development and application are usually dealt with by scientific research and design institutes, construction bureaus and other energy organizations and enterprises, while solving certain organizational tasks for achieving optimal reliability is left to certain specialist departments with a rather vague aim and function. For these reasons, economic and technical tasks are still not solved systematically, but are usually given as constraints in solving the tasks of optimal design of equipment.

Insufficiently elaborated models and methods for a more detailed budget-based assessment of reliability indicators and a still insufficiently formulated budgetary-normative basis provide a wide range of opportunities for work and research. It can also be noted that the development of theoretical bases for achieving optimal reliability and their possible use in practice is delayed in relation to the development of the energy as a whole, with the uneven distribution of the necessary funds.

The prognosis and assessment of the reliability of the newly designed power plants, for the given project-operational conditions, should provide and comprise the following parts of the budget: assessment of the probability of non-failure operation of the elements, subsystems and plants as a whole, depending on the project-constructive, technological and repair-exploitation conditions; assessment of the structural connection between the plant and the plant as a whole in the given reliability of the elements and the pre-defined repair and maintenance program in function of the possible variants of the technological scheme of the plant, including the creation of reserves in elements and connections; assessment of the reliability of the plant at time average intervals during the lifetime in function of the variant of the cycle of repair and maintenance of the plant in the given scheme and given indicators of potentially possible or temporary durability and overhaul of its elements.

The harmonization and increasing of the exploitation level of the reliability of the plant and its constituents is not possible only on the basis of experience data, and it is similar with the rational distribution of costs according to the stages of achieving reliability.

In order to carry out this complex analysis, the following elements must be taken into account: the changes in the operation of the elements and plants completely without failure, the dependence of maintenance and overhaul (i.e. the convenience of maintenance and overhaul) from the functional scheme, the determination of the optimal cycle of preventive overhauls, tolerance and limiting elements in processes of manufacture, assembly, exploitation, maintenance and overhaul, operating mode without crossing and washing of boiler heating surfaces, changeability of stated values in relation to thermodynamic, technological and other constructive parameters, life of use, wear, etc. It is necessary to emphasize that in practice there are still norms of technological design and development of standard solutions based on acquired experience, which, together with norms related to the control and diagnosis of individual equipment elements, are not based in detail on experimental and budgetary arguments relating to reliability indicators.

The subject of research and achieving optimal reliability of technical systems as a whole is very complex, both in the inter-branch environment (fuel and energy complex), and from a global point of view to achieve optimal reliability of the higher hierarchical system in general. Development of mathematical modeling methods in the preparation, development and design of new technical systems; elaboration

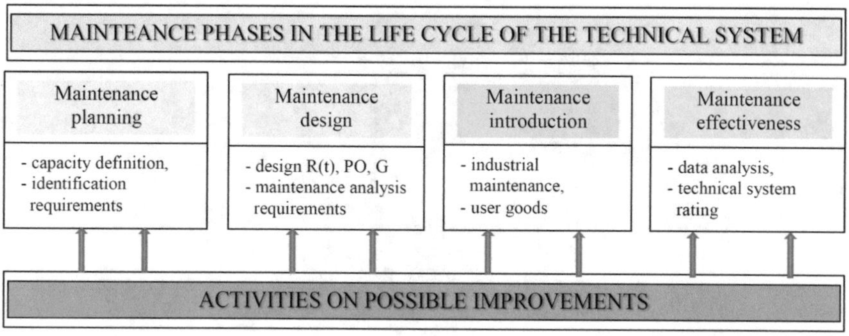

Fig. 12 Overview of important activities in the process of maintaining the technical system within its lifetime of use

of prognozing algorithms, optimization and standardization of reliability indicators in function from technical, economic and global factors; development of methods and algorithms for achieving optimal reliability at all stages of the life cycle of the technical system; tests carried out on existing facilities; collecting and processing of experiential data and forming a normative technological base—the elements are required for solving the problem of reliability optimization.

The achievement of the optimal reliability of the technical system during its lifetime, with lower maintenance costs (Fig. 12), is solved by applying the optimization theory in two ways: the first way, which leads to the tendency of the overall capture of all systems and all optimization tasks in them, and the second way is a method that implies the development of methods, techniques, and optimization algorithms designed to solve particular groups of systems and optimization problems within them. Both methods result in subordinating variable sizes in the system in such a way that they fulfill certain relations between them, as well as limits that are set, provided that the chosen optimization criterion obtains an extreme (minimum or maximum) value, [1].

Depending on the stage of the life cycle of the technical system, in the literature there are a large number of possible classifications of optimization tasks. From the aspect of the formal description, the optimization tasks of design, management tasks, as well as tasks of planning and terminating activities are distinguished.

Starting from the adopted optimization tasks, appropriate optimization methods are developed: mathematical, stochastic and heuristic optimization methods, as well as optimization methods based on the application of artificial intelligence, Figs. 13 and 14.

In the theory of reliability for the earliest stages of development, elaboration and design of the thermoelectric power system, statistical and logical probabilistic models are developed for assessing the change in the reliability indicators of complex technical systems, such as the thermal power plant. Figure 15 gives a graphic representation of the above procedures, with the most important models and methods.

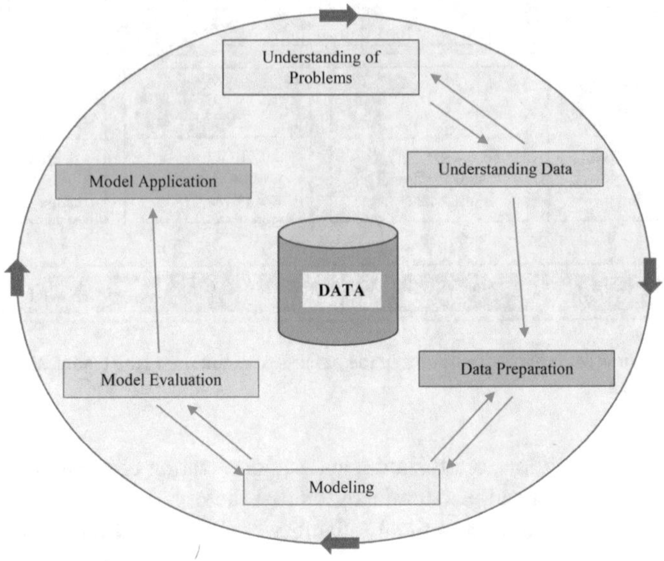

Fig. 13 The basic phases of the process of a intelligence data analysis according to the CRISP-DM standard, [3]

The methods used today in the prognosing and assessment of the reliability of complex installations, such as, for example, Thermal power plants are based on general methods of estimation of elemental reliability developed in the reliability theory based on probabilistic calculated strengths, along with their simultaneous reliance on modern methods of deterministic project-constructing calculations of elements, subsystems and systems of thermal power plants (strength, heat, hydraulic, aerodynamic and other budgets), [50].

The method of Ganttograph (Henry L. Gantt), as one of the first methods of line planning and management in production and maintenance, was aimed at rationalizing production in machine processing by reducing failure caused by waiting processing between adjacent technological operations. The ganttograph is a rectangular coordinate network with a horizontal (registered time unit: hour, day, week or month) and vertical scale (the registered entries from which the project consists), [1].

Ganttographs have their advantages (a joint presentation of the project plan and its realization, getting a easy overview of the status of the project, easy drafting of ganttographs for multiple levels of leadership through disassembling the project into components/activities), but they also have disadvantages in relation to some other techniques used in controlling the flow of projects (joint elaboration of the project plan and its termination, making it difficult to produce a quarterly plan, showing the relationship within the project in the form of terminal rather than technological plans, with the difficulty of developing an alternative plan, difficult to see the impact of delay on the project, the need to develop new gantographs, due to their

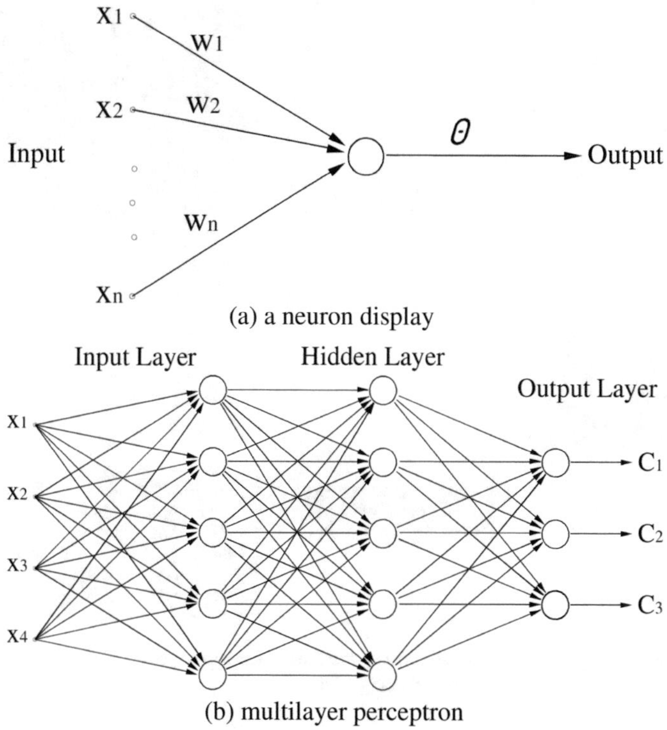

Fig. 14 Elements of the neural network, [3]. **a** A neuron display, **b** multilayer perceptron

frequent updates, difficult tracking or completely disabled following of monitoring for complex projects, etc.).

On the basis of the Gantograph method, some new ways of termination have been developed, of which the following are highlighted, [1]:

- **Overview of events**, which represents a list of events that mark the beginnings and endings of individual project phases/events, with introduced standard labeling of planned and realized events, which allow very effective control of the deadlines, and thus defining the critical stages of the project in advance (with disadvantages: a pre-elaborated project plan is needed, you cannot see the impact of an event on the whole project, etc.);
- **Wall panels and Kardex charts**, as organizational means, composed of several elements embedded on panels or on maps (in the case of brick slabs in the grooves most often slide many colored sliders into which the riders are inserted as carriers of information);
- **Special graphic forms**, referred to as *Transplan*, as a special form of the Gantt chart, then the *Line of Balance* (LOB), as an auxiliary method for presenting the total time on the project realization, tracking the time taken and the readiness of the main components for a certain time interval, as well as the *Gray-Kidd* algorithm

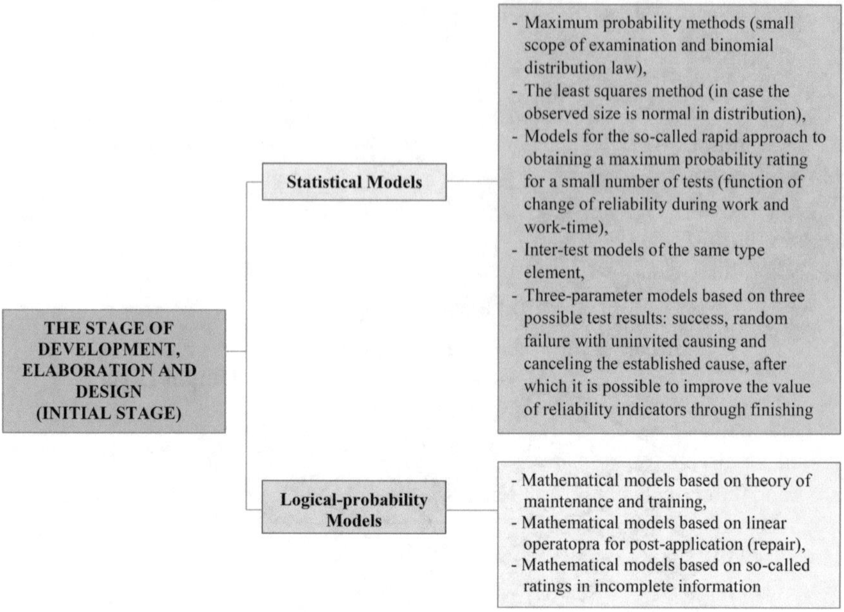

Fig. 15 Methodological bases of reliability assessment for type-specific details, subsystems and systems at the stage of development and design, [50]

(the first method of line diagrams with the included scheduling depending on the resource capacity to perform the planned activity) and cyclographs (graphic representations of work planning, organized on the basis of the application of the chain method).

Today, more than twenty modified network planning methods have been applied with the use of computer technology, in order to obtain as reliable information as possible for the preparation and timing of deadlines and the production of the appropriate product. The ultimate effect of their use can be seen in the identification of activities that can shorten the total duration of works, as well as the creation of all assumptions for more objective insight into the possible shortening or delays in completing the planned works, as well as the causes of their occurrence. In classical network diagrams, the links between individual activities are accomplished by linking the final event of the previous one with the initial event of the next activity, with the fact that at PRECEDENCE network link diagrams are possible also between parts of the activity, which is an even greater possibility, [3]. Based on the previous methods, two separate graphic representations of network plans, also known as *"i-j network activity-oriented diagrams"* and *"event-oriented PDM network diagrams"*, have been developed.

Activity-oriented network diagram has been developed as a *critical path method* (CPM) and PERT *method*, with the presentation of a straight line activity over which the description or activity code is entered, and the duration of the activity is listed below the line. The arrow shows the direction of the progress of the works. Linking of activities is done by numbers entered into circles, which represent an event at the beginning and at the end of the activities, symbolically marked with *"i-j"*. Fictive activity is shown as a dashed line and represents an activity for which the realization does not require engagement of the costs, reserving certain capacities and time. A network diagram is obtained by linking all events, while realizing the conditions by which it defines that the beginning of the next activity of an event requires the completion of all activities that precede that event. Activities that have the same initial and final events are commonly called *"coffee grinding"* and *"water heating"* activities. They can only be distinguished if they do not have the same initial and final event, which is achieved by introducing a zero-duration fictitious activity necessary to distinguish between the two activities, [3].

On the other hand, in order to avoid the occurrence of common differences between the set requirements for reliability and their dependence on the fulfillment of operational requirements, special attention should be paid to defining analytical expressions and numerical values of reliability parameters. In order to accomplish this task, it is necessary to form an appropriate database, related not only to the system as a whole, but also to the components of the system, as the basic links in the reliability chain. The intensity of the failure of some of the system components depends on many factors (mechanical and thermal overload, environmental impact, conditions of exploitation, the way of repairs or replacement, the influence of human factors, etc.). Thus, the reliability assessment, depending on the purpose and phase of the life cycle of a complex system such as a thermal power plant, is generally realized in three basic ways, [50]:

- Assessment of reliability on the principle of similarity of equipment, based on its typification or retrospective analogue information, with correction for new prognosing project conditions,
- Assessment of reliability by method of component listing or so-called a "rough" calculation of reliability, with the formation of appropriate statistical methods and logical-stochastic models, as well as rating for the incomplete determination of information and estimation of reliability by the straining analysis method, or the so-called " "fine" reliability calculation (characteristics of possible relations of working parameters and loads),
- Assessment of the durability parameters probability and possible deviations of structural elements, expert correction of the characteristics of sustainability and resource details with the participation of harmful effects.

The intensive development of stochastic security analysis methods resulted in the formulation of a set of probabilistic methods for analyzing the security of technical systems. There are different methods of realization of the mentioned methods, which, when observing the system as a complex unit, can be classified as follows:

- use and reduction to the probability model of failure, frequency and duration, which corresponds to the law of the "breaker" Boolean algebra with two basic states: *full working ability* or *complete failure*,

- methods based on the use of Markovian or Semi-Markovian security models, that are distinguished by multiple states (including the state of the reserve), and on the function of the time dependence of the state probability,
- the use of Weibul's distribution, both for the elements, the subsystems and the system as a whole and its testing.

Methods of calculating that deviate from the classical directions in reliability theory, where the distribution of the failure is not explicitly stated, but is dependent on the system of planned overhauls, after which the assessment of the existing state and exploitation possibilities for the next period is carried out.

Further progress in improving the reliability assessment, except in the adaptation of classical methods to the specificities of a given compound technical complex, lies in the need to shorten the test time of one or more factors by selecting an optimal short-cut plan, by automating (*on-line*) reliability assessment procedures and optimizing it on a base of selected criteria (most often economic criteria). It is also necessary, taking into account the very structure of the technical system and the reliability characteristics of certain elements, to give a measure of importance and to rank the elements from it in terms of rational distribution of resources in raising the reliability of each of them. As a result of solving the problem, a list of criticality of the consequences (effects) of the failure is determined. The conditions that need to be present, in order to get to the list of criticality, are the knowledge of the operating conditions of the system, its structure and the possession of the database of elements failure, [4].

6.2 Mathematical Problem Optimization for the Efficiency of Energy Systems

Optimization is defined as a process that determines the "best" solutions of a particular, mathematically defined, problem. Having the best solution or closest solutions saves the material and energy (whose resources are limited) or achieving financial profit or achieving the highest reliability or safety in the operation of energy systems, the least impact of energy systems on the environment, with the maximum achievement of sustainable development through strategic planning and similarly. Optimization aims to minimize negative effects (effort, costs, environmental pollution, etc.) or maximize positive effects (profit, community development). Optimization can be roughly divided into experimental and mathematical optimization, [1]. Experimental optimization is based on statistical data, or on long-term measurements, it serves to optimize a very complicated process without creating a very complex mathematical model, which would ultimately be very inaccurate. Other suggestions for the classification of optimization methods are [3, 51]:

- The first optimization method is based on conventional methods such as non-linear programming, weighting methods and ε-constraints.

- The second optimization method is based on evolutionary techniques such as the NPGA method—Niched Pareto genetic algorithm, NSGA—Non-dominated Sorting Genetic Algorithm, SPEA—Strength Pareto Evolutionary Algorithm and SPEA2—Strength Pareto Evolutionary Algorithm 2, [52].

Optimization methods can generally be divided into three groups:

(a) Conventional optimization methods;
(b) Intelligent Search Methods;
(c) Methods for solving the uncertainty of aims and constraints.

Conventional methods have certain limitations, challenges and uncertainties in performing optimization. Some of them are: a long time for performing optimization, a very good knowledge of the problems and variables that affect the outcome. Also, classical methods are very complicated, which additionally makes it difficult to obtain an optimal solution. All of these factors make conventional methods unsuitable for many problems, and especially applications in the power system. In the history, these methods were used by some of the more prominent scientists Lagrange, Fourier, Newton, Gauss, etc., and later these methods were improved and became the foundation of further development of algorithms and optimization methods.

A group of conventional methods and algorithms includes [3, 51]: unconstrained optimization, linear programming, nonlinear programming, quadratic programming, Newton's method, interior point method, mixed-integer programming, network flow programming.

The optimization approach without limitations is the foundation of optimization algorithms with constraints. Most optimization problems with constraints in the distribution power system can be transformed into optimization problems without limitations. The main optimization methods without limitations used in the optimization of distribution power systems are: gradient method, line search, Lagrange multiplier method, Newton-Raphson method the quasi-Newton method, etc., [53].

Linear programming is a reliable technique to solve a large number of optimization problems with linear constraints and aims. Software packages for the optimization of power systems that are used to linearize nonlinear optimization problems of the distribution power system, contain powerful linear programming algorithms, [3]. Linear programming uses a simplex method, a revised simplex method, and an internal point technique to solve linear problems. The most famous is the *simplex method*. In practice, engineering results are satisfactory, in solving the problems of the distribution power system, such as optimization of power flows and reactive power, management of the electric power system, etc. This approach has several advantages, some of which are, [3, 51]: reliability, especially when it comes to convergence properties, fast identification of non-performance, and the compilation of a large selection of constraints on the distribution system, including very important unpredicted constraints. The disadvantages of this approach are: inaccurate estimation of system losses, as well as insufficient precision in finding the correct solution.

The nonlinear programming method is used for nonlinear problems, such as problems in the distribution system, in which the function of the goal and the constraints is nonlinear. The first step in applying these techniques is the selection of the search direction in the iterative procedure, which is determined by the first partial derivation of the equation (reduced gradient) [2]. Therefore, these techniques can be classified as first order methods. Nonlinear programming methods have much greater accuracy than linear programming methods and also have global convergence, which means that such a convergence guarantees a solution that is independent of the starting point. However, in certain cases, the convergence to the solution may be slow due to the zigzag movement in the search space [51].

Square programming is a method for solving non-linear limited problems. This method is based on the generalization of the Newton method for unlimited minimization problems. The aim function is square while the constraints have a linear form. This method is more accurate than linear programming, and in practice it is used to optimize the cost of generators, the cost is an aim function that is generally square, [2, 54].

The Newton method has the requirements of computing partial derivatives of the second order in the equations of flows of forces in which there are certain constraints, also called the second order method. The necessary conditions of optimum are the Kuhn-Tucker conditions. The Newton method is widely used because of its square convergence properties [53].

According to the author Frano Tomasevic, the internal point method has been developed with the aim of solving linear programming. This method has more advantages over the conventional simplex method of linear programming, the advantage of this method is the speed of computation. It is used to calculate optimal power flows and optimum distribution of reactive power, [2, 54].

The method of mixed integer programming is based on the assumption that the optimization problems are mixed integer, and that the variables of the problem are integer. In the electric power system, some of these integer variables are the angle of phase shift, the transformer ratio, the status of involvement of some of the elements, and the like. This method is very computationally demanding, and is therefore performed for a long time because it possesses a large number of discrete variables. In order for the optimization process to accelerate, the problem is split into continuous and integer problems using decomposition techniques such as, for example, Benders' decomposition. Decomposition results should significantly reduce the number of iterations, as well as the timing of the calculation execution. This method can be used to solve optimal power flows, determine the driving schedule, optimize the reconfiguration of the distribution network.

The method of programming network flows is a special form of linear programming. It is primarily used on linear models, and later, the method of nonlinear convex programming is used to optimize the network flows of the electric power system. Programming of network flows is characterized by simplicity and speed of execution of algorithms, [51].

Intelligent search methods belong to heuristic (experiential) methods and are now applied in solving a wide range of optimization problems. This method is a good choice for solving problems that are difficult to be mathematically formulated. Also, they are suitable for solving problems that could otherwise be dealt with by various complicated mathematical methods, however, due to the lack or unreliability of the available data, they are not suitable. These methods are suitable for solving problems with a well-known mathematical algorithm, sometimes even simple, but its application would require an unacceptably long time due to a large number of unknown sizes, [54]. Intelligent Search Methods Artificial Intelligence seeks to avoid solutions that are very likely not an extreme function, and it can be concluded that they do not even contain the optimum at the time. This way of solution is obtained at a relatively high speed, but it is difficult to say with certainty that this solution is the best, therefore this solution can not be considered an optimal solution because we can not prove it. For such solutions, it can only be claimed, with a high probability, that it is better than a solution that would be obtained without applying a particular method. A list of these methods is quite long, but some of the most widely used in solving optimization problems are, [55]: Monte Carlo method, Expert systems, Neural networks, Genetic algorithms, Evolutionary programming, Simulated annealing, Tabu search algorithm and Particle swarm.

Any method that solves problems by generating a large number of random numbers and observing part of those numbers that show the desired properties is called the Monte Carlo method. The method is useful for solving numerical solutions that are more complex for solving analytical procedures, [4].

Expert systems are used by experienced (heuristic) rules according to which they imitate live experts. In principle, rules are determined by the cooperation of a large number of experts from the observed area, they can also automatically be determined by other methods. These systems have proven effective in many practical applications, [2].

Neural networks are based on biological neural networks, i.e. nervous system. Networks comprised of programmatically derived neural networks are called artificial neural networks, and hence the terms artificial neurons or neural nodes. By programming, an artificial mimic of biological neurons or the function of the human brain is achieved. They solve problems of linear optimization, problems of nonlinear programming, and many more problems. They alter the solution of the optimization problem at the equilibrium point of nonlinear dynamical systems and change the optimal criterion in the energy functions of the dynamic systems, [2].

Genetic algorithms are flexible heuristic search algorithms based on an example of natural selection. The fundamental concept of the genetic algorithm is to design a process model based on a natural system of evolution, guided by the principle of the survival of the strongest. This algorithm represents the intelligent development of random search within a defined space to find an optimal solution to the problem. It is superior over classical methods and is remarkable for solving problems from the real world, and is therefore well accepted in engineering, computing, electrical engineering and other technical but also economic and social branches of science, [5].

Evolutionary programming is a stochastic method of optimization similar to a genetic algorithm, which emphasizes the behavior of relations between parents and their offspring, but it should try to imitate specific genetic operators as observed in nature. Evolutionary programming is similar to evolutionary strategies, although two approaches have been developed independently and singly. This method is a useful way of optimization when other methods such as a gradient of decreasing or a direct analytical solution are not applicable. Combinatorial and real-valued functional optimizations, in which the optimization surface or the target function environment is rough (*fitness landscape*), i.e. they have many local optimal solutions, well adapted to evolutionary programming. Evolutionary programming depends more on mutation and selection operators while the genetic algorithm relies mainly on crossover, [1].

The simulated hardening algorithm belongs to a group of approximation algorithms. The algorithm is based on stochastic technique, but includes many principles of an iterative improvement algorithm. The algorithm is known as simulated hardening, but in the literature it is also called Monte Carlo hardening, stochastic cooling, stochastic relaxation, and random algorithm, [2]. The results obtained by this algorithm are very close to the optimum, and do not depend on the initial configuration. In this way, the method of simulated hardening does not show the disadvantages of the iterative improvement method, it gives better results, but it is somewhat slower. In its original form, the simulated hardening algorithm is based on the analogy between the metal hardening simulation and the combinatorial optimization problem, [3].

Tabu search is a powerful optimization process that has been successfully applied to a set of combinatorial optimization problems, [4]. Taboo search has the ability to avoid the solution in a single local minimum. The Tabu algorithm performs multiple searches, and due to its ability to memorize already published searches, it does not search the same areas, which increases the speed and the finding of a global optimus. The best solutions are obtained if combined with other methods.

Optimizing by method of particle swarm optimization is a stochastic method based on the population. The method is inspired by the social behavior of flocks of birds or fish. It possess similarities to programming with a genetic algorithm. By this method, the system is started by random population and search optimization by updating generations. In contrast to the genetic algorithm, particle swarm has no evolutionary operators, such as crossing and mutations [56].

In methodology of optimization by particle swarm intelligence, potential solutions, called particles, pass through the solution space following the current optimal particles. The particle swarm combines an individual and social model of respecting the best solutions of the population. A social model suggests that individuals neglect their own experience and are guided by the successful behavior and beliefs of individuals in the neighborhood. On the other hand, the individual model treats an individual as isolated beings. The particle using one of the above models changes its position in the space. Each particle remembers its coordinates as well as the neighbors' coordinates to find the best solution in the problem area. If the particle takes the entire space of the solution as a neighborhood then the best solution is also the global best value, [50].

The optimization process of a particular system can be formally divided into two levels. At the first level it is necessary to uniquely define the problem and establish the exact relations of the influence parameters and solutions under the conditions under which the system should function. The second level is the choice of some of the known or the development of a new methodology for solving the problem, where the defined problem must satisfy the formal and mathematical limitations of the chosen method, and the chosen method must enable reliable and accurate determination of optimal solutions in the simplest and quicker way. The process of solving optimization tasks consists of the following five phases: problem formulation, creation of a mathematical model representing the real system, selection and application of the method, choice of algorithms and programs for the computer (possibly modification or development of a new method, algorithm development or computer program development) testing of models and solutions obtained and implementation.

The formation of a mathematical model involves defining the function of the objective $f(X)$, the set of constraints $G(X)$ and systematically collecting, verifying and arranging the necessary data in order to complete one or more variants of the model. The mathematical model is usually the result of the original synthesis of the interdependence of variable optimization. It is necessary to find the extreme value of the function aim: (min/max) $f(X)$ with the satisfaction of the restriction $G(X)$.

Aim function is a function that defines the criterion i.e. aim of optimization. In a mathematical sense, the function of an aim is expressed by a function $f(X)$. The optimization task is to determine its extreme value, i.e., the variable optimization vector X_{opt} for which aim function reaches the extreme value. How each system functions and achieves aims under certain conditions defined by the constraints. A set of constraints is defined by a system of m equations and/or inequality variables in which variable optimisations are figured out:

$$g_i(X) = 0; \quad i = 1, 2, \ldots, m; \quad g_i(X) \geq 0; \quad i = m_1 + 1, m_1 + 2, m_1 + 3, \ldots, m \tag{18}$$

In the set of restrictions G, the following cases can occur, [52]:

- A set G is contradictory, which means that there is no solution that meets all limitations;
- The set of constraints G is not contradictory, but domain D (domain of permissible solutions) determined by the set G is unlimited. The optimization task will have a solution if the target function is not unlimited in a tolerable domain D;
- The G restriction set is not contradictory and the scope of the permissible solutions is limited. In this case, the solution can be found if the aim function has a finite value in the restricted domain D.

The division of the mathematical optimization method is given in Table 3, [50].

Figure 16 gives an overview of the classification of operational survey methods. In addition, the properties of the aim function can be: the function of one variable, linear function, the linear/nonlinear function of the smallest squares, the square function, the continuous/interrupt nonlinear function, the non-differential nonlinear function, etc., with certain properties of the constraint function (without constraints, simple boundaries, linear functions, differential/non-differential binary functions, interrupt linear functions, etc.), [1].

Nonlinear programming is a wider set of methods of operational research for solving optimization tasks whose mathematical model contains at least one, and

Table 3 Division of mathematical optimization methods

Division criterion	Optimization methods	Short description of the method
Kind of a function problem	– Linear programming – Non-linear programming – Dynamic programming – Heuristic programming – Network programming – Optimum reservation – Theory of games	If all the functions of the problem are linear then it is a linear programming. For linear programming tasks, there is a general method of solving (Simplex) developed by *Dantzing* in 1947 Nonlinear programming is a wide set of methods for solving problems in which there is at least one non-linearity in the aim function or constraint functions. In addition to linear and non-linear programming, there are a number of methods applicable in some special optimization tasks Dynamic programming is a mathematical programming method that solves the problems of optimizing time-dependent or multi-path processes. *Bellman* published first work on dynamic programming in 1957, and that is a pure numerical method, and its application is practically inconceivable without the use of a computer Heuristic programming is a set of combinations of different heuristics used to solve incomplete defined optimization tasks. Heuristics are defined as a procedure that, based on elementary facts, ensures the finding of complex truths Network programming is a special area of operational research that has been developed and is most widely used in the field of production organization. There are a number of network programming methods, and the best known are CPM (*Critical Path Method*) and PERT (*Program Evaluation and Review Technique*) Optimal reservation solves the reliability problems of complex technical systems. The optimal scheduling task can be set up in the following way: to achieve maximum reliability of a system with limited resources, or to achieve a certain reliability or other parameter of the functioning of the system with minimal resource consumption (financial or technical) Game theory refers to the tasks of operational research that solve situations in which two or more subjects attempt to achieve their own, usually completely opposite objectives

(continued)

Table 3 (continued)

Division criterion	Optimization methods	Short description of the method
Limitations presence	-Tasks without constraints – Tasks with constraints	The existence and form of constraints affect the choice and efficiency of the method. Restrictions may be in the form of equality or inequality. Restrictions in the form of inequality are defined as hyper-superficial in *n* dimensional space. Points on one side of a surface are permissible, and on the other side they are unacceptable. If a point is found on this surface it is said that the restriction is active at that point. Methods for solving problems without limitations can also be used to solve constraint problems by limiting in some way the transformed aim function or not transforming the aim function, instead of that the problem is considered and resolved as constrained at any time, which means that at no time permissive domain is left
Method of resolution	– Analytical methods – Numerical methods	Analytical methods are based on the mathematical determination of stationary function points. The condition that a point is stationary is: $\partial f/\partial X = 0$ In case of simple and rather rough models, the solution can often be determined analytically. As good models are most common and very complex, finding solutions are often not possible with analytical methods. Then the methods of numerical mathematics are used. Many prominent mathematicians, such as Newton, Euler, Gauss, Lagrange, Hermite and others, have contributed to the development of numerical methods. Especially intensive development of this field of mathematics is experienced by the appearance of computers (1940), with the possibility of implementing a large number of computational operations in a short time. A special importance was the numerical solution of new classes of tasks, especially those described by partial differential equations
Type of variable optimization	– Optimization problems with continuous variable – Optimization problems with discrete variable – Optimization problem with integer variable – Optimization problem with "0–1" variable – Optimization problem with mixed variable	There are several methods for solving problems with mixed variables: *methods of branching, approximate methods, Ad hoc methods*. Numerical methods for solving mixed discrete nonlinear programming are divided into six groups: methods of branching and fencing, methods of hardening simulation, integral programming, sequential linearization methods, methods based on retributive functions, *Lagrange* relaxation methods and other methods

(continued)

Table 3 (continued)

Division criterion	Optimization methods	Short description of the method
Number of criteria	– One criterion optimization – Multi-criterion optimization	If the quality of the solution of a concrete construction can be expressed by only one criterion, it is one-criterion optimization. When more non-compliant aims are taken into account then it is a multi-criteria optimization
Number of variable sizes	– One-dimensional optimization problem – Multi-dimensional optimization problem	The number of variable sizes can be one or more, which determines the optimization problem as a single or multidimensional problem for solving
Testing and ranking of various optimization algorithms	– Simplicity of implementation – Calculation complexity – The speed of convergence – Reliability – Applicability in practice (usage experiences)	Defining criteria for evaluating alternative methods of energy system optimization, testing and ranking

usually more, nonlinear connections. The stationary points of the nonlinear function—points 2 and 4 are the points of the minimum,—points 1, 3 and 5 are the points of the maximum, and—the point 6 is the transverse (sediment) point, Fig. 17, [57, 58].

Convex functions are functions for which between any two points 1 and 2 functions get values less than or equal to the values on the straight line segment that connects the fluxes 1 and 2, Fig. 18.

Nonlinear programming methods are usually classified according to the order of the used statements in the following way, [1]:

- the methods of the null derivatives (the test point coordinates in r iterations can be mathematically defined in the following way: $X_r = X_{r\text{-}1} + s_r \cdot P_r$), with characteristic representatives of these methods: *Hookee—Jeeves*, *Powell, Rosenbrock*, etc.);
- methods of first derivatives (methods of first derivatives or gradient methods use the gradient direction as the direction of change and improvement of acceptable solutions: $Xr = X_{r\text{-}1} + s_r \cdot f\ (X_r)$, with a characteristic representative of gradient methods: method of the fastest ascent and its modification: method of variable metrics);
- methods of other (higher) derivatives or *Newton's* methods (used in the process of optimizing the value of other statements of the aim function at the observed point to which the optimization process arrived);
- the method of a flexible polyhedron (the basic idea of this approach is to derive from the n—dimensional polyhedron, which includes the point of the optimum, so that in sequential iterations the narrowing of this domain and approximation to the optimal value is carried out), with the characteristic representatives of the flexible polyhedron method: the *Nelder—Mead* and the *Complex Box* method, and

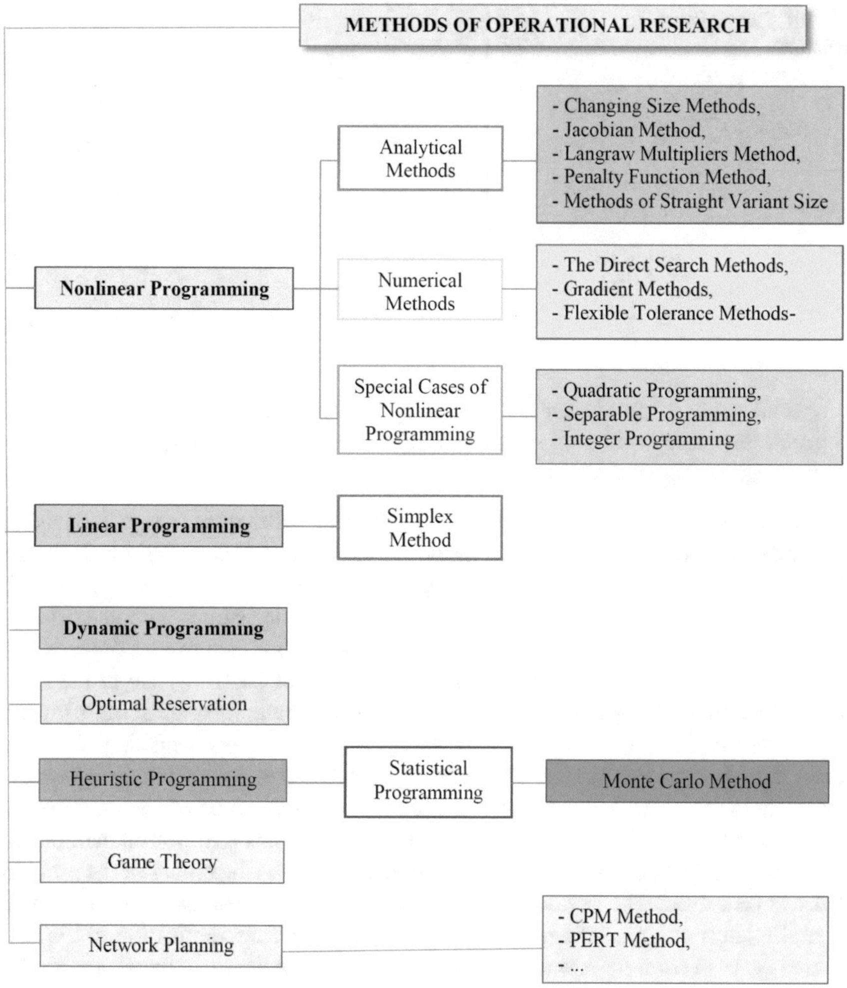

Fig. 16 Classification of operational research method, [2]

Fig. 17 Stationary points of non-linear functions

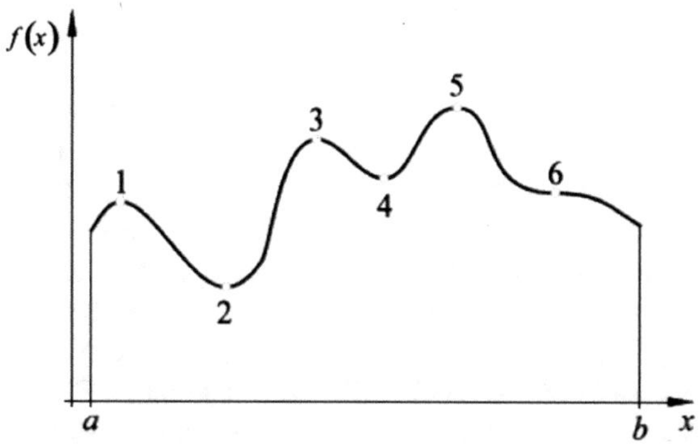

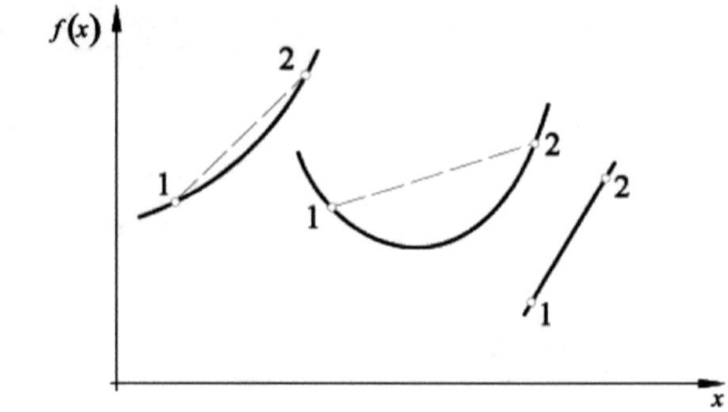

Fig. 18 Convexity of functions

- special cases of non-linear programming (square programming, separable programming, stochastic programming, integer linear programming, geometric programming, aim programming, etc.).

Although different methods and decision-making techniques are used, decision-making issues are an important activity, which has become increasingly important in the course of time (the intensive development of decision-making theory, as a new scientific discipline). According to *Schermerhorn*, the decision is a choice between several alternative problem solving options, which aim to achieve an objective, as it usually responds to a specific need. It is the result of choosing one of the many alternatives that are available to the decision maker (individual or most often group). The notion of more criteria decision-making refers to decision-making situations when there is a greater number of conflicting criteria. It is this fact that represents a significant step towards the reality of the problem, which can be addressed by methods of multiple criteria decision making, unlike the classic optimization method that uses only one decision criterion, which drastically reduces the reality of the problem that can be solved. The development of new and improvement of existing methods of multiple criteria analysis has influenced their increasing application in the field of energy, both from theoretical and practical aspect. In theory, a multiple criterion analysis involves solving insufficiently structured problems, while in practical terms it is an essential tool in solving everyday tasks of choosing decisions, management actions, both at the design level and in the exploitation of complex energy and process systems. A specific segment of the application of the method of multiple criteria analysis is the adoption of strategic or operational decisions in resolving multidisciplinary with the dominant technical or predominantly economic content whether it is to observe a part or the energy system as a whole. The application of these methods in the selection of the right solutions in the tasks of deciding management in the design, maintenance and exploitation of thermal power plants is based on the development of information technologies and computer technology. Today, many methods are applied, and the following should be emphasized: PROMETHEE (Preference Ranking Organization METHod of Enrichment Evaluation), AHP (Analytic Hierarchy Process), IKOR, ELECTRE (ELimination Et Choice Translating REality), MAX-MIN, MAX-MAX, Hurwiczova (combination of max-max and max-min methods),

SAW (Simple Additive Weighting Method), TOPSIS (Technique for Order Preference by Similarity to Ideal Solution), a conjugate method, a disjunctive method, and so on, [1, 3].

Depending on the method used as a solution, a ranking alternative or criterion is obtained, the best alternative or a criterion or a set of alternatives or criteria that meet certain conditions. Problems that can be considered using multiple criteria decision-making have certain common characteristics: a greater number of criteria, which must be created by the decision maker, high likelihood of conflict between criteria, incomparable (different) units of measure (as a rule, each criterion has different units of measure) as well as designing or choosing the most optimal concepts in a predetermined area. As solutions, there are either designing the best alternative or selecting the best action from a set of previously defined final actions. Within this chapter, a brief description of the most important methods is given, with an analysis of their advantages and disadvantages, as well as the possibility of applying it to solve the specific problem of the selection of microlocation for power plants. In the process of valorization and selection of acceptable potential microlocation for energy plants within the selected macrolocation, it is necessary to apply a certain procedure, which will be uniform in all its aspects. In order to achieve this goal, it is necessary to define the general criteria for selection and mutual comparison of the selected micro locations within the previously determined macrolocation using the multi-criteria optimization method. The basic principle when selecting the criteria is that they can be measurable, that is, the available location data, based on them, can be valorized. However, in contrast to that, there are a number of criteria that are not measurable, that is, their influence in relation to micro-location can not be accurately valorized and determined, but only through certain indirect indicators. The level of development where the methods of multiple criteria analysis are located can be used without major problems to solve the real problems of selecting variant microlocation solutions for a specific thermal power plant.

When selecting and evaluating conceptual variant solutions in the process of comparing and selecting optimum microlocation and energy facilities within the selected macrolocation are usually used within the following criteria and conditions: the space required for the accommodation of facilities and plants of the thermoenergetic block, the occupancy of microlocation with industrial or other objects, topographic conditions, seismological conditions and engineering-geological characteristics of the soil, conditions of transport and delivery of plants and equipment, conditions for shipments and waste disposal, conditions for transportation and delivery of coal, conditions for dispatch and landfilling of ash and slag, conditions and possibilities of water supply, conditions and manner of connection with the electricity grid, conditions and manner of connecting with public roads, environmental conditions, economic conditions, social justification, population and development, reliability, maintenance and risk assessment, placement of electricity, as well as other conditions, which include possible other conditions or criteria which are not covered by the previously specified ones, which may be of significance for the considered microlocation within the macro-location (radioactive areas, areas treated by special legislation, areas that are the subject of special attention of the investor, planned reserved area within already built plants, etc.).

Macrolocation for each of the coal mines from the observed region is defined by certain strategic documents at the republic or local level. Sometimes these solutions are very expensive and do not follow the trend of developing new technologies and the use of new equipment in the field of thermoenergetics, so it is necessary to additionally perform evaluation of the variant solutions according to certain criteria and conditions, which will be applied in the procedure of comparison and selection of selected micro locations for the realization of the energy plant within the pre-selected macrolocation. These criteria, beside their diversity, can be mutually opposing, so it is necessary for evaluation of the alternatives in the multi-criteria analysis to use a method that would allow their simultaneous processing, while respecting the inherent possibilities of each technological approach and their relative importance to each other according to the criteria.

7 Racionalization of Energy Consumption and Prevention for Their Implementation

The tendency for rational use of energy in developed countries implies that for the same quantity of products less energy is consumed, [2]. The energy and process industry has always used energy more rationally for economic reasons. When considering the total energy required, for a particular production process, it is necessary to take into account not only energy consumers, but the whole production system as a whole. One such analysis uses different following criteria, of which the following should be emphasized: assessment of the use of energy and raw materials, reduction of environmental pollution, improvement of working conditions, improvement of product quality, as well as the production process. Thus, the implementation of the rationalization of energy consumption within the energy and process company will have a double effect (both the micro-economic effect for the work organization that implements it, and the macroeconomic effect for the whole country). On the one hand, the company will reduce its energy costs, with the volume and size of potential savings increasing with the age of the plant, as well as the quality of the conducted maintenance, and on the other hand, the benefit for the state lies in the fact that energy consumption will remain at the same constant level or will be reduced with the increased industrial production and construction of new factories. The need for investment in new energy sources is also being reduced. Therefore, the state should stimulate investments in energy efficiency projects and rationalization of energy consumption with its policy (primarily tax and credit policy). A special effect in rational energy consumption is also from the aspect of environmental protection. As increased energy consumption disrupts the maintenance of the existing ecological balance, any reduction in energy consumption has an impact on the micro climate (from improving working conditions to reducing pollution closer to the environment, but also to the entire humanity, because pollution of nature does not recognize state boundaries). The rationalization of energy consumption is a set of measures that, with changes in the organization of work, the way of using plants and devices with accompanying equipment, environmental and material management, taking into account work safety, health and environment protection, achieves optimal productivity, the required level of product quality, as well as profitability and economy, with the accompanying reduction of the energy consumption per unit of product, Table 4.

Rationalization of energy use and obstacles to their implementation can be grouped into four groups, Table 5.

As the rationalization of energy consumption in an industrial plant is a complex problem that touches upon different scientific fields of science, that is, it has an interdisciplinary character that requires cooperation between experts of various profiles (energetics, electrical technicians, electronics, mechanics, experts of mechanics, chemistry, economy, etc.) the implementation of certain measures sometimes requires the participation of specially trained specialists in this area. It is important to note that the implementation of measures to increase efficiency and rationalization of energy covers the whole life cycle of the technical system, starting from the development of the idea and the development of an investment project (basic working life), as well as the reconstruction, revitalization and modification of existing equipment on the technical system (extended working life), as well as in the eventual upgrading and maintenance of existing systems. Thus, the assessment of all potential energy management measures (except for technical justification) requires an economic analysis, which should include, with pure economical indicators justification of investments in some measure or measures related to energy efficiency and financing, with precise identification of the source of funds, which today can pose a major problem for countries in transition like BiH, [3].

Depending on the financial resources needed for financing, potential measures can be classified into measures that do not require the investment of financial resources, but mainly a certain effort and changing habits of people, then into the measures whose realization needs small investments that are carried out once and for which the costs are not large (but not negligible), as well as measures requiring significant (large) financial resources, which are carried out once during the working life of the technical system, Table 6.

8 Conclusions

Starting from the definition of energy efficiency as the relationship between the achieved results in the production of electricity and the energy consumed by energy sources and losses related to their own consumption on the power plant, it is necessary to use the existing methodology for assessing the operation of the power plant (PP) within the electric power system (EPS) calculate certain efficiency indicators, such as: coefficients of time and power utilization, energy efficiency coefficient of

Table 4 Review of optimization result of parameters from experimental teams at TPP

Expert team	Project assignment	Recommendations and data contained in the regime card	Note
Team for preparing, monitoring and analyzing the quality of fuel	Regime card for optimal delivery of coal delivery	1. The method of receiving coal—monitoring of the quality and quantity from Rudnik 2. The method of disposal and mixing on coal landfill of power plants 3. Disposal of coal on landfill by quantity and quantity 4. Plan for taking coal from the landfills and loading boiler bunkers 5. The method of sampling coal from the landfills and the type of analyzes to be carried out 6. Cooperation with other teams and exchange of results 7. Cooperation with operating personnel	A. By creating and applying the registry card, it is impossible for coal of unknown quality to reach the boiler furnace B. The regime chart is elaborated in detail with the exploitation personnel of the delivery and preparation of coal, with strict adherence to the prescribed recommendations C. The application of the registry card makes operation easier for the staff members of the block boiler unit and has a positive effect in the operation of the boiler within the thermal power plant as a whole

(continued)

Table 4 (continued)

Expert team	Project assignment	Recommendations and data contained in the regime card	Note
Optimization team of parameters for technological boiler plant process	Parameters and sizes contained in the regime card of optimum control of the block boiler plant of the thermal power plant	1. Flow and temperature of the supply water 2. Media temperature in the drum (for boilers with a drum) 3. The temperature of the medium at the exit from the furnace part (if it works with flow boilers with supercritical parameters) 4. The medium temperature at the outlet from the individual steam heaters 5. The medium temperature and pressure of the primary steam at the boiler output before the turbine 6. The temperature of the inter heated steam at the inlet and outlet from the boiler (if the boiler has interheaters) 7. Number of fans: flue gases, fresh air and recirculation of cold flue gases and hot air (if it is built in boilers) during operation 8. Fan speed 9. Number of mills in operation 10. The height of the coal layer in the dispensers, with a note of the height of each dispenser, with the possibility of regulating the dosing of the coal (prescribed values), the prescribed values of the position of the device for the arrangement of coal powder on burners and the fineness of the melt (separate mill valves) 11. Subpressures in front of mills and temperatures aero compound behind mills 12. The content of oxygen in the flue gases behind the individual heating surfaces of the boiler, with the flow of air to the boiler 13. Percentage of openness of the air valves on secondary air for coal burners and valves during the addition of primary and tertiary air 14. Percentage of openness of the valves during the addition of recirculated cold gases (if it is used in the process) and the openness of the valves during the addition of air for drying	A. Uses the results of the team for the preparation, monitoring and analysis of the fuel quality in the development and the analysis of the results for adjustment and researching of the boiler plant, and making of technical report and regime card of the boiler plant technological process B. The regime card is made after the finishing optimization of the regime for maximal, nominal, minimal power and three powers between the nominal and the minimum C. The regime card should contain the names of parameters, mark and power with values and instructions D. There is also a regime for the cleaning of heating surfaces for existing systems (if the boiler has them) E. Recommendations are given for some of the plants and incidents that occurred during the adjustment and testing phase F. Team leader with specialists represents regime card to personal staff and explains its application G. The regime card is placed in a prominent place in the thermal command of the block so that operators can use it in managing, guiding, comparing sizes and analyzing parameters H. Exploitation personnel send remarks on the application of the regime card (if any illogicality appears) to the leaders of the optimization team, who, with their associates, investigate, re-examine and testify it

(continued)

Table 4 (continued)

Expert team	Project assignment	Recommendations and data contained in the regime card	Note
Team of parameters optimization for technological process of boiler plant	Regime card for maximum, nominal, minimum power and three powers between nominal and minimum	1. Maximum temperature in the furnace for boiler mode without slag 2. The flue gas temperatures along the boiler flue (behind the hot surfaces of the steam preheater) and the flue gases behind the air heaters 3. Pressure at the top of the boiler furnace 4. Underpressure behind the water heaters 5. Resistances in air heater 6. Underpressure in front of smoke fans 7. The air temperature behind the calorifier before the input in the air heater 8. The minimum temperature of the walls of the heat exchanger tube at the inlet (if the heater is tubular and if the coal has a high sulfur content) 9. The content of compliant in slag and ash in%	
Type of parameter optimization for turbine plant	Regime card and accompanying technical reports with the presentation to the exploitation staff	1. Pressure and temperature of overheated steam before turbine 2. Pressure and temperature of the lubricating and regulating oils 3. Axial displacement of the turbine rotor 4. Pressure in the chamber of turbine regulation grade (with regeneration turned on and off) 5. Maximum power (with regeneration turned on and off) 6. Vacuum in a turbine condenser 7. The flow of demi water into the condenser 8. Media level in the condenser 9. Steam pressures and temperatures at the subdued parts from the turbine towards the consumers of steam 10. Level in high and low pressure heaters 11. Pressure in the water tank 12. Pressures and media temperatures in regenerative devices (LPP, HPP etc.)	A. The parameters and values, achieved through the settings and tests, processed and optimized are entered into the regime card B. The regime card contains examined regimes

(continued)

Table 4 (continued)

Expert team	Project assignment	Recommendations and data contained in the regime card	Note
Parameters optimization team of the technological process for generator and transformer plant	Regime chart and accompanying technical report, with presentation to the exploitation personnel	1. Parameters and sizes obtained by the adjustment settings and tests processed and optimized for generator plant and supporting equipment 2. Parameters and sizes obtained by adjustments and tests processed and optimized for transformer plant and supporting equipment 3. Parameters and sizes obtained by the settings and tests processed and optimized for the transmission network	A. The results of the optimization of the generator process and the transformer plant are given in the form of a technical report and represent to the exploitation personnel of this part of the installation and management of the plant B. Parameters and sizes obtained by setting and tests processed and optimized are entered in the regime card C. The regime card contains values for examined regimes

LPP low pressure feed water heater, *LHP* high pressure supply water heater

Table 5 Measures of rationalization of energy use and prevention for their implementation

Ordinal number	Name of the group	Description of the measures for action	Obstacles to their implementation
1	Improving the level of action (technical and organizational measures)	– increase the level of action in the production, transformation, accumulation and transport of energy – for a certain amount of energy consumed, primary energy consumption should be reduced (burning or burner change etc.)	– technical nature (lack of information on appropriate technical solutions and lack of experience in energy management, inability to analyze energy consumption due to the lack of special equipment and measuring instruments) – financial nature (lack of money, high interest rates and the lack of a simple and acceptable method of financing the purchase of equipment) – economic nature (energy prices, which are not real compared to the price of the product, lack of knowledge about the exact share of energy costs in the price of the product)
2	Improving efficiency of use (measures of reconstruction, revitalization and modernization)	– existing useful energy to be used more efficiently (e.g. better insulation, increased load, reconstruction of obsolete plants, etc.)	
3	Managing and directing consumption (forced measures)	– direct impact on the needs of useful energy (e.g. lower room temperature, exclusion of excessive lighting, higher technological discipline) – their implementation depends largely on the human factor, but they can be accepted as a completely normal state over time	
4	Long-term structural reduction of needs (planning measures)	– energy consumption is reduced by the change in the structure of production – high-energy products are abandoned in favor of high-productivity products with low energy consumption per unit of product (application of new technologies) – depends on the company's business policy and requires long-term planning	

Table 6 Division of potential measures of energy management related to financial funds

Ordinal number	Group of measures	Alternative name	Short description	Dynamics of realization
1	Measures without financial	Good management or Household	– do not require the investment of financial resources but mainly a certain effort and changing habits of people – they do not require the use of new technologies	They are delivered regularly, at least
2	Measures for whose implementation insignificant (small) investments are	Balance management	– for which costs are not large, but not negligible – the costs of implementation include the costs of purchasing certain devices, installation and running costs of devices and people – mainly existing technologies are used, so the requirements for new technology are small	At least once a year (regular overhaul)
3	Measures requiring significant (large) financial resources	Strategic Management	– which are carried out once during the working life of the technical system – require a detailed technical and economic analysis of justification of the investment and sources of means	The implementation of these measures must be closely related to the planning of a company's business policy so as not to implement a measure in the production process that is relatively quick to leave (before the funds are returned)

constituent plants and block as a whole, method and quantity of produced electricity and spent fuel energy, amount of used heat and electricity for own needs, etc. Proper calculation of the achieved block characteristics enables evaluation of energy efficiency. The energy efficiency of the power plant should follow the activities and actions that in normal circumstances lead to a controlled and measurable increase of the energy efficiency of the block facility, technical block systems, production processes and primary energy savings. These activities are based on the application of energy-efficient technologies, i.e. processes that achieve energy savings and others that have positive effects, and can include proper handling, maintenance and adjustment on a thermoelectric power plant, improved efficiency of existing equipment and systems without any changes in any production the process of a given plant, or in an energy supply system. The process of engineering optimization represents a systematic search for the optimal solution of a given engineering problem, taking into account the defined criteria of optimum and in the conditions of meeting the given limits. The assumption is, of course, that the problems are sufficiently "excess" internal degrees of freedom (those that can be decided by the designer because they are not determined by the physical legitimacy of the problem) which become variable of the optimization process. Goals of predicting reliability, i.e., of process of determining numerical values for the ability of the structure to meet the established reliability requirements are: feasibility assessment, comparison of possible solutions, identification of possible problems, provision and maintenance planning, determination of data defects, harmonization in cases of interdependence of parameters, reliability allocation, and measurement of progress in achieving established reliability.

References

1. Milovanović NZ (2003) Optimization of power plant reliability. University of Banja Luka, Faculty of Mechanical Engineering Banja Luka, Banja Luka, 323 pp
2. Papić RLj, Milovanović NZ (2007) Maintenance and reliability of technical systems, DQM monograph library Quality and reliability in practice, Book 3, Prijevor, 501 p
3. Milovanović NZ, Papić RLJ, Dumonjić-Milovanović RS, Milašinović NA, Knežević MD (2017) Sustainable energy planning: technologies and energy efficiency, DQM monograph library Quality and reliability in practice, Book 9, Prijevor, 779 pp
4. Milovanović NZ (2011) Monographs: "energy and process plants" tom 1: thermal power plants—theoretical foundations. University of Banja Luka, Faculty of Mechanical Engineering Banja Luka, Banja Luka, 431 pp
5. Milovanović NZ (2011) Monographs: "energy and process plants" tom 2: thermal power plants—technological systems, design and construction, exploitation and maintenance. University of Banja Luka, Faculty of Mechanical Engineering Banja Luka, Banja Luka, 842 pp
6. Dueck G, Scheuer T (1990) Threshold accepting: a general purpose optimization algorithm appearing superior to simulated annealing. J Comput Phys 90:161–175
7. Rechenberg I (1965) Cybernetic solution path of an experimental problem. Royal Aircraft Establishment. Library translation No. 1122, Farnborough, Hants., UK

8. Holland J (1975) Adaptation in natural and artificial systems. The University of Michigan Press, Ann Arbour, ISBN 10: 0-262-58111-6
9. Dorigo M, Maniezzo V, Colorni A (1996) The ant system: optimization by a colony of cooperating agents. IEEE Trans Syst Man Cybern B 26(1):29–41
10. Moscato P (1989) Report No. 826 (C3P). Caltech concurrent computation program, Pasadena, CA, USA, pp 158–179
11. Pareto V (1896) Cours d'economie politique professe a l'universite de Lausanne, vol 1, 2, F. Rouge, Laussanne, 97 pp
12. Saaty TL (1980) The analytic hierarchy process: planning, priority setting, resource allocation. McGraw-Hill, New York
13. Chan FTS, Kumar N (2007) Global supplier development considering risk factors using fuzzy extended AHP-based approach. Omega 35(4):417–431
14. Chang D (1996) Applications of the extent analysis method on fuzzy AHP. Eur J Oper Res 95(3):649–655
15. Van Laarhoven P, Pedrycz W (1983) A fuzzy extension of saaty's priority theory. Fuzzy Sets Syst 11(1):199–227
16. Buckley JJ (1985) Fuzzy hierarchical analysis. Fuzzy Sets Syst 17(3):233–247
17. Boender CGE, De Graan JG, Lootsma FA (1989) Multicriteria decision analysis with fuzzy pairwise comparisons. Fuzzy Sets Syst 29(2):133–143
18. Feng DZ, Chen LL, Jiang MX (2005) Vendor Selection in supply chain system: an approach using fuzzy decision and AHP. In: Proceedings of ICSSSM'05 services systems and services management, pp 721–725
19. Haq AN, Kannan G (2006) Fuzzy analytical hierarchy process for evaluating and selecting a vendor in a supply chain model. Int J Adv Manuf Technol 29(8):826–835
20. Chamodrakas I, Batis D, Martakos D (2010) Supplier selection in electronic marketplaces using satisfying and fuzzy AHP. Expert Syst Appl 37:490–498
21. Onut S, Kara SS, Isik E (2009) Long term supplier selection using a combined fuzzy MCDM approach: a case study for a telecommunication company. Exp Syst Appl 36(2):3887–3895
22. Krajnc D, Glavič P (2005) A model for integrated assessment of sustainable development resources. Conserv Recycl, Elsivier, Amsterdam, pp 189–208
23. Sirikrai SB, Tang JCS (2006) Industrial competitiveness analysis: using the analytic hierarchy process. J High Technol Manage Res 17:71–83
24. Nagesha N, Balachandra P (2006) Barriers to energy efficiency in small industry clusters: multi-criteria-based prioritization using the analytic hierarchy process. Energy 31:1969–1983
25. Singh RK, Murty HR, Gupta SK, Dikshit AK (2007) Development of composite sustainability performance index for steel industry. Ecol Ind 7:565–588
26. Ugwu OO, Haupt TC (2007) Key performance indicators and assessment methods for infrastructure sustainability-a South African construction industry perspective. Build Environ 42(2):665–680
27. Tseng ML, Lin YH, Chiu ASF (2009) Fuzzy AHP based study of cleaner production implementation in Taiwan. PWB manufacturer. J Clean Prod 17(14):1249–1256
28. Chatzimouratidis AI, Pilavachi PA (2009) Technological, economic and sustainability evaluation of power plants using the analytic hierarchy process. Energy Policy 37(3):778–787
29. Perez-Vega S, Peter S, Salmeron-Ochoa I, Nieva-de la Hidalga A, Sharratt PN (2011) Analytical hierarchy processes (AHP) for the selection of solvents in early stages of pharmaceutical process development. Process Saf Environ Prot 89:261–267
30. Hsu CW, Hu AH, Chiou CY, Chen TC (2011) Using the FDM and ANP to construct a sustainability balanced scorecard for the semiconductor industry. Expert Syst Appl 38(10):12891–12899
31. Bottero M, Comino E, Riggio V (2011) Application of the analytic hierarchy process and the analytic network process for the assessment of different wastewater treatment systems. Environ Model Softw 26:1211–1224
32. Hobbs BF, Meier PM (1994) Multicriteria methods for resource planning: an experimental comparison. IEEE Trans Power Syst 9(4):1811–1817

33. Huang J, Poh K, Ang B (1995) Decision analysis in energy and environmental modeling. Energy 20(9):843–855
34. Landhelma R, Salminen P, Hokkanen J (2000) Using multicriteria methods in environmental planning and management. Environ Manage 26(6):595–605
35. Afgan NH, Carvalho MG (2002) Multi-criteria assessment of new and renewable energy power plants. Energy 27:739–755
36. Cormico C, Dicorato M, Minoia A, Trovato M (2003) A regional energy planning methodology including renewable energy sources environmental constraints. Renew Sustain Energy Rev 7:99–130
37. Pohekar SD, Ramachandran M (2004) Renew Sustain Energy Rev 8(377):365–381
38. Roh S (2012) The pre-positioning of humanitarian aid: the warehouse location problem. A thesis submitted in fulfilment for the requirements for the Degree of Doctor of Philosophy in Cardiff University, Transport and Shipping Research Group, Logistics and Operations Management Section of Cardiff Business School, Cardiff University, 373 pp
39. Roh SY, Jang HM, Han CH (2013) Warehouse location decision factors in humanitarian relief logistics. Asian J Shipping Logistics 29(1):103–120
40. Roh S, Pettit S, Harris I, Beresford A (2015) The pre-positioning of warehouses at regional and local levels for a humanitarian relief organization. Int J Prod Econ 170(Part B):616–628
41. Agyei W (2015) Project planning and scheduling using PERT and CPM techniques with linear programming: case study. Int J Sci Technol Res 4(8):222–227
42. Sharma JK (2007) Operations research; theory and application, 3rd edn. Macmillan India Ltd., pp 525–557
43. Soroush HM (1994) The most critical path in a PERT Network. J Oper Res 45(3):286–300
44. Hajdu M (2015) Point-to-point versus traditional precedence relations for modeling activity overlapping. Proc Eng 123:208–215
45. Francis A, Miresco ET (2010) Dynamic production-based relationships between activities for construction project planning. In: The international conference on computing in civil and building engineering, Nottingham, United Kingdom, June 30–July 2
46. Hajdu M (2015) One relation to rule them all: the point-to-point precedence relation that substitutes the existing ones, In: 5th international/11th construction specialty conference, Vancouver, Canada, June 8–10
47. Tangen S (2005) Demystifying performance and productivity. Int J Prod Perform Manage 54(1):34–46
48. Aruldoss M, Travis ML, Venkatesan VP (2014) A survey on recent research in business intelligence. J Enterp Manage 27(6):831–866
49. Aruldoss M, Lakshmi TM, Venkatesan VP (2013) A survey on multi criteria decision making methods and its applications. Am J Inf Syst 1(1):31–43
50. Milovanovic Z (2000) Modified method for reliability evaluation of condensation thermal electric power plant (in Serbian), Ph.D. thesis, University of Banja Luka, Faculty of Mechanical Engineering Banja Luka, Banja Luka
51. Soliman SA, Mantawy A-AH (2012) Modern optimization techniques with applications in electric power systems. Springer, New York, p 414
52. Corne DW, Knowles JD, Oates MJ (2000) The Pareto envelope-based selection algorithm for multiobjective optimization. In: Proceedings of the parallel problem solving from nature VI conference, Springer, Berlin, pp 839–848
53. Sumathi S, Surekha P (2010) Computational intelligence paradigms: theory and applications using MATLAB. CRC Press, Taylor and Francis Group, Boca Raton, 821 pp
54. Zhao B, Guo CX, Cao YJ (2005) A multiagent-based particle swarm optimization approach for optimal reactive power dispatch. IEEE Trans Power Syst 20(2):1070–1078
55. Haykin S (81998) Neural networks: a comprehensive foundation, 2nd Edn. Prentice Hall, 842 pp
56. Rao SS (2009) Engineering optimization—theory and practice, 4th edn. Wiley Inc, New Jersey, p 813

57. Mučibabić DS (1995) Multicriteria aspects of decision making in conflict situations (in Serbian), Ph.D. thesis, FON, Beograd, 213 p
58. Mučibabić DS (1997) Multicriteria approach to solving real conflict situations. Math Moravica 1(1):73–85

Planning Methods for Production Systems Development in the Energy Sector and Energy Efficiency

Z. N. Milovanović, Lj. R. Papić, S. Z. Milovanović, V. Z. Janičić Milovanović, S. R. Dumonjić-Milovanović and D. Lj. Branković

Abstract Economic development of a country requires the consumption of appropriate energy resources. Any deviation from the timing as a result may have the appearance of development restrictions in the other economic activities, so the development of energy consumption should be continuously monitored. On the other hand, energy consumption itself is linked to certain influential factors (population growth, science and technology development, economic development, standard, etc.), which intensity of activity changes over time. Choosing the optimal structure for covering consumption is very important for the development of energy. Planning of development in the field of energy is important as for the dependence of the development of the society on safe, sufficient and appropriate quantities of necessary forms of energy, as well as for the engagement of large financial resources in this field. When planning the development of energy, the following criteria should be followed: the security

Z. N. Milovanović (✉)
Department of Hydro and Thermal Engineering, Faculty of Mechanical Engineering Banja Luka, University of Banja Luka, Stepe Stepanovića71, Banja Luka, Bosnia and Herzegovina
e-mail: zdravko.milovanovic@mf.unibl.org

Lj. R. Papić
DQM Research Center, Prijevor, Poštanski fah 132, Čačak, Serbia
e-mail: dqmcenter@mts.rs

S. Z. Milovanović
Department of Construction Project Organisation, Technology and Management, Faculty of Architecture, Civil Engineering and Geodesy, University of Banja Luka, Stepe Stepanovića 77/3, Banja Luka, Bosnia and Herzegovina
e-mail: snjezana.milovanovic@aggf.unibl.org

V. Z. Janičić Milovanović
Routing Ltd., Prvog Krajiškog Korpusa 16, Banja Luka, Bosnia and Herzegovina
e-mail: valentina.mil@live.com

S. R. Dumonjić-Milovanović
Partner Engineering Ltd., Kralja Nikole 25, Banja Luka, Bosnia and Herzegovina
e-mail: svetlanadm@ymail.com

D. Lj. Branković
SHP Celex, Veljka Mlađenovića bb P.O. Box 142, Banja Luka, Bosnia and Herzegovina
e-mail: dejan.brankovic@shpgroup.eu

M. Ram and H. Pham (eds.), *Advances in Reliability Analysis and its Applications*, Springer Series in Reliability Engineering,
https://doi.org/10.1007/978-3-030-31375-3_3

of consumer supply with minimal costs, the rational use of domestic sources, with proper evaluation of imported energy forms, maximum prevention of monopolistic and single forms of energy, and achieving acceptable conditions for environmental protection and sustainable development. Planning of the development of the electric power system includes all activities from the first assumptions about the possibility of building an object until its entry into operation. But in terms of terminology, the term planning refers primarily to the planning of power generation plants (hydro, nuclear and thermal power plants). When planning the development of the electro-energy system (EES), the goal and criteria are uniquely determined: the settlement of the predicted consumption of electricity at a minimum cost and assuming that the specified limitations are met, such as, financial, technical, ecological, limitations on the availability of primary forms of energy, etc. Satisfying the limitations is imposed as a primary task, regardless of the method of planning, which is the reason that a particular limitation is often the decisive factor in deciding on the final strategies of development. Although long-term plans bear a great deal of uncertainty, such planning is necessary primarily for two reasons: the first is the basic and extended lifetime of the production plants (e.g., 25–30 basic plus 15–20 years of extended service life after revitalization, reconstruction and modernization of thermal power plant), while the second is the time necessary for the preparation of the construction and the construction itself (3–6 years, not counting the possibility of a delay in the project). Planning of the construction of production plants should be determined by the necessary construction to meet the future consumption (volume and capacity), the time of entry of a particular production plant into operation and the possibility and improvement of technologies for production of electricity (improved and cleaner technologies, cogeneration and trigeneration systems, hybrid systems, …). Only the making of development studies can be divided into two parts, which include simulation of the legality of work in the system (system operation), as well as the economic evaluation of production facilities or entire development plans. The first part requires the development of a system model, i.e. it is necessary to describe the system in mathematical equations and to approximate it with the inevitable neglect and simplification. In the second part, the energy contribution of each plant should be evaluated and by economic methods to do its valorization. The methods used in planning the development of the electric power system differ with respect to: optimization technique (linear programming, nonlinear programming, etc.), type of approximation (linear, nonlinear) and economic valorization (with inflation, without inflation). None of these methods has proven to be absolutely acceptable for all problems so far, so a large number of methods are being developed that are intended to solve partial problems, i.e. or they serve to optimize the self-production facilities, or optimize only the transmission/distribution network, or the process of optimization refers to the whole energy (with very simplified relationships in the given system). The basic assumption of applying optimization models is the high reliability of the parameters on which the budget is based. Namely, if the parameters are not reliable enough, the question arises of the need for optimization. Sometimes, even the sensitivity analysis, usually used in the final phase of any optimization model, cannot eliminate effective reliability and input data. Furthermore, it is never possible to include all the limitations within the model, so that only an approximately objective picture of the condition in the system is obtained. However, today, it is impossible to rely solely on the intuition of planners in selecting the most favorable development option, so the application of the model is inevitable (more criterion evaluation methods). Energy development planning is based on the security of consumer supply with minimal costs, with the accompanying rational use of domestic resources, which implies correct evaluation of imported energy forms, maximum prevention of monopolistic behavior (the only form of energy available) and achieving satisfactory environmental protection conditions. When planning the development of the electricity sector of any country, the goal and criteria are uniquely determined through the settlement of the estimated electricity consumption, with minimum costs and assuming that certain financial, technical and environmental limitations on the availability of primary forms of energy are met. The new value of own consumption of a thermal power plant, which can be achieved by applying the proposed measures, is still above reference values in relation to similar systems in the world, which means that there is still a serious work on analyzing possible savings and reducing losses as a whole. It should be noted that it is expected that in the first period of implementation of the energy management system, several short-term measures will be identified, which do not require financial investments and are more of an organizational nature. After the introduction of the BAS ISO 50001 standard and the application of appropriate standardized procedures in practice, the number of short-term measures will be reduced, which also indicates an organized monitoring of the implementation of energy efficiency measures.

Keywords Energy sector · Production systems · Development planning · Planning methods · Optimization criteria · Optimization models · Energy management · Energy efficiency

1 Introduction

Due to its light transformation into other forms of energy, as well as possible transport to a greater distance by the transmission network (voltage levels 110, 220 and 400 kV), electricity is extremely important for the economic development of each country, [1]. Empirically it has been established that consumption of electricity per capita is one of the key parameters, which indicates the development of the national economy and the living standard of people of a particular country. Most predictions today show that in half a century the population of the country will double, before it is halted to 12 billion, which requires the provision of additional energy, [2].

All previous energy projections in the future have shown that fossil fuels will be an important source of energy in the upcoming period. Predictions in 2004 about directions of world energy in the future have shown that energy demand in the developing world will double in the next two decades, as the population and economies in developing countries grow faster than in industrialized countries. Finding ways to enable the development and ways to meet the growing world energy needs should at the same time mitigate the potential impacts of energy supply and use on the environment, while ensuring long-term quality for life on earth, [3].

Requirements for the continuous supply of energy needs in sufficient quantities for industrial plants, traffic and standard of human life require the development of new technologies based on fossil fuels (fluidized bed combustion plants, combined cycles with gasification, combined cycles with natural gas as fuel, burning cells, technologies with external heat energy—Stirling machine, thermal photovoltaic conversion, thermal-electric converters with alkaline metals), as well as increasing energy efficiency (saving and rational use of energy, reduction of distribution and other losses).

By analyzing today's current forms of conventional (thermal power plants, heating power plants gas power plants, nuclear power plants and large hydropower plants) and unconventional energy sources (such as small hydropower plants, active thermal and photovoltaic use of solar radiation, wind, ambient heat, biomass and waste, geothermal energy), it can be determined in each case that there are certain deviations from the general properties (in the form of certain improvements or deterioration), [1, 4–7]. The connection between the use of certain energy sources with savings and more efficient use has a double significance. On the one hand, rational use of energy is a direct means to reduce total energy consumption and environmental impact, and on the other hand, the special designs of devices and facilities for the use of energy increasingly enable the participation of non-conventional or new renewable energy sources.

The energy system in sustainable development is defined in accordance with the six compliance criteria: environmental compliance, intergenerational compliance, compliance of consumption, socio-political coherence, geopolitical compliance and economic compliance. The aim of the research on sustainable development is to integrate ecological, economic and social dimensions into the socio-ecological system, which is being continued, while maintaining the required state of balance (sustainability). As sustainable development is not limited to a clearly defined balance situation, it already corresponds to a more dynamic process, in which priorities and actions are continually redefined according to the needs and wishes of the energy consumption observed. The primary function of the indicator of sustainable development is the evaluation, assessment and condition of the three dimensions of the socio-ecological system (society, economy and the environment), in which sustainable development must more strongly reflect on inter-dimensional relations than on intra-dimensional relations.

Starting from the main components of sustainable development in the field of energy, such as availability, affordable price, energy security, energy efficiency, ecological acceptability and possible risks, it is necessary to further stimulate the production of useful energy produced and used in the same way to assist the development of mankind through longer period of time. Energy must be available and acceptable as a supply service, but also available and reliable as an energy service. On the other hand, there is no risk-free technology that has no waste and does not affect the environment, so it does not make sense to speak isolated about a particular better or worse technology for transforming the primary into useful forms of energy. It is better to compare the characteristics of one technology or energy service with possible alternative solutions. Every economic development of the country requires the consumption of appropriate energy resources. The occurrence of a deviation from the time interval for the result may have the appearance of restrictions in the development of other economic activities, for example, the lack of electricity leads to major disruptions in industrial production, which in turn entails significant losses in the final economic balance of the given country. For these reasons, it is necessary to monitor the development of energy consumption, which is related to certain influential factors (population growth, development of science and technology, economic development, standard, etc.), whose intensity of activity changes in time. Choosing the optimal structure for covering consumption is very important for the development of energy. It is a very complex problem, the solution of which depends not only on the energy sources of the country, but also on the mutual dependencies between primary and transformed forms of energy, then forms of energy that cannot be replaced by other types of energy, as well as the possibilities and needs of importing certain forms of energy, location consumers, consumption characteristics, etc. It is understandable that the situation and development in the field of energy are constantly being studied today.

2 Review of Past Research

Energy consumption is a key factor that causes social stratification to those that can and cannot fully afford it. As energy becomes more expensive in the future, and the impact of energy activities on the environment becomes more pronounced, rational use of energy will lead to less energy consumption per unit of national product. As, regardless of the changes in the structure of world and national economies and on energy saving measures or rational use of energy, energy consumption in the world is constantly increasing at a rate of about 1–3% per year over the past thirty years, with an increase in energy consumption more pronounced in those countries that recorded and economic growth, [8]. According to [9–16], the use of energy to a certain limit allows a social progress. Above this limit, energy consumption is growing at the expense of equality, i.e. of total well-being, since fewer people can afford the benefits of using it. In that case, the growth of a gross national product can only be an illusion of increased welfare (only a small number of people enjoy this welfare).

Starting from the correlation between gross national income and energy consumption per capita, from the economic point of view, the prosperity of a nation can be identified with high energy consumption per capita. On the other hand, from the thermodynamic point of view, prosperity could be characterized by high efficiency in energy transformations, while from the social aspect prosperity would result in the least possible use of the mechanical energy of the most eminent part of the society. Many authors consider that large amounts of energy are destroying the social structure [8–10, 12], Illich [10]. Thus, Ivan Illich and Howard T. Odum, looking at the same problem (development of civilization) through the same indicators (energy consumption) and at different times (time difference from 30 to 40 years), using different methods (Odum used system models and developed special methods of calculating energy for those models, Ivan Illich came to philosophical conclusions by thinking), they came to practically the same conclusions (Ivan Illich believes that with a low energy consumption, human civilization would have a wide choice of different lifestyles and cultures, but in a society based on high energy consumption, technocracy is the one that dictates social relations, and such a society collapses wether it was called capitalism or socialism), [10–12].

Starting from the fact that more than 70% of the world's energy needs are currently met on fossil fuel (combustion of coal, oil and natural gas), and that the resulting harmful products of combustion create major problems for the possible consequences of global climate change on a global scale (accumulation of greenhouse gases in the atmosphere, largely as a result of combustion of fossil fuels), it is clear that from the ecological point of view there must be a limit to the destruction of natural resources (claims and tests for energy products, their extraction or extraction from the soil or their use by combustion). Paris conference and previous ones on a world-level try to agree on greenhouse gas emissions, primarily carbon dioxide, [8].

The less accepted fact is that there are limits to the growth of energy consumption from a social point of view, beyond which a greater consumption of energy destroys the social structure, which does not have to coincide with the limit of acceptable physical destruction of the environment, [1].

The phenomenal growth of the world economy has so far been based on the use of fossil fuels (coal, oil and natural gas), which had a very high energy return on energy investment, even above 50 (*Energy Return On Energy Investment*, EROEI), while renewable energy sources have significantly smaller EROEI, [2–4]. It is also a fact that such growth would not be possible with renewable energy sources (EROEI for fossil fuels decreases over time, which has the effect that energy is becoming more and more expensive).

On the other hand, the steady growth of the economy and gross national income requires more and more energy. As for the energy sources used so far, there are limitations in quantity (the reserves, i.e., the physical quantities of coal, oil and gas in the country are final) or in the inflow (solar radiation comes to the earth's surface with a maximum power of 1 kW/m^2, about 200 W/m^2), the alternatives grow and are steady state or oscillation, [1, 8]. The constant level, as well as the mean value of the oscillation, can be at some level, greater than or equal to the current level. The key question posed by analyzes is whether the economy can be maintained at a high level with available resources based on renewable energy sources or the future is reduced to less energy inevitable. In addition, the future with less energy does not have to be catastrophic, so instead of sustainable development one should begin to think about the so-called prosperous decline, as described by the famous ecologist Howard T. Odum in his book A Prosperous Way Down, [12]. There are also thoughts about the aspirations towards the so-called steady state economy, [13] whereby sustainable development is still possible in such an economy, but, of course, development does not have to mean growth. In an economy that stagnates or decreases the key values should become: less instead of more, more efficient instead of faster, more economical rather than more costly, and cooperation instead of competencies. On the other hand, the key influencing factors in future planning, which positively and negatively influence the choice of solutions, are [17]:

- Restriction of greenhouse gas emissions for the post-Kyoto period, as a global agreement to mitigate climate change, which will result in a very strict obligations by the EU and its members in terms of emission reduction;
- Increasing energy demand at global and European level, as well as in the region and in the countries of the Balkans (raising the general standard at the expense of increasing energy consumption, while ensuring development in both countries in transition and countries with a weaker level of development in the EU);
- Increased demand for energy in industry, services, transport and households (mitigating the growth of energy demand at the expense of greater energy efficiency, technological development, legislative norms, standards, organization of business activities and the economic strength of an individual, company, and every country as a whole);
- The development of the energy market, the establishment of unique rules of market functioning, and the effectiveness of the mechanisms of compulsion to respect the uniform rules;
- Technological development; (the development of new or accelerated technologies that reduce greenhouse gas emissions, nuclear power plants, renewable energy sources and energy efficiency, and new devices that are needed for citizens and businesses with lower energy consumption);
- Build-up and construction of network infrastructure, interconnection of national networks and construction of transnational networks (analysis of the impact on the structure of sources and delivery directions, with associated material and non-material costs);
- Coherence (global) energy policies with other policies (primarily food, science and technological development policies);
- Citizens perception, acceptability and marketing of particular technologies;
- Energy price for the last consumer, which includes realistic environmental protection prices;
- Development of international relations, in particular the development of institutional relations in the EU and the process of EU enlargement.

Strict restrictions on greenhouse gas emissions in the production, transformation, transport, distribution and energy consumption to reduce their concentration in the atmosphere are directly reflected on the price of energy (the cost of reducing greenhouse gas emissions). Global emission policy will also increase energy costs, and the cost of reducing greenhouse gas emissions is the result of all the above-mentioned impact factors.

3 Basic Planning Schemes in the Field of Energetics

Planning of development in the field of energetic is important because of the dependence of the development of the society on safe, sufficient and appropriate quantities of necessary forms of energy, as well as because of the engagement of large financial resources in this field, [3, 5, 18]. When planning the development of energy, the following criteria should be considered: the security of consumer supply with minimal costs, the rational use of domestic sources, with proper evaluation of imported energy forms, maximum prevention of monopolistic and unique forms of energy, and achieving satisfactory conditions for environmental protection and sustainable development. Planning of the development of the electro energy system includes all activities from the first assumptions about the possibility of building an object until its entry into operation. But in terms of terminology, planning is primarily meant for the planning of power generation plants (hydro and thermal power plants).

When planning the development of the electric power system, the goal and criteria are uniquely determined: the settlement of the anticipated consumption of electricity with minimum costs and assuming that certain restrictions are met, such as, for example, financial, technical, ecological, limitations on the availability of primary forms of energy, etc. Satisfying the constraints is imposed as a primary task, regardless of the method of planning, which is the reason that a particular constraint is often the decisive factor in deciding on the final development strategy. Although long-term plans carry a great deal of uncertainty, such planning is necessary primarily for two reasons: the first is the basic and extended lifetime of the production plants (25–30 basic plus 15–20 years of extended service life after the revitalization, reconstruction and modernization of the plant) and the other is the time needed for the preparation of the construction and the construction alone (3–6 years, not counting the possibility of a delay in the project). It means that the analysis of the development plans of the electric power system should be carried out for a period of 15–50 years in advance. When planning the construction of production plants, the necessary construction for satisfaction of the future consumption (volume and capacity) should be determined, also the time of entry of an individual production plant into operation and the possibilities of improving the technologies for electricity generation (improved and cleaner technologies, cogeneration and tri generation systems, hybrid systems, etc.). Making of development studies can be divided into two parts, which include simulation of the legality of work in the system (system operation), as well as the economic evaluation of production facilities or entire development plans. The first part requires the development of a system model, i.e. it is necessary to describe the system with mathematical equations and to approximate it with the inevitable neglect and simplification. In the second part, the energy contribution of each plant should be evaluated and by using economic methods its valorization should be carried out.

Energy efficiency represents the relation between the achieved result in the production of electricity and spent primary energy for its production (from energy sources and for its own needs), [19–26]. With the notion of "efficacy", the terms "effectiveness" or "efficiency" are often used in the Serbian language for description of the action that produces the desired result, regardless of its characteristics and its related losses. Energy efficacy and energy efficiency always mean the same—to use the minimum energy for the same activity or for the same desired result, while retaining the achieved comfort (convenience) on the side of the energy user. Therefore, energy efficiency is a sum of planned and implemented measures, which aim is to use the minimum possible amount of energy, so that the level of comfort and level of production will remain constant, i.e. energy efficiency will result in less use of the amount of energy (energy products) for performing the same effect. In addition, energy efficiency does not imply energy savings. Namely, saving implies certain disclaimers, while efficient use of energy never disturbs the existing working and living conditions. Improving the efficiency of energy consumption does not only imply the implementation of technical solutions. Energy efficiency primarily represents the level of consciousness of people and their will to change established habits towards more energy efficient solutions. Satisfying the 3E form, which includes energy, economics and ecology, is a challenging task that is placed before the designers and the construction sector in the energy sector. On the one hand, we meet with the problem of new construction in line with the modern standard of living and sustainable development, and on the other hand with the problem of modernizing the existing construction, which in large percentage does not meet today's standards, consumes enormous amount of energy and becomes a significant problem and a big pollutant of the ecological system.

In order to achieve an ambitious level of energy efficiency by 2030, the European Commission has started to establish funds and instruments that treat energy efficiency as a separate source of energy. In 2015, the Commission proposed as a first step a revision of the Energy Efficiency Labeling Directive, which would increase the effectiveness of the existing legal settlement on the designation of energy efficiency and the strengthening of the bases for its realization. Also, in 2015, a number of measures for ecological design and energy efficiency labeling were introduced, which could further reduce household energy consumption and bills as well. Energy efficiency plays an important role in the European Investment Strategic Fund, which is already supported by strategic projects in the field of energy efficiency in France and Italy. There are many other projects that are pending an approval. This will complement investment from the European Structural and Investment Funds. The progress report on achieving the goal of increasing energy efficiency by 20% by 2020 indicates that, despite considerable progress, only 17.6% of primary energy savings have been achieved by joint efforts of member states compared to projections for 2020. Nevertheless, the Commission believes that the 20% aim can be achieved if the existing EU legislation is implemented properly and on the whole. Member States should become more ambitious and improve investment conditions so that energy efficiency in Europe can continue to grow. One of the goals of the energy union strategy is further progress from the economy based on fossil fuels. In 2015, progress has been made in three areas that are key to this transition—emissions trading, renewable energy sources and further investments in low carbon technologies and energy efficiency. An agreement on the introduction of market stability reserves, which will be available from 2019, will intensify the EU Emissions Trading System

(ETS) system. In mid-2015, the EU Commission presented a proposal for a revision of the EU's emissions trading system. This is the last step in adapting the EU's emissions trading system to be able to act as full force as the main European instrument for achieving greenhouse gas emissions targets for 2030. In support of the ambition of the EU to take the lead in the world in the field of renewable energy sources, the Commission also announced in 2015 an advisory communication on a new electricity market model, whose main goal is to prepare markets for an increasing share of renewable energy sources. Renewable energy sources become a common source of energy. They already meet the energy requirements of 78 million Europeans, while the EU is on the right track to reach the target of 20% of final energy consumption from renewable sources by 2020. Thus, in order to move to a low carbon economy (economy), significant investments are needed, especially in electricity networks, energy production, energy efficiency and innovation. The EU budget contributes to the achievement of this move by including climate goals in all relevant political initiatives and ensuring that at least 20% of the EU budget for the period 2014–2020 is linked to the climate. For this period, from 2014 to 2020, about 180 billion Euros will be allocated. More than €110 billion has been made available through European Structural and Investment Funds (ESIF). Furthermore, projects in the field of sustainable energy sources were one of the first projects approved for the guarantee of the European Fund Strategic Investment (EFSI), especially in Denmark, Finland, France, Spain and the United Kingdom. Based on the Commission's energy and climate framework by 2030, the EU defined as the binding goal of reducing domestic greenhouse gas emissions at the level of the entire economy for at least 40% by 2030 compared to 1990 levels, which was their starting point for a climate conference in Paris, which was a willingness to negotiate an ambitious, binding and transparent global climate deal that would clearly set the path to restricting the growth of the average global temperature to less than 2 °C. To achieve these goals, further decisive measures are needed at the local level. Bearing this goal in mind, on October 15, 2015, the Commission called representatives from cities to launch a new agreement together with the mayor to include mitigation of climate change and adaptation initiatives. The launch of a global agreement by the mayor will encourage the action of local authorities around the world, including in regions that have not been involved so far. Also, more than 4000 companies have committed themselves to taking action until the start of the Paris conference. Fulfilling these obligations on the grounds offers significant business opportunities for innovative EU companies, creating jobs and growth in the EU.

Energy efficiency is an integral part of the development guidelines of all sectors of the energy system. Particular attention should be paid to energy efficiency in the sectors of direct consumption, buildings, industry and traffic, as there are the greatest possible effects here. Energy efficiency measures reduce the increase in energy consumption, which reduces the need for building of new capacities, energy imports and they increase security of supply. Increasing energy efficiency with energy savings has contributed to the reduced use of fossil fuels and the reduction of emissions of harmful gases into the environment, the development of the economy, the increase in the number of jobs, and increased competitiveness. In discussing the possibilities and potentials of improving energy efficiency and rational energy management in all sectors of consumption in the Republic of Srpska, the current situation in the building sector—housing and non-residential buildings, transport and industry buildings—was examined, and on the basis of the collected data and the findings the possible concrete measures are identified which will increase energy efficiency in individual sectors of consumption, with an impact analysis and the consequences of their implementation. The emphasis was placed on the need to build an institutional and legislative environment as one of the basic preconditions for a successful implementation of energy efficiency measures. All available experiences from other countries are unanimous in concluding that without the incentive measures it is extremely difficult, almost impossible, to launch the implementation of energy efficiency measures (EE), which are seeking greater investments.

According to the Energy Sector Development Strategy of the Republic of Srpska until 2030, energy consumption in industry in the Republic of Srpska in the past period accounted for 12–25% of the total energy consumption in the Republic of Srpska, or from 18 to 35% of the share in energy consumption in the BiH industry, [23]. At the beginning of this decade, this share fell from 34 to 18%, to rise later to 35%. In absolute terms, energy consumption in the Republic of Srpska industry has been increasing since 2002. Given the share in energy consumption in industry, the most important industries are: metal production, chemical industry, non-metal minerals industry, mining and quarrying, food, beverage and cigarettes production, textile and leather production, paper and graphics production, industry. The consumption of final energy in the largest part, over 50%, is a waste of the metal production sector. In this sector, the Zvornik alumina factory has by far the largest share. For the same reason, natural gas is most involved in consumption among fuels. The next large-scale sector is the production of food, beverages and cigarettes, while the other individual sectors are significantly less represented. In the consumption of electricity, the largest percentage is spent on the food, beverage and cigarettes sector.

By comparing the form of final energy consumption, the most commonly used energy source is natural gas, whose consumption has varied in the previous period. Then the most consumed are liquids fuels, primarily oil fuel. Electricity follows, while wood waste, coal and other solid fuels are less represented. The scenario of the development of energy consumption in industry predicts a doubling of final energy consumption by 2030 according to scenario S1, by approximately 3% lower growth in scenario S2, and by about 12% lower growth in scenario S3, [23]. It is assumed that the processing industry has 4–5 times higher share of agriculture, construction and mining. Electricity in total consumption so far has less than a fifth. It is assumed that this percentage will grow, but for measures of increasing energy efficiency it is therefore advisable to concentrate on the rationality of using heat energy. Natural gas will retain the leading share among energy products, so the area of application of measures should focus on the efficiency of heat aggregates, reduction of distribution losses and efficiency of heat processes. In electrical energy, the greatest potential for the application of measures is in the efficiency of electric motors. Appropriate organizational measures—energy audits, sectoral analyzes and practicability studies to increase energy efficiency—are indispensable in the proper energy policy for industry. According to complexity and investment intensity, the applicable measures of increasing energy efficiency can be divided into measures: fundamental rationalization of energy consumption (behavior change, load management, etc.), revitalization of electrical and thermal infrastructure, compensation of reactive energy, consumer interventions (replacement, repair, etc.), the work on energy aggregates (repair, upgrading, etc.), the construction of a new energy system (new energy and energy infrastructure), as well as the introduction and improvement of complete regulation and automation. Improvements should be planned in such a way as to take measures according to the above schedule, due to the fact that the former imply smaller investments, simpler projects and faster cost-effectiveness. Taking more complex and expensive measures takes full meaning only when simpler and cheaper measures are implemented.

The stages of realization of energy efficiency projects in industry can be presented in the following hierarchy: preliminary audit (knowledge of potentials), full energy audit (realistic situation recorded and known necessary measures), practicability study (technical and economic analysis), project (technical details and financial plan), as well as implementation (savings and rational use of energy), [20–22]. The representation of all these elements depends on the complexity of the projects, but basically the results of one phase enable the quality preparation of the other. Accordingly, where specific stages of implementation are insufficient or missing, they need to be promoted, financially and organically supported.

Potential savings in electricity consumption can effectively be grouped into savings for: electric motors, plants, appliances, lighting, air conditioning and ventilation, as well as heating. These groups are mostly present when observing all groups of consumers. Specific technological processes in the industry that consume larger amounts of electricity should be analyzed separately. The use of thermal energy in industry is related to the specificity of the technological process of individual industries. Water is the most commonly used medium. Only the process industry in its technology uses other fluids as heat transfer media. Here, one should observe the efficiency of heat aggregates—mostly boilers, depending on the fuel used, then the losses in the distribution of the media, and the rationality of using heat in the production process itself.

The implementation of energy audits (examinations) in industrial plants has proven to be an extremely effective measure of increasing energy efficiency. Elements of the implementation of energy audits of production plants, which are based on monitoring and analysis of the use of heat and electricity for various categories of consumers, most often refer to these areas of energy use: lighting, electric motors, fans and pumps, compressed air systems, steam systems and other manufacturing processes characteristic of individual industrial plants (cooling, drying, specific heat processes, other private industrial processes). It should be taken into consideration that the production and processing of aluminum is a particularly energy intensive industry, where approximately 22.3 GJ of energy per ton of alumina/clay is specifically spent. The position of this industry emphasizes special attention in considering energy rationality.

Energy consumption in the food, beverage and cigarette industry in the next sector according to the amount of consumption is difficult to specify because it is a matter of a lot of different processes, and until recently, it was considered that energy consumption is not so significant in this branch of industry. However, analyzes show that about 10% of the total amount of energy needed for the entire industrial sector is spent in the developed countries on the food industry. Here are measures of heat energy saving especially important.

Previous analyzes show that, among industrial enterprises in the Republic of Srpska, in most cases, there is no need for efforts to increase the rationality of energy use. Especially with heat energy, over 90% of the situation it is considered that there is no need for that, while in the case of electricity in just over 15%, the situation it is considered that improvements are necessary. In order to improve the situation in that area, the Government of the Republic of Srpska will take concrete measures to raise awareness of the importance of rationalizing energy consumption in industrial subjects, [21, 22].

The energy efficiency of the energy blocks should be followed by activities and doings in normal circumstances leading to a verifiable and measurable increase in the energy efficiency of the plant unit, then the technical systems of the block, production processes and the cost of primary energy fuel. These activities are based on the application of energy-efficient technologies, i.e. processes that achieve energy savings and other accompanying effects, and may include appropriate handling, maintenance and settings on a block of thermal power plants. Energy efficiency has several components: economic-financial, technical, human and organizational component, [20].

In the tendency for lower operating costs and greater competitiveness, it is necessary to prepare energy-saving projects that would reduce energy costs, improve competitiveness, reduce greenhouse gas emissions, and increase energy supply security. In order to achieve energy savings, the best way is to establish energy management in their organizations and prepare action plans for energy efficiency, [24].

Investing in the energy efficiency project primarily has a need for the realization of a quick return on investment, then for the security of investments, as well as achieving the expected results of the project (energy savings, lower operating costs for energy, higher reliability in operation, higher efficiency and lower losses in the process transformation of primary into useful forms of energy—electric and thermal energy). In addition to these expectations, it is necessary to eliminate possible risks in the realization of the project (hidden costs of exploitation and maintenance of the system, the impact of technical shortcomings during exploitation in the volume of production, rationalization of one's own consumption, etc.). In order to achieve the goals of reducing operating costs through energy savings, it is often necessary to deliver and install more energy efficient equipment (often combined with its management and maintenance), monitor energy consumption, and introduce the increase in company competitiveness (Asset Management) and ESCO principles (contracting performance of energy savings), [20–25]. Starting from the tendency of rational energy consumption, it is necessary for the production process, besides energy consumers, to observe the entire production system as a whole, using the previously defined criteria, such as: estimating the use of energy and raw materials, reducing environmental pollution, improving work conditions, product quality and improvement of the production process. Using the ISO 50001 standard and the Law on Efficient Use of Energy, an insight into the energy state of an object or plant is made in order to determine the potential for increasing energy efficiency, [25, 26]. By inspecting the previous measurements, based on the process of statistical data for the results of previous exploitation on certain production, transmission and distribution companies within the electricity system and reconstructions, modernizations and improvements, the preliminary method of energy management is defined, which can be integrated with the company business and strategic planning. The success of projects in the field of energy efficiency can be evaluated through several criteria. One of the most important criteria is the economic effects of the project, [27–32].

4 On Energy Safety, Diversification of Energy Sources and Quality of Energy

Although energy security is spoken and written about a lot, there is still no single and generally accepted definition of security of energy supply. Monitoring through history, it can be concluded that energy security has received different definitions, continuously expanding it, which also involved the introduction of more and more elements and criteria for fulfillment, in order to achieve a satisfactory level of energy security. The International Energy Agency (IEA) gives a definition of energy security as "adequate, affordable and reliable energy supply" (International Energy Agency, 2010), that is, determines the reliability and continuity of supply, its price and acceptability of that price for economy, and an adequate form and amount of energy needed. A somewhat broader concept of energy security has been developed and used by the Asia Pacific Energy Research Center (2007), which includes availability, which signifies geological aspect, approachability or accessibility, which implies geopolitical relations, price and economic affordability, as well as environmental and social acceptability.

The security and stability of energy supplies is an objective necessity for the continued functioning of the economy at global and regional levels. Control over energy sources, that is, the construction of strategic partnerships, therefore, are increasingly becoming a strategic factor.

During the 1990s, when most of the national electricity and natural gas markets were still monopolized, the European Union began the process of their gradual opening to the market competition. The first liberalization directives (the first energy package) were passed in 1996 (electricity) and 1998 (gas), and into the legal systems of the member states they were to be transmitted by 1998 (electricity) and 2000 (gas). The second energy package was adopted in 2003, and the directives from that package were supposed to be transposed into the national law of the member states by 2004, while certain provisions came into force only in 2007. The basic characteristic of the adopted rules is to make free choice of suppliers of gas and electricity among a large number of competitors by industrial and private consumers. In April 2009, a third energy package was adopted, aimed at further liberalization of the internal electricity and gas market, which in principle presents the change of the second package and the basis for the implementation of the internal energy market. In February 2015, the EU Commission published a communication on the Energy Union Package entitled "A Framework Strategy for a Resilient Energy Union with Advanced Climate Policy" (COM (2015) 0080), which explicitly states that the goal of the energy union is "to provide consumers in the EU—to households and businesses—safe, sustainable, competitive and affordable energy." In order to achieve these goals, the package describes five closely related and mutually reinforcing dimensions: energy security, solidarity and trust, a fully integrated European energy market, energy efficiency that reduces demand, decarbonization of the economy, and research, innovation and competitiveness. As a result of this process, there are a series of legislative proposals on the new model of the EU energy market, presented by the EU Commission in late 2016. The aim of the "Clean Energy for All Europeans" package (COM (2016) 0860) is the implementation of the EU's policy and objectives, including energy efficiency, renewable energy, a model of the electricity market, security of electricity supply and governance rules for the energy union.

In order that EU's internal market become completely revived, in the energy sector, it is necessary to eliminate numerous obstacles and trade barriers, harmonize tax and price policies and measures regarding compliance with norms and standards, and environmental and safety regulations are required. The goal is to create a functioning market which guarantees everyone fair access to the market and a high level of consumer protection and a satisfactory level of connectivity and capacity for energy production. Although the European Council has set a deadline for the full establishment of the internal energy market by 2014, this goal has not been fully met, with some progress being made in diversifying energy suppliers and in terms of cross-border energy trade (COM (2014) 0634). Diversification (lat. Diversification–changing, diversity, change, making changes) of energy products, sources of energy and sources that include the selection and supply of energy products that will ensure the safety of consumers' supply and spatial distribution of resources in order to ensure the stability of the supply of each area. In addition, the use of energy means the acquisition, conversion, transmission, distribution and use of all types of energy.

Analysis of the quality parameters of electricity according to the current standard BAS EN 50160 (BiH) represents the basis for optimizing the consumption of electricity, which allows to increase the reliability of supply, maintain quality of production, and help maintain a stable power system. Under the term quality of electricity, it is understood that the quality of the frequency (related to maintaining the frequency at the prescribed value and representing the measure of the performance of the electricity system in relation to consumers), the voltage quality (the measure of the deviation of voltage and the shape of the voltage wave from ideal references, and similar to the quality of frequency it represents a measure of the performance of the electric power system in relation to consumers,) the power quality (complementary to the concept of voltage quality and refers to the deviation of the wave currents from an ideal reference, which, like the voltage reference, is a sinusoidal, of constant amplitude and frequency, and in a phase with a voltage wave); the quality of the electricity (characterizes the performance of the consumer in relation to the system), the quality of power or energy (a combination of voltage and current quality and characterizes the interaction between systems and consumers, where it should be distinguished that the technical terms "system" or "manufacturer"—"consumer" from commercial ones "supplier" or "seller—"buyer"), the quality of the supplier (includes all technical and commercial aspects of the quality of frequency and voltage that are affected by the customer), as well as quality of consumers (includes all technical and commercial aspects of the quality of voltage, current and energy, from the influence on the supplier), Table 1.

Indicators of the quality of electricity supply to consumers, which are subject to normalization and standardization, are usually related to the two most important variables that characterize the operation of each power system, which is the frequency and voltage. Although these sizes do not have direct commercial value, their indirect impact on the quality of delivered electricity to consumers is obvious since

Table 1 Energy quality parameters according to EU standard EN 50160, [33]

Characteristics of voltage events	Acceptable limits	Interval of measurements	Period of monitoring	Acceptance percentage (%)
Network frequency	49.5–50.5 Hz 47–52 Hz	10 s	1 week	95 100
Slow voltage changes	230 V ± 10%	10 min	1 week	95
Voltage failure (≤1 min)	10–1000 times per year (below 85% of the nominal value)	10 ms	1 year	100
Short breaks (≤3 min)	10–100 times per year (below 1% of nominal value)	10 ms	1 year	100
Accidental, long interruptions (>3 min)	10–50 times per year (below 1% of nominal value)	10 ms	1 year	100
Temporary voltage leap	Mainly < 1.5 kV	10 ms	N/A	100
Transient overvoltage	Mainly < 6 kV	N/A	N/A	100
Voltage asymmetry	Mainly 2% but occasionally 3%	10 min	1 week	95
Harmonic voltage	8% Total harmonic distortion (THD)	10 min	1 week	95

all disturbances or abnormalities in the system, through the frequency of the system and supply voltage, are copied as effects whose material impact on end consumers is obvious, [33].

With the advancement of the economy, in addition to the continuity of supply, voltage and frequency as still important parameters, the group of characteristics of electricity supplied to the consumer is increasingly being emphasized. In addition, the technical parameters of electricity, i.e. the voltages observed in the network are: network frequency, size and voltage changes (oscillation or flickering of voltage, collapse, short-term and long-term voltage failures), temporary network overvoltage between phase conductors from the ground, impulse overvoltage between phase conductors from the ground, voltage unbalance, voltage of higher harmonics, voltage of inter harmonics, and signal voltages. The essence of the electricity quality is the voltage at 0, 4 kV. The quality is ideal if the voltage is 230 V, the frequency is 50 Hz, and has a sinusoidal waveform. The quality is satisfactory if the voltage in the network is within the limits given in Table 1, or within the European Norm EN 50160. Outside these limits, the voltage quality is unsatisfactory [34]. With the quality of electricity at 0, 4 kV, voltage, frequency, amplitude, and waveform are monitored.

The European standard EN 50160 has the most important voltage characteristics for consumer connections in the public low-voltage and medium-voltage networks under normal operating conditions. The purpose of this standard is to give a description and determine the distribution voltage characteristics, without describing the average value of the observed parameters, but the maximum deviations of the individual parameters, which can be expected in the distribution network for the given measurement period, determined by the recommendations of EN 50160 (seven days, without stopping).

4.1 Energy Policy

With the basis of the energy policy of any country an overview of the situation is given and the basic priorities, directions and guidelines in certain sectors are defined in the energy sector. Specificities of different types and forms of energy are taken into account, and as such are individually considered, and also in the energy balance. Most often, through the basis of energy policy, its strategic goals are defined: ensuring security of energy supply, promoting sustainable development in the context of energy production and use, ensuring gradual introduction of competition in the energy market, creating conditions for accessibility and uniform use of networked energy throughout the geographical area of the country, compliance between the above objectives and etc. In addition, it is necessary to create legal preconditions for collecting more precise data on energy statistics in accordance with the EUROSTAT methodology, [2, 3].

In the oil industry of the Republic of Srpska in the coming period, domestic production of petroleum products for the needs of the market is planned, and the surplus will be exported to neighboring countries. The shipping of oil derivatives according to the present conditions is directed to the transport of derivatives by road, and partly by rail. In the following period, activities will be carried out that will ensure better and safer shipping of petroleum products. Privatization of the oil industry will provide the improvement of the organizational structure and a stronger interconnection of production and transport organizations with the aim of providing both refineries with the most successful placement and export of finished products through transport organizations.

The security of natural gas supply is conditioned by the diversification of sources, the existence of a natural gas storage, long-term supply contracts and debts to Gazprom. Since the natural gas infrastructure in the Republic of Srpska and Bosnia and Herzegovina is markedly underdeveloped and since there are no specified elements to ensure the security of supply, the situation in the supply of natural gas consumers will be monitored and timely measures will be taken for the continuous supply of consumers with these energy products, i.e. intervention measures for switching to others types of energy (use of fuel oil instead of natural gas in heating plants).

There is also a commitment to include alternative sources and energy from renewable sources in the production structure, such as: bio diesel, hydropower, biogas, biomass, solar energy, wind energy, etc. Since the most important renewable energy sources in the Republic of Srpska are biomass energy and hydropower, stimulating measures should be directed to these two types of renewable energy sources, but increased use of biomass residues for the production of heat energy as well as the production of liquid motor bio fuels should be encouraged.

It is understood that an integral part of the energy management policy should be a mandatory environmental policy in line with the European Union standards. In light of the current energy position of the European Union countries, two aspects of energy policy are of particular importance: energy independence and energy security. In any case, the unavoidable need for sustainable development of the energy must be taken into account (till today, around 170 countries have ratified the Kyoto Protocol), that is, the economic aspect and the aspect of environmental protection through reducing fossil fuel consumption. In this context, the most important mechanisms for achieving these energy policy goals are a constant increase in the share of renewable (non-fossil) energy sources, and others, an increase in energy efficiency. This current and wide initiative gives a special importance to the more detailed management of energy balances through which measurement and recording of progress made by energy saving activities and reduction of fossil fuel consumption are carried out.

In the years of the economic crisis, it is necessary to recognize the energy efficiency philosophy and its potential that can permanently achieve significant savings (without loss of standards) not only in the households' budgets, but even in the public budget, if adequate education, programs and technical and technological measures are applied in the administration, administrative institutions, public institutions, municipalities, hospitals, judiciary, educational institutions, public heating plants, public transport, etc.

Energy independence is an important part of the independence and stability of a country, and therefore it has to be given special significance for each sector separately (natural gas sector, oil and oil derivatives, coal and electricity, bio fuels and biomass sectors). The unique market of electricity and gas, signed by the Treaty establishing the Energy Community, established in 33 European countries, determines a part of the energy policy between the European Union and the countries of Southeastern Europe (Croatia, Bosnia and Herzegovina, Serbia, Montenegro, Macedonia, Bulgaria, Romania, Albania and the UN Interim Administration in Kosovo). The signing of the Energy Community Treaty has created an obligation to respect the European Union's legislation in the field of energy, and the ultimate goal of this arrangement is to create a legal and institutional framework for the free transfer and trading of energy products, as well as a greater obligation to protect the environment and the rights of the end customer, i.e., consumers.

All this requires harmonization of regulations in line with good practice and EU directives. Therefore, BiH's energy policy is based primarily on ensuring safe, high-quality and reliable energy and energy supply, ensuring optimum development of energy infrastructure, introducing modern technologies, providing conditions for improving energy efficiency, creating conditions for stimulating the use of renewable energy sources and improving protection environment. Reasons for the implementation of projects in the field of energy efficiency improvement can be: economic (energy costs), energy (local and global), legislative (Republic of Srpska and BiH as a whole, Southeastern Europe and the EU), environmental protection (local and global) and competitiveness (local, EU and global).

Work in the field of energy efficiency training should be endless (continuous activity). The increase in energy efficiency is the result of the use of new technical knowledge and the application of modern experience through staff training with the aim of achieving energy conservation policy, while maintaining the achieved quality level of energy users. The training program emphasizes the transfer of knowledge and experience through consideration of energy on an international scale to draw benefits from the experiences of other countries. Mostly, in order to provide an understanding of the situation in the world energy sector and its complexity, especially on the geopolitical, economic and ecological plan. The aim of the training is to understand, formulate and implement a policy of ensuring the efficiency of energy use. The training process concerns all sectors of economic and non-economic activity and is directed towards all present and future specialists in these fields. The training is intended for engineering and technical personnel, economists, architects, heads of economic organizations, managers of public utility companies, etc., with the aim of acquiring modern knowledge in the field of standardization in the field of energy saving, energy efficiency and energy management.

Energy management system is a set of measures that guarantee the efficient use of available energy resources in the company, treating energy savings and rational use, treating replacement of certain types of energy sources with cleaner ones, and those that can be more efficiently used. The introduced energy management system is essentially development of strategic and targeted activities to improve energy efficiency, which are most often integrated into business philosophy and planning.

The environmental protection management system directs organizational activities in ways that reduce their adverse effects on the environment, while continuously improving business performance in relation to the environment. The energy management system allows organizations to expand their environmental responsibility, reduce energy costs and CO_2 emissions. Since the energy management system is introduced most often as the last, that is, after the established standards ISO 9001 and ISO 14001, it is not necessary to introduce procedures that are common to all standards. On the other hand, the documentation and reporting system that relies on the applicable international standard ISO 50001: 2011 should be adapted to existing management systems, which together lead to the introduction and implementation of the Integrated Management System (IMS) within the production TPP, as a constituent unit of the EES. In addition, very simple measures in the form of small changes related to work procedures can have a significant impact on reducing energy consumption. Other important integral parts of the energy management system are: energy reviews, respectively Energy Audits, as well as monitoring the realization and setting goals, [3, 25, 26].

The interest of the company in the introduction of energy management and, consequently, the increase in its energy efficiency, is accompanied by the introduction of the international standard ISO 50001, which gives them real support in these activities. In many organizations, elements of energy management are already being applied, for example, programs, plans or projects for energy saving are being prepared and implemented, procurement of equipment with higher energy efficiency, analyze possibilities for increasing the energy efficiency of individual processes and/or production or the like. Therefore, using the international standard ISO 50001, an organization can discover that it already fulfills some of those requirements that are included in it. Thus, the ISO 50001 international standard is a special point of view by which it is possible to evaluate and improve management methods that are already applied in the organization, focused on energy savings. They are the starting point in the implementation of the international standard ISO 50001 which enables the system approach to energy management to be provided. The ISO 50001 international standard should supply every organization, regardless of its size, the proper working mode, in both the management and technical aspects so that it can realistically increase its energy efficiency, increase the use of renewable energy sources and reduce the greenhouse effect. Obviously, the introduction of an energy management system is an innovative solution, which is related to the modernization of existing production and management based on the use of the best world practice in the field of energy saving.

Preparation of the Energy efficiency study of the energy plant shall be carried out in accordance with the given Terms of Reference. In doing so, account should be taken of the basic principles on which the efficient use of electricity generation capacities are based, the most important of which are: energy security, competitiveness of products and services, sustainability of energy use, organized energy management, economic cost-effectiveness of energy efficiency measures, as well as minimum energy efficiency requirements. The Energy efficiency study aims to process and demonstrate: the analysis of energy efficiency in the previous work of the observed energy increase, more efficient use of its production capacities, increased competitiveness in the electricity market, reduction of the negative impact of the energy sector on the environment, encouragement of responsible behavior towards energy production based on the implementation of policies for the efficient use of electricity, heat and cooling energy (and technological steam) and energy efficiency measures in the electricity generation sector.

Starting from the definition of energy efficiency as the relation between the achieved result in the production of useful (final) forms of energy and for the energy consumed from energy products and losses related to their own consumption at the plant, it is necessary, using the existing methodology for evaluating the operation of power plants (EPs) within the electric power system (EES), to calculate certain performance indicators, such as: time and power utilization coefficients, the energy efficiency of the constituent installations and the block as a whole, the method and quantities of the produced electricity and the consumed heat from the fuel, the amount of heat and power used for their own needs and etc.

The correct calculation of the achieved characteristics of the block can help to evaluate the energy efficiency. The energy efficiency of the thermal power plant block should be followed by activities and actions that in normal circumstances lead to a verified and measurable increase in the energy efficiency of the plant unit, the technical systems of the block, the production processes and the savings of primary energy. These activities are based on the application of energy-efficient technologies, i.e. procedures for achieving energy savings and other accompanying positive effects, and may include appropriate handling, maintenance and adjustments on a block of thermal power plants, improving the efficiency of existing equipment and systems, without modification in any production process given plant, or in the energy supply system. Energy efficiency studies are most often conceived from several parts: energy efficiency analysis of the plant unit, techno-economic assessment of the possibility of increasing the energy efficiency of the plant and proposals for measures to improve the energy properties of the block and the production of electric, heat and cooling energy (and technological steam) without change equipment, improving the efficiency of production capacities for supplying useful energy by introducing new equipment or dismantling old ones and replacing new energy-efficient equipment (this type of measures is characterized by the necessary investments, which, in a short period, can significantly increase energy efficiency, which indirectly, through realized savings, secure financial resources for return on investment).

The first group of measures include initiatives and measures in terms of managing and reducing energy consumption and losses without changing process and waste energy use, such as: stopping the operation of the equipment in neutral, lowering unnecessarily high temperatures in the flows of the process, limiting the use of hot water for cleaning and rinse, monitoring of power consumption, remote or on-line monitoring and management of electricity consumption to avoid a large factor of simultaneity and reduce peak loads, systematic and planned maintenance of equipment, elimination of steam, water, compressed air and vacuum leaks, improved operation planning, better thermal insulation of pipes, atomic temperature control, introduction of centralized control of air conditioning systems, introduction of predictive maintenance, controlled and rational use of lighting in workspaces, installation of highly efficient lighting (replacement of standard bulk "energy saving"), frequent "guiding" of large electric-motor consumers, for improvement the power factor of the electric motor, compensating reactive electricity, preheating of waste working materials, returning of condensate, and installing additional heat exchangers.

The second group of measures relates to changes in the production process and the introduction of technological and technical innovations: replacement of parts or the entire production plant with more modern and more efficient, reconstruction of hot water and steam installations, installation of modern high-efficiency boiler plants, replacement of indirect drying with heated air by direct drying of warm gases from the process of combustion of natural gas, the application of new or improved technologies, such as cogeneration (coupled production of thermal and power energy from one source, thus achieving a factor of utilization of primary fuel over 85%), and tri generation or combined steam-gas installations.

Energy Saving is the sum of planned and implemented measures aimed at utilizing the minimum amount of energy, regardless of preserving the level of comfort and rate of production. Unlike energy efficiency, energy conservation usually reflects certain disclaimers (often as well as the result of the economic position of the user), while the efficient use of energy never distorts the working and living conditions. If the concept of savings is considered through two possible meanings, one of which refers to devices and the other to measures and behaviors, then energy-saving device is one that has small losses when transforming one type of energy into another. Regarding energy saving measures, under energy savings, all measures are meant which are applied to reduce energy consumption in the context of energy efficiency (replacement of non-renewable energy sources by renewable energy, introduction of tariff systems by distributors that will stimulate energy savings, replacement of worn-out carpentry in heating spaces, the installation of measuring and regulating devices for energy consumers, the isolation of the space that is heated, etc.), as well as technical or non-technical measures, or behavioral modifications from the aspect of savings that result in a decrease in the level of comfort and rate of production, which implies the most commonly lower step comfort and standard (reduced energy consumption through reduced consumption, rarely with the optimization of energy costs, etc.). Potential that he has energy saving is enormous both on the producer's side (processing of coal, oil, gas and production, transmission and distribution of electricity), as well as on the side of consumers (use of energy in households, transport and production).

4.2 *Planning in Energetics*

Every economic development of the state requires the consumption of appropriate energy resources. Any deviation from the time schedule for the result may have the appearance of restrictions in the development of other economic activities. For example, the lack of electricity leads to major disruptions in industrial production, which in turn entails significant losses in the final economic balance of the country. For these reasons, it is necessary to monitor the development of energy consumption, which is related to certain influential factors (population growth, development of science and technology, economic development, standard, etc.), whose intensity of activity changes in time. Choosing the optimal structure for covering consumption is very important for the development of energy. It is a very complex problem, the solution of which depends not only on the energy sources of the country, but also on the mutual dependencies between primary and transformed forms of energy, then forms of energy that cannot be replaced by other types of energy, as well as the possibilities and needs of importing certain forms of energy, location of consumers, consumption characteristics, etc. That is why it is understandable that the situation and development in the field of energy are constantly being studied today.

Planning of development in the field of energy is important because of the dependence of the development of the society on safe, sufficient and appropriate quantities of necessary forms of energy, as well as because of the engagement of large financial resources in this field, [10, 13, 18]. When planning the development of energy, the following criteria should be considered: the security of consumer supply with minimal costs, the rational use of domestic sources, with proper evaluation of imported energy forms, maximum prevention of monopolistic and unique forms of energy, and achieving satisfactory conditions for environmental protection and sustainable development. Planning of the development of the electric power system includes all activities from the first assumptions about the possibility of building an object until its entry into operation. But in terms of terminology, planning is primarily meant for the planning of power generation plants (hydro and thermal power plants).

When planning the development of the electric power system, the goal and criteria are uniquely determined: the settlement of the anticipated consumption of electricity with minimum costs and assuming that certain restrictions are met, such as, for example, financial, technical, ecological, limitations on the availability of primary forms of energy, etc. Satisfying the constraints is imposed as a primary task, regardless of the method of planning, which is the reason that a particular constraint is often the decisive factor in deciding on the final development strategy. Although long-term plans carry a great deal of uncertainty, such planning is necessary primarily for two reasons: the first is the basic and extended lifetime of the production plants (25–30 basic plus 15–20 years of extended service life after the revitalization, reconstruction and modernization of the plant) and the other is the time needed for the preparation of the construction and the construction itself (3–6 years, not counting the possibility of a delay in the project). It follows that the analysis of the development plans of the electric power system should be carried out for a period of 15–50 years in advance. When planning the construction of production facilities, the necessary construction should be determined to meet the future consumption (volume and capacity), the time of entry of a particular production plant into operation and the possibilities of improving the technologies for electricity generation (improved and cleaner technologies, cogeneration and tri generation systems, hybrid systems). The preparation of development studies can be divided into two parts, which include simulation of the legality of work in the system (system operation), as well as the economic evaluation of production facilities or entire development plans. The first part requires the development of a system model, i.e. it is necessary to describe the system with mathematical equations and to approximate it with the inevitable neglect and simplification. In the second part, the energy contribution of each plant and economic methods should be evaluated.

4.3 Integrated Planning and Energy Policy

The objectives of the integrated energy development planning are realized through the application of the conceptual scheme given in Fig. 1. It implies the search for the optimal balance between the work programs for ensuring energy efficiency and work programs in the field of energy supply through a continuous comparison of their impact on the economy and the environment as well as on the stable development in general. As a result of the implemented integrated planning according to this scheme, a database is provided, which includes: defining a program of work on the side of energy consumption, determining the level of energy requirements and possibilities for their satisfaction with existing systems of production and distribution of fuel resources and energy, continuous comparison of the variant for preventive action, applied on the energy supply side and on the energy consumption side.

Figure 2 shows the distribution of potential interested parties for energy savings in two groups, i.e. macro and micro participants.

The belonging of each particular participant to a particular group practically does not depend on the amount of capital invested, since the main determining criterion is the degree of motivation of the participants in achieving the stimulating goal for its participation in the development of energy savings. Of course, the state and its structures (funds and agencies) have the greatest motivation. Their aspiration is to ensure a stable long-term development of the economy, strengthen energy security and reduce the speed of spending of national reserves of primary energy resources at the expense of reducing energy consumption per unit of national income. In addition, the problem of preventing the increase of the negative ecological effects of production energy facilities on the environment, which was and will remain in the predictable future, will be the basis of the world energy balance.

All energy planning models have a simplified description of reality and are based on certain assumptions, which are more or less secure and reliable (stochastic character). Depending on the chosen criteria, there are also ways of classifying energy models. Most commonly, energy models are classified according to eight criteria (general and specific model applications, analytical approach (top-down versus bottom-up), methodology used, available mathematical background, level of coverage—global, regional, national, local or project level, it also includes certain sectors, time coverage—short-term, medium-term, long-term, requirements related to the volume of required data).

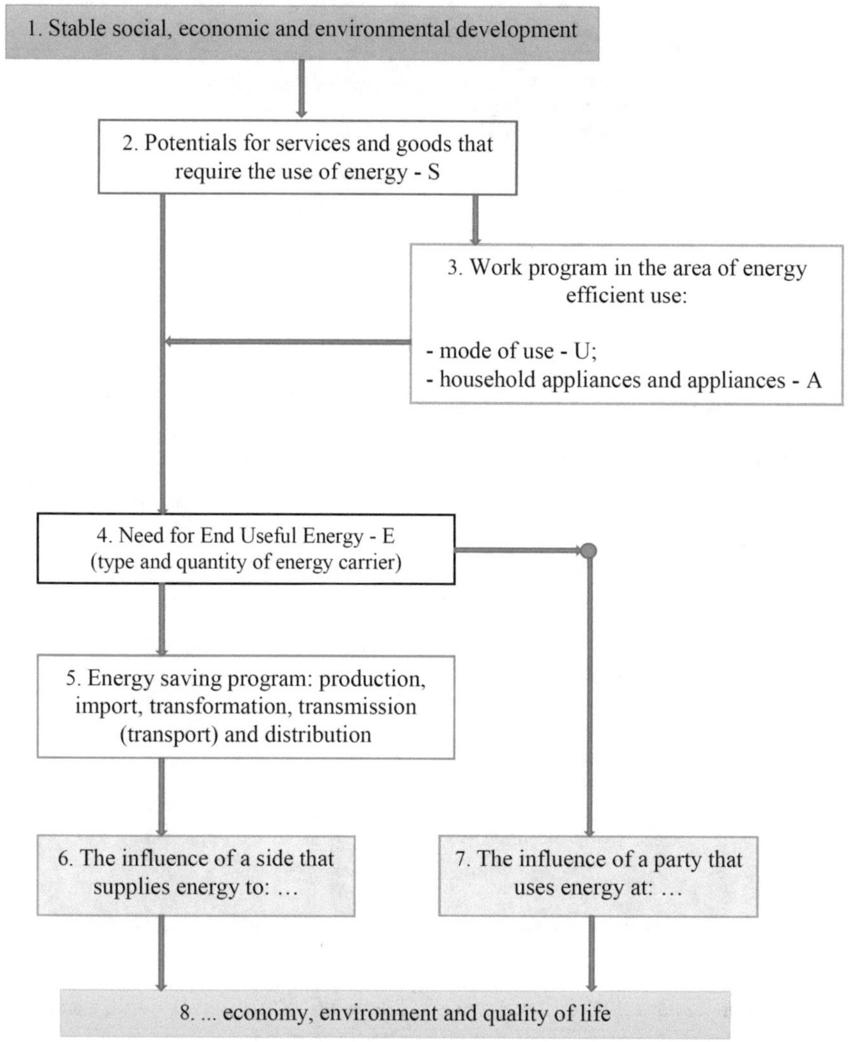

Fig. 1 Integrated energy development planning, [25]

Models are typically designed for specific applications, and any mismatched application by the model user can lead to wrong conclusions.

The general characteristics of energy models relate to the way of treating the future (predicting a future situation based on past trends, analyzing development scenarios or modeling development based on a given future goal). Predicting or forecasting the future, since the extrapolation of past trends, as well as models for the forecast of future developments, are usually short-lived, with the fulfillment of a key precondition for their application that key parameters, such as elasticity, remain constant. The scenario analysis explores the future by analyzing different

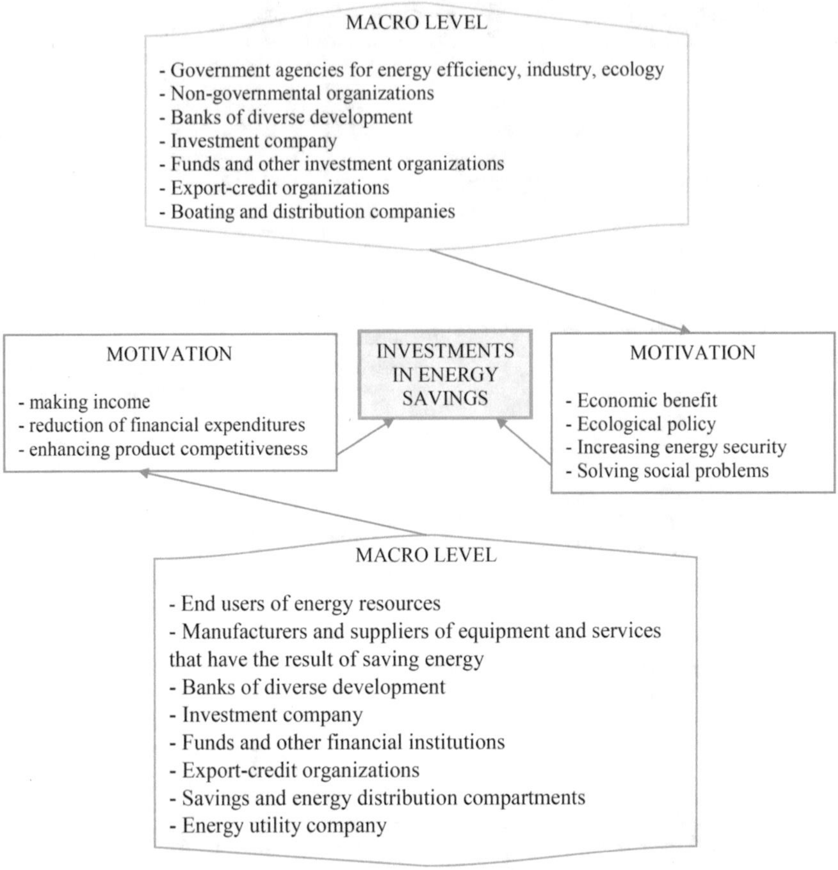

Fig. 2 Motivation of potential investors in energy saving, at micro and macro level, [25]

development scenarios, with a number of "intervention" scenarios being compared with a reference scenario. The reference scenario assumes the continuation of the previous practice (the so-called Business as Usual). In alternative scenarios, certain economic behavior (most often on the basis of satisfaction function), the need for resources, technical progress, and economic and demographic development must be assumed. Of crucial importance in these models is the analysis of the sensitivity of changing default assumptions. Scenario analysis can also be applied in bottom-up and top-down models, [25, 26]. Modeling the development based on the desired goal (modeling backward) is applied in constructing the desired future according to the expert's instructions, identifying and analyzing the steps that will lead to the desired goal. Such a method can also be used to analyze the long-term economic consequences of alternative energy scenarios. In this way, bottom-up and top-down models can be connected. More specific applications are supply and demand analysis, impact analysis and evaluation of different options.

Energy demand models are directed either to the whole economy or to its particular sector, while looking at energy demand as a function of population change, income and energy prices.

Energy supply models are mostly focused on technical issues related to energy supply, but can also use financial aspects analysis. Energy supply models are most often associated with demand models into a common model.

Impact models determine the consequences of choosing a particular energy variant or option or energy policy measure. Impacts can be financial-economic, social (welfare distribution, employment) or health and environmental impacts (emissions, solid and liquid waste, biodiversity, etc.).

Evaluation models evaluate different scenarios with respect to the default criterion or group of criteria. The most common criteria are technical and economic efficiency, as well as the impact on the environment. Evaluation models are often combined with impact models.

Models differ in the form of energy that is being analyzed; sometimes it is only electricity, and sometimes electricity, technology, and heat.

5 Straight Planning of Electricity Production

All energy planning models have a simplified description of reality and are based on certain assumptions, which are more or less secure and reliable (stochastic character). Depending on the chosen criteria, there are also ways of classifying energy models. Most commonly, energy models are classified according to eight criteria (general and specific model applications, analytical approach (top-down in relation to bottom-up), methodology used, available mathematical background, level of coverage—global, regional, national, local or on the level of project, includes certain sectors, time coverage—short-term, medium-term, long-term, requirements related to the volume of required data).

5.1 Integrated Planning and Energy Policy

The general characteristics of energy models relate to the way of treating the future, i.e. to predict a future situation based on past trends, to analyze development scenarios, or to model development based on a given future goal. Predicting or forecasting the future, since the extrapolation of past trends, as well as models for future development forecasts, are usually short-lived, with a key precondition for their implementation (key starting parameters should remain in the given range of change).

The scenario analysis explores the future by analyzing different development scenarios, with a number of "intervention" scenarios being compared with a reference scenario. In the reference scenario, the continuation of previous practice (the so-called Business as Usual) is assumed, while in alternative scenarios certain economic behavior (most often on the basis of satisfaction function), as well as the need for resources, technical progress, and economic and demographic development must be assumed. Of crucial importance in these models is the analysis of the sensitivity of changing default assumptions. Scenario analysis can also be applied in bottom-up and top-down models. Modeling the development based on the desired goal (modeling backward) is applied in constructing the desired future according to the expert's instructions, identifying and analyzing the steps that will lead to the desired goal. Such a method can also be used to analyze the long-term economic consequences of alternative energy scenarios. In this way, bottom-up and top-down models can be connected. More specific applications are supply and demand analysis, impact analysis and evaluation of different options. Energy demand models are directed either to the whole economy or to their particular sector, while looking at energy demand as a function of population change, income and energy prices. Energy supply models are mostly focused on technical issues related to energy supply, but can also use financial aspects analysis. Energy supply models are most often associated with demand models in a common model. Impact models determine the consequences of choosing a particular energy variant or option or energy policy measure. Impacts can be financial-economic, social (welfare distribution, employment) or health and environmental impacts (emissions, solid and liquid waste, biodiversity, etc.). Evaluation models evaluate different scenarios with respect to the default criterion or group of criteria. The most common criteria are technical and economic efficiency, as well as the impact on the environment. Evaluation models are often combined with impact models. Models differ in the form of energy that is being analyzed; sometimes it is only electricity, and sometimes also electricity, technology, and heat.

Analytical approach: top-down in relation to bottom-up approach. Regardless of the same input assumptions, top-down and bottom-up approaches can give different results. The difference in results comes from the fact that these models differently approach the choice of technologies, the method of decision making of economic entities and the way of functioning of the market in a given period of time. The top-down approach is based on a pessimistic economic assumption, while the bottom-up uses an optimistic engineering approach. Namely, economists consider that technology is a means of transforming input resources (labor, capital and energy) into useful products. The "best" or optimal technology (the one that leads to maximum market efficiency) is determined by the so-called the production limit, which can be constructed by observing market behavior, with no investment beyond the production boundary being cost-effective. Technological progress can shift the production limit closer to the starting point. The complete economic model does not display explicitly the technology, but uses the elasticity that implicitly means the application of certain technologies, where the technology is treated as a black box, which makes it difficult to demonstrate technological development in the production function of economic models. In contrast, engineering models do not depend on the behavior of the market, describe techniques, properties and direct costs of all technological options in order to identify the possibilities for their improvement.

In practice, the technological potential that is calculated in engineering studies differs from the "best" technologies, which represent the limit of production in economic models. The difference comes from the fact that the engineering approach generally neglects the existing restrictions on the market and only considers technological potential, while an economic approach is viewed as part of the given market conditions. Restrictions on which engineering approach is not taken into account are hidden costs, costs of introducing certain technologies, market irregularities, macroeconomic relations (i.e. prices), as well as macroeconomic indicators (GDP, employment). Behavior on the market is the consequence of these restrictions. As a rule, models that reflect consumers' behavior on the market automatically take into account all these restrictions, and are therefore to a certain extent pessimistic.

Supporters of the opposite bottom-up approach claim that the application of appropriate measures could allow the market conditions to change and mitigate the existing restrictions, and that, therefore, top-down access leads to poorer results than they are in reality. Another feature of the top-down model is to use aggregate data to examine interactions between the energy and other sectors of the economy, as well as to examine the complete macroeconomic picture of the economy as a whole. This is done by adopting past consumer behavior patterns, as well as by their extrapolation into the future. That's why top-down models are only suitable for short-term predictions, prognoses in which past behavior patterns still apply. By contrast, engineering, bottom-up models usually concentrate only on the energy sector and use detailed (non-aggregate) data to describe the final energy consumption and technological options. Bottom-up models are suitable provided that the structural development of one sector (energy) does not significantly affect the development of the entire economy, i.e. if their interaction is negligible. And finally, the differences between top-down and bottom-up approaches can be most commonly shown as differences between aggregate and non-aggregate access, or the difference between the maximum and the minimum predefined consumer behavior. Top-down access is generally used for forecasting, and bottom-up for research, with the differences given in Table 2.

Methodological background. Depending on the applied methodology, models can be econometric, macro-economic, economic balance models, optimization, simulation, tabular, with modeling backward and multicriteria models. In practice, there is no strict division between models; the more common approach is to combine elements of several models.

Econometric models. Econometric models are defined as "the application of statistical methods in solving economic problems", [1, 15, 16, 35]. It is about the application of statistical methods for the extrapolation of past market behavior into the future. They are based on measured aggregate data from the past, which are used for short-term and mid-term future predictions of labor, capital and other inputs. Econometric models are also often used to analyze energy-economic relations. Generally, their purpose is to accurately predict the future using measured parameters. Although early energy models (demand models) were purely econometric, today the econometric principle is used mainly in the context of macroeconomic models. The disadvantage is that in econometric models no specific technological option can be assigned. In addition, since the variable (variables) based on market relations in the

Table 2 Comparison of top-down and bottom-up models, [1, 25, 26]

"Top-down" model	"Bottom-up" model
Economic approach	Engineering approach
A pessimistic evaluation of the "best" characteristics	Optimistic rating of the best feature
Cannot give an explicit description of the technology	Provides a detailed description of the technologies
Reflects market-acceptable technology	Reflects technical potential
The "most efficient" technologies are determined by the production limit (which is conditioned by the behavior on the market)	The best technologies can be more efficient than those suggested by an economic model
Use of common (aggregate) indicators for predicting	Using disaggregated data for analysis and research
They are based on observed behavior on the market	They do not take into account behavior on the market
They do not take into account technically most efficient technologies; therefore, they underestimate the potential of increasing efficiency	They do not take into account market constraints (hidden costs and others), therefore they overestimate the potential of increasing efficiency
Energy demand is determined by aggregate economic indicators (GDP, price elasticity) but there is a different treatment of energy supply	Energy supply technologies are presented in more details, but energy consumption is differently treated
Consumer behavior in the market is one of the starting assumptions	Directly include the costs of individual technologies
During modeling past trends are assumed	It is assumed that interactions between energy and other sectors are negligible

past, it is required that market relations be more or less permanent. It should be noted that econometric models require a fairly large amount of data and a fairly experienced modeling person. Long-term planning is only possible with a higher degree of aggregation, which reduces fluctuations over time.

Macroeconomic models. Macroeconomic models are focused on the complete national economy, as well as on cross-sectored interaction. Often they are applied in energy demand analyzes in which production (output) is assumed to be conditioned by demand. The input-output tables are used to describe interactions between sectors and serve in the analysis of energy-economic relations. The input-output method is justified only if there are assumptions about constant relations and the possibility of perfect aggregation. Macroeconomic models evolve as a rule for research purposes rather than predicting because they use input parameters and assumptions that do not necessarily reflect the real situation. Similar to econometric models, the lack of a macroeconomic method is that it does not describe specific technologies and requires a high degree of expertise. It does not take into account inter temporal preferences and long-term expectations, which results in a fairly static representation of technological development.

Models of economic equilibrium. In contrast to econometric and macroeconomic methods, which mainly describe short-term and medium-term effects, the economic equilibrium method is intended for long-term analyzes. It is used for the analysis of the energy sector as a part of the economy and it focuses on the relations between the energy sector and the rest of the economy. Models of economic equilibrium are sometimes referred to as resource allocation models. The models of partial and general balance are distinguished. Partial Equilibrium models assume the balance of only one part of the economy, e.g. balance of supply and demand of energy. General Equilibrium models analyze the conditions in which a simultaneous balance is achieved in all markets, allowing for backward effects between individual markets. Economic equilibrium models are used to simulate long-term economic growth. They are based on the assumptions of the perfect market, the output is determined by supply (production) and there is no unemployment. The disadvantage of these models is that they do not provide complete information about the time path to the new balance and underestimate the transitional costs.

Optimization models. These models are used to optimize investment decisions in energy, with the results directly dependent on input parameters. The result is the best solution for given variables and with given constraints. Optimization models are often used by electric power companies and utility companies to achieve an optimal investment strategy. They are also used in the planning of national energy. The basic assumption of optimization models is that all market participants act optimally in the given conditions. The disadvantages are that these models require a good knowledge of mathematics and that all processes must be analytically defined. Optimization models often use linear programming techniques.

Simulation models. These models are descriptive models representing a particular system, with the often simplified mode of reproduction of the actual system. If the simulation model describes the operation of the system in only one-time interval, it is a static model. If the results in a given interval are conditioned by system behavior (expansion, development) at previous intervals, this is a dynamic model. Simulation models are especially useful in cases where it is impossible or too expensive to conduct experiments on the system. The disadvantage is that they are quite complex. They are often used in scenario analysis.

Table models. Table models are bottom-up models, highly flexible. These are more program packages rather than models as such. Usually these are tools that contain a reference model (for example, a reference database model) and simply enable changes in accordance with the needs of users.

Modeling backwards. The reverse modeling method is used to construct the desired goals in the future and then find the way to achieve the given goal. This method is often used in the analysis of alternative energy scenarios. For example, a Dutch research program called "Sustainable Technological Development" uses backward modeling to research the technological requirements needed to achieve the target.

Multicriterial models. The multi criteria method in the analysis also introduces other criteria, not only economic efficiency. It enables the inclusion of quantitative and qualitative criteria. This approach is not yet widespread in energy models. The concept of multi-criteria decision-making refers to decision-making situations when there are a greater number of conflicting criteria. In theory, the multi-criteria analysis involves solving insufficiently structured problems, while in practical terms it is an essential tool in solving everyday tasks of choosing decisions, management actions, both at the design level and in the exploitation of complex energy and process systems. The development of new and improvement of the existing methods of multi-criteria analysis has influenced their increasing application in the field of energy, both from theoretical and practical aspect. A specific segment of the application of the multi-criteria analysis methodology is the adoption of strategic or operational decisions in solving multidisciplinary problems with dominant technical or predominantly economic content, regardless of the part or energy system as a whole. The application of these methods in the selection of the right solutions in the tasks of deciding management in the design, maintenance and exploitation of thermal power plants is based on the development of information technologies and computer technology. Problems that can be considered using multi-criteria decision-making have certain common characteristics: a greater number of criteria, which must be created by the decision maker, high likelihood of conflict between criteria, incomparable (different) units of measure (as a rule, each criterion has different units of measure) or the choice of optimal concepts in a predetermined space. As solutions, either designing the best alternative or selecting the best action from a set of previously defined final actions appears. Within this paper, a brief description of the most important methods is given, with an analysis of their advantages and disadvantages, as well as the possibility of applying it to solve the specific problem of the selection of micro location for power plants. In the process of valorization and selection of acceptable potential micro location for energy plants within the selected macro location, it is necessary to apply a certain procedure, which will be uniform in all its aspects. In order to achieve this goal, it is necessary to define general criteria for selection and comparison of selected micro locations within the previously determined macro-location using the multi-criteria optimization method. The basic principle when selecting the criteria is that they can be measurable, that is, the available location data, based on them, can be valorized. However, in contrast, there are a number of criteria that are not measurable, that is, their impacts in relation to micro-location cannot be accurately valorized and determined, but only through certain indirect indicators. The use of multi-criteria decision-making methods should provide assistance to decision-makers when there is a wide choice of alternatives to the problem being addressed. In addition, the decision-making process and choosing an optimal solution when designing power plants is of a multi-criteria type, taking into account a number of factors and interests of different groups and levels of the social community (often mutually opposing), with the participation of several interest groups in the decision-making process. The question arises as to how to reconcile all these criteria, from the perspective of different preferences and often mutually opposing interests. When choosing variant solutions for power plants, the ideal case would suit a situation in which all the criteria in the problem could be classified into two categories: the profit category, where the criteria are maximized, where the criterion does not necessarily have to be a profit and cost category, where the criteria minimize. The ideal solution would be one that maximizes all profits and minimizes all cost criteria, which in practice is very difficult to achieve. As an ideal solution cannot be achieved, it is necessary to look for the so-called non-dominant solutions (the solution is dominated if there is at least one other solution, which is at least one better than the one observed, provided that it is at least equivalent to other attributes). The third category consists of the so-called satisfactory solutions, as a reduced subset of possible solutions. Finally, there are desirable solutions that are non-dominated, and best meet the expectations of the decision maker. In principle, two groups of methods are distinguished: simple no compensation methods for simpler decision making and more precise compensatory methods for more complex decision making. Starting from the three most common types of multi-criteria tasks, it is possible to extract multi-criteria optimization (one solution), multi-criteria ranking (ranking multiple solutions) and multi-criterion separation of good from a set of bad solutions (a subset for further troubleshooting). The final set of alternatives is observed (decisions, actions, potential solutions)—$A_i \in A$ (this number in practice does not exceed 4), where each of the alternatives is described with more criteria (attributes, indicators)—$PK_j \in PK$. For a specific case of selecting the location of an energy plant, alternatives and criteria are regulated in a matrix form. Each of these criteria is either maximization (max) or minimization (min) type, while xi is given the value of the i-th alternative according to the j-th criterion and with Wj the weight of the criterion (evaluation of its importance). Based on a large number of published papers, it can be concluded that AHP is a convenient and highly flexible system for supporting decision-making in various problems with multiple choice criteria, which on the other hand is realized as a PC platform software with complete technical support in the field of multi-criteria decision making.

Certain uncertainty in decision-making may arise in resolving some of the problems, so the so-called "fuzzy multi-criteria decision" making is used, [20, 36]. This uncertainty can be caused in two ways: first, in the case where the decision maker is not 100% certain when making some subjective assessment, which is called the uncertainty of a subjective assessment, and the other, when information about some criteria is not complete or not available at all, and is called uncertainty due to lack of data or incomplete information. In principle, the use of common multi-criteria decision methods is used to solve such problems, which are intended for problems defined with a certain uncertainty (fuzzy variant methods, such as: fuzzy AHP, fuzzy TOPSIS, etc.), [36]. Gray theory is a method used to study uncertainty, especially in the realization of mathematical analyzes of technical systems with input uncertain information, in particular, problems of uncertainty caused by discrete data and incomplete information. The original concept of this method was developed by professor Denga, in 1982. It is used to solve decision problems that are defined with relatively little input information and data, known on a specific scale and unknown on a specific scale, [36]. Gray theory deals with high mathematical analysis of systems, which are partially known and partially unknown, and are defined with insufficient data and insufficient knowledge. When the decision process is not clear, the gray theory examines the interaction analysis between a large number of mutually different and incomplete input data. In addition to multi-criteria decision-making, this method is also used in computer graphics, predictions, and control of the system as a whole, [36]. Gray theory method was used, among other things, in works by Ozcan and in the selection of storage locations, and under certain conditions it is also applicable in the analysis of macro and micro location choices for complex energy objects, [23, 36].

Mathematical approach. The usual mathematical methods used in energy models are linear programming, mixed integral programming, and dynamic programming. Of course, combinations are also possible. It is worth mentioning the latest mathematical methods, which have only just started to apply in energy planning, which are multi-criteria methods and the method of fuzzy logic.

5.2 Level Includes—Global, Regional, National, Local or Project Level

The level of model coverage reflects the scope of the analysis, which is an important factor in determining the model structure. Global models describe the world economy or the situation in the world, while regional models often include continents or subcontinents. National models treat conditions in world markets as external parameters, but cover all major sectors of the national economy at the same time, including back links and relations across sectors. Examples of national models are econometric models for short-term analyzes, i.e. models of general market balance for long-term analyzes. The local level includes regions within the country. The level of the project usually relates to the concrete location of the project. If the project is national or international, the level of the project would have a different meaning, but such projects are rare. Comprehensiveness of global, regional and national models usually requires highly aggregated data, i.e. top-down approach, covering all major sectors and their macroeconomic relations, which inevitably follows a fairly simplified view of the energy sector. In contrast, local models usually require bottom-up access and use disaggregated data, [37].

Covered sectors. The model can be directed to only one sector, as is the case with many bottom-up models, or involving multiple sectors. For analysis, it is crucial that the system of economy is divided between sectors. Multi sectored models can be used internationally, nationally and sub nationally, and analyze interactions between these sectors. Single-sector models provide information only on one sector (for example, energy) and do not take into account the macroeconomic relations between these and other sectors. Other economic sectors are presented in a very simplified way. Almost all bottom-up models are sectored, but the opposite is not true—not all sectored models use only bottom-up approach. For example, the partial equilibrium model, which is focused on the long-term development of only one sector, is a top-down type model.

Time scope—short-term, medium-term, long-term. There is no standard definition of short, medium and long deadlines. But it is usually considered that the period or period is up to 5 years in the short term, 5–10 years in the medium term, and more than 10 years in the long term. The observation period plays a crucial role as economic, social and environmental because aspects vary depending on the time frame. Therefore, the observation period determines the structure and goals of the energy model. For example, in long-term analyzes, one can assume the economic equilibrium in the market (resources are fully allocated, no unemployment), while short-term models should include transitional and non-equilibrium effects, e.g. unemployment.

Required volume of data. Most models require quantitative data, which often have to be expressed in monetary units. However, sometimes data do not exist or are unreliable (e.g. for developing countries), in which case the model must be concerned with qualitative data. Data can be aggregated or disaggregated. Long-term global and national models will require highly aggregated data and few technical details. A more detailed overview of the energy system (e.g. supply and demand) will only be possible in models that are viewed exclusively by the energy sector.

5.3 General Considerations of Optimizing the Operation of the Electricity Energy System (EES)

In every country, electricity generation is a vital part of its economic system. This is also the reason for imposing the need for a more economical production that is using existing capacities in the most efficient way. The main characteristics of the electricity system (EES) are: continuity of production, rapid transfer of energy to consumers and inability to create a reserve of electricity. For these reasons, production aggregates must, at all times, ensure a total power corresponding to consumption, while acting elastically, i.e. adapting to changes in network load. These tasks are performed on three levels, [38]:

- Presently establishing and maintaining a balance between consumption and production (this task is performed by automatic regulation);
- Creating the conditions for automatic regulation to fulfill its task even over a longer period of time with increased system load, which is achieved by ensuring primary energy reserve, i.e. filling the hydro-storage accumulation or storing sufficient fuel in the thermal power plant dumps;
- Long-term planning for the construction of new production capacities, which will follow the growth of consumption, is conditioned by the growth of industrial capacities, the increase in communal needs, and the growth of individual electricity consumption.

Optimal use of electric power aggregates comes down to including their most favorable number and determining their load, as well as the choice of most suitable dynamics, duration and volume of planned overhauls, elaboration of methods for adapting to unplanned and excessive situations, etc. All of the above is solved by enabling the production of the necessary energy with the least cost, i.e. achieving maximum production of energy from the available potential in hydro power plants and minimal fuel consumption for the production of the required energy in thermal power plants and installations based on renewable energy sources. Although it seems simple, solving this task is very complex due to the diversity of power plants in the system: storage and flow hydroelectric power plants, steam or gas turbine power plants using different fuels, including nuclear, and in addition, reversible plants with pumping and turbine modes, as well as installations based on renewable energy sources. In addition, some plants have been built as thermal power plants for heat and power generation. It is evident that in such a complex system, it is impossible to achieve a situation in which all production aggregates would work under the most favorable conditions. Optimum system is by no means a superposition of the optimum of all its parts. It is determined on the basis of adopted and precisely defined criteria, which are often in contrast with each other. Therefore, optimum is often a compromise solution. The long-term development plan and also the strategic plan represent a framework for the company's operation in the long term for sustainable development both in the power industry and in the development and sustainable operation of the mines in the production of thermal power plants. The goal is to create a document that will be the basis for the future development of the company, which will define the position of this production part that operates within the company's electric power company, as well as the established policies, business strategies and strategic projects necessary for sustainable growth of the company in the most economically efficient and environmentally friendly way. One of the objectives is to define a base for the establishment of standards applied in the EU, i.e. fulfilling the obligations of BiH and Serbia within the Energy Community of South East Europe, with the ultimate goal of effective integration into the EU.

Starting from the general goals defined at the level of individual power companies, which include, among other things: safe and reliable production and supply of electricity, quality of electricity supply in accordance with quality standards, increased business efficiency, competitive prices of electricity reflecting economic and financial costs, increasing energy efficiency in all segments of the system of production, distribution, supply and use of energy, optimal use of local energy sources in order to achieve economic growth and social stability, use of opportunities in the regional market with the aim of exploiting export opportunities, energy management in an environmentally acceptable manner, satisfaction interests and needs of electricity customers in a liberalized environment and restructuring in order to improve business and fulfill the requirements within the Energy Community and EU standards, the individual initial objectives of future policy development are defined. Individual initial objectives, grouped into three categories, are shown in Fig. 3, [23].

When making a development plan, while respecting existing regulations, it is necessary, in the fields of energy, environmental protection, competition and renewable energy sources, to take into account future regulation after harmonization of regulations and practice with European legislation.

A. Electricity balance	B. Environmental aspect	C. Financial position
Ensure continuous production growth and product portfolio improvement Satisfy the future demand of end customers (security of supply and energy independence) Provide balance surplus / reserve (market participation / security) Ensure continuity of coal production	Meet EU standards regarding limit values of emissions from thermal power plants Improve efficiency (primarily in the production of thermal power plants), while reducing its own consumption Increase capacities based on renewable sources (sHP, photovoltaic panels, wind generators, etc.)	Continuously positively operate with the gradual growth of capital returns Satisfy liquidity and indebtedness indicators Increase investment ability Ensure sufficient reserves for the new investment cycle, for more strict environmental requirements, as well as the risks associated with the mine operations in the TPP

Fig. 3 Categories of individual initial objectives of future development policy, [23, 39]

The essence of the relevant legal settlement is security of supply, with focus on: consumer interest and protection, reduction of energy dependence and mitigation of environmental impact. To this end, the EU has defined for the establishment of a framework that will lead to: creating a competitive and unified electricity market, limiting emissions, increasing the share of renewable resources and increasing efficiency, [40, 41]. From the point of view of market and competition, the key changes relate to: ensuring the right for buyers to choose a supplier, separating network activities that are a natural monopoly and to be regulated (transfer, distribution) from other activities where competition is possible (production, supply) [42].

5.4 Energy Efficiency Management System

In order to create the conditions for further improvement of the energy efficiency aspect of the company, and in the efforts to institutionalize energy efficiency and energy efficiency in all segments of the company's business, within the framework of electricity companies, it is necessary to establish an appropriate energy efficiency management structure. By designating the structure of managers who are competent to conduct energy efficiency policies and preparing plans for increasing energy efficiency and energy savings, as well as the structure of specialized experts in the company in order to implement plans and measures for increasing energy efficiency and energy savings in the company, will ensure the implementation of the set goals of the energy management system efficiency in power companies, as a systematic approach to addressing energy efficiency issues with the structure of employees and with clear responsibilities and competence (ensuring active and permanent participation of all employees on promotion and realization of tasks of increasing energy efficiency and energy savings and improving the image of the company), with more efficient use of energy products and reducing fossil fuel consumption, reducing costs in the company, reducing greenhouse gas emissions (primarily CO_2) and reducing acid emissions (SO_2 and NOx) and solid particles (contributing to the development of a national policy to increase energy efficiency and reduce emissions, as well as the global fight against climate change), [39].

By introducing the energy efficiency management structure, companies will establish a solid foundation for the establishment of an energy efficiency management system in accordance with EN 16001 (Energy Management System), that is, the EN ISO 50001 standard, all in the function of company's commitment to continuously improving energy efficiency and energy savings in the coming period, [37].

5.5 *Energy Efficiency Regulations in the EU and Republic of Srpska and BiH as a Whole*

From the current EU energy efficiency regulation that influences the determination of the power sector of the Republic of Srpska (as well as the electric power industry in the Federation of BiH) in terms of long-term development, one can distinguish: directives related to energy efficiency, EU standards/norms for energy efficiency, directives related to renewable energy sources and greenhouse gas emissions. As part of the obligations from the Energy Community Treaty, the transposition of the legal settlement on energy efficiency into the legislation of the Republic of Srpska—Bosnia and Herzegovina is also carried out. Based on the Directive 2006/32/EC on ultimate use of energy efficiency and energy services, the Law on Energy Efficiency of the Republic of Srpska was prepared. The First National Action Plan for Energy Efficiency was also drafted, as well as a preliminary report on its implementation (done by the Energy Community). National Energy Efficiency Plans define quantitative targets for increasing energy efficiency in terms of savings in final energy consumption, bearing in mind households, transport and industry. Based on the effects of Directive 2006/32/EC on energy efficiency in energy consumption and energy services in EU Member States. The EU Member States and the EC Member States should make the Second National Action Plan for Energy Efficiency, with the aim of improving the effects and state of energy efficiency. This Second NEEAP should have been adopted in 2013. BiH is late with these activities, but it is inevitable that they will have to be implemented in the near future, and should be taken into account through the guidelines for the long-term development of the electric power industry of Republic of Srpska and the Federation of BiH in the domain of energy efficiency. In November 2012, the new Energy Efficiency Directive 2012/27/EU came into force in the EU, which replaced the Directive on the promotion of cogeneration 2004/8/EC, as well as the Energy Efficiency Directive 2006/32/EC, and it intensified and expanded activities to increase energy efficiency and energy savings, now in the entire chain of production, through distribution and supply, to the final energy consumption. The directive has ordered the allocation of public funds for such projects and it predicts the development of plans for energy saving companies, the performance of energy audits as regular activities in companies, the introduction of the Energy Management system into companies, the installation of smart metering, etc. The directive is currently mandatory for the EU Member States, and for the Energy Community countries, reports on the effects of the directive in the member states will be awaited before it becomes binding on the countries of the Energy Community. However, it is expected that this will happen sooner, so this directive, and all that it carries in terms of plans, goals and obligations for energy efficiency, should be taken into account when defining the development plans of the EP in Republic of Srpska—BiH.

5.6 Guidelines for Energy Efficiency in Production

Although significant progress has been made in the production of MH Electric Power Industry in the Republic of Srpska in recent years in raising energy efficiency, it is still lagging behind compared to developed countries, primarily due to the obsolete existing production part of power plants and associated technologies. In this sense, MH Electric Power Industry in the Republic of Srpska is facing the challenge of introducing into the production portfolio of efficient and clean technologies—power plants on renewable energy sources and thermo-blocks with clean coal technologies, in order to further increase energy efficiency and reduce emissions. Concerning the existing production facilities within the Republic of Srpska Energy Sector, the value of specific heat consumption for the existing part of the reconstructed and modernized blocks of TPP Gacko 1 and TPP Ugljevik 1 from about 30 to 33%, with planned maintenance and investment measures, will be maintained until their exiting from the operation, with a proactive increase in energy efficiency with the planned entry of new more efficient thermal units of the new generation into the company's production portfolio. In order to implement such commitments it is necessary to develop detailed plans for the implementation of projects such as operational plans for increasing energy savings by reducing their own consumption in production units that will be continuously implemented, performing energy audits and monitoring the effects in production facilities, then activities and projects in order to optimize the load distribution of production units of MH Electric Power Industry in the Republic of Srpska, based on the actual (on-line) specific heat consumption of thermal blocks, all with the goal of managing production units (load distribution) in the function of the goal of production efficiency. There is also the continuous implementation of energy audit activities in the mines, with the aim of identifying and implementing measures for increasing energy efficiency in mines and improving the quality of coal delivered (coal separation projects on mines to improve and equalize the quality of coal delivered—1st degree of homogenization of coal). One should not forget the realization of development activities and projects for the expansion of cogeneration from TPP Ugljevik 1 and TPP Gacko 1: Heating of the Gacka region from TPP Gacko 1 (cogeneration)—realization of the project in accordance with the recommendations of the Study with preliminary design and improvement of the level of heating of Ugljevik from TPP Ugljevik 1. Making of Study on the implementation of partial increase in energy efficiency in existing power generation plants should be predicted, as well as the implementation of development activities and projects that imply reducing CO_2 emissions. When it comes to new production facilities, from the aspect of net efficiency of thermal power plants, the European Commission document of July 2006 is important: IPPC8—Reference Document on Best Available Techniques for Large Combustion Plants, which lists the net efficacy of the thermo blocks according to the best available techniques in relation to the type of plant and type of fuel, Table 3.

The important thing for production portfolio of Power Industry in the Republic of Srpska, or planned new thermal power projects in the forthcoming planning period, from this document, it can be seen that: for the lignite block 2 in TPP Gacko, for the predicted FBC combustion technology, the minimum level of net efficiency is 40%, for block 3 in TPP Ugljevik, which will use brown coal, with combustion technology in the flight (PC), the net efficiency would be at least 43%, and with technology combustion in the fluidized layer (FBC) higher than 41%. Furthermore, for cogeneration (CHP), for all types of coal, a net efficiency of 75–90% is indicated, with the stated values for the net efficiency of cogeneration plants (CHP) conditioned by the share of heat power in the total power of such plants, i.e. the net efficiency is It is also dependent on the heat consumption. Finally, for gas-fired power plants with gas and steam turbine (CCGT), the net efficiency in the condensation regimen (regime only for the production of electric power) is 54–58%. The mentioned reference document also contains the emission limit values, which are included in the Industrial

Table 3 Efficiency levels according to BAT for thermal power plants (TPP), [1]

Fuel	Burning	Net efficiency, %	Net efficiency, %
		New TPP	New TPP
Lignite/brown coal	Cogeneration (CHP)	75–90	75–90
Lignite	FBC PFBC PC (DBB)	>40 >42 42–45	Specific for each power plant: Indicative: 36–40% or 3%
Brown coal	FBC PFBC PC (DBB and WBB)	>41 >42 43–47	

PC Pulverized Combustion, *DBB/WBB* Dry/Wet Bottom Boiler, *FBC* Fluidized Bed Combustion, *PFBC* Pressurized Fluidized Bed Combustion

Emissions Directive 2010/75/EU (IED), and thus became an integral part of the binding legislation in this field for EU Member States, and in the perspective of the EC countries, when it comes to new thermal power plants. Unlike emissions, values for net efficiency are not further regulated by appropriate directives or laws, and it is difficult to estimate when and whether this will happen at all. However, Directive 2010/75/EU on industrial emissions includes energy efficiency among the criteria for determining BAT for reaching ELV, which is also underlined in the preamble of the new Directive 2012/27/EU on energy efficiency.

In the new energy efficiency directive, the promotion of efficient heating and cooling through cogeneration and tri-generation is done to the extent that EU member states are obliged to identify the potential for high efficiency cogeneration and/or efficient remote heating and cooling, and to implement such projects, and that in cases where investment costs exceed the benefits of such projects, Member States are obliged to take adequate measures to implement such projects. Also, this directive distinguishes between high-efficiency cogeneration and cogeneration; in a way that high-efficiency cogeneration needs to meet the criterion of achieving primary energy savings of at least 10%, with such facilities having the potential for appropriate incentives. This is important in the development of future cogeneration projects in MH Power Industry in the Republic of Srpska in the next period, as well as in the design of planned new thermal units for MH Power Industry in the Republic of Srpska (block 2 TPP Gacko, solution of block 2 problem in Ugljevik). Finally, in the EU 2020 strategy, Priority 1—Energy Efficiency, Activity 3—which refers to electricity companies, it is mentioned that net efficiency should be among the criteria for issuing permits for power plants (authorization), which implies that only thermal power plants with the corresponding net efficiency can be obtained permits.

6 On Example of the Adopted Policy for the Strategy of Development of the Electric Power Sector

Sustainable development on the principles of economic growth, with the imperative of preserving the environment and respecting the social aspect, is the essence of the EU's energy and development policy. Rising energy prices and the dependence on energy imports threaten the stability of energy supplies, as well as the competitiveness of Europe. In addition, the negative impacts on the environment and the gradual drain of fossil fuel reserves are key problems in the EU energy today. An additional problem is the imperative of reducing emissions, as well as the fight against climate change. Therefore, the central objectives of the EU energy policy are: stability of supply (reduces dependence on energy imports), competitiveness (enables economic growth) and sustainability (enabling environment and social acceptability). The European Union's energy market is liberalized for all consumers who can choose electricity suppliers that appear on the market. This means that all legal and administrative barriers to entry into the market of electricity and gas companies are abolished. New suppliers have the ability to provide services to consumers at the prices that meet the competition. The open market should help to achieve real competitiveness in the European market and to improve security of supply. It should also help to protect the environment as companies need to innovate in the field of renewable energy.

Climate preservation policy assumes a radical reduction in CO_2 emissions and other environmental impacts. Therefore, in addition to the limitations used so far, that come from the characteristics of energy/technology/location of the plant, a dominant limitation is introduced—the cumulative right to greenhouse gas emissions that has a downward character. By 2030, greenhouse gas emission rights can be expected to be at least halved in relation to the early 1990s, which will affect structural changes in production and energy consumption. The obligations of radically reducing CO_2 emissions and other greenhouse gases additionally require an increase in the use of non-fossil fuels, primarily renewable sources, such as water, wind, sun, biomass, and energy efficiency and the application of new technologies.

The EU 2020 Strategy (20-20-20) provides solid EU energy policy frameworks based on five basic groups of action measures (priorities), such as: increasing energy efficiency and energy savings, integrating the internal energy market, ensuring quality, safety and stability of delivery for consumers, intensifying research and development activities in order to achieve the set goals and increase competitiveness, as well as further strengthening of the external position of the EU as the largest regional energy market in the world. This strategy sets ambitious targets for the energy sector, including the fight against climate change, which was set up by 2020: reducing greenhouse gas emissions by 20%, increasing the share of renewable sources to 20% of total final energy consumption and increasing energy efficiency by 20%. The European Parliament continuously supports these goals. The European Council also announced a long-term commitment to de-carbonizing the energy sector with the aim of reducing the EU's emissions from 80 to 95% by 2050 by the EU and other industrial countries. Energy efficiency is particularly emphasized as the most economically efficient way to reduce emissions, improve energy stability and competitiveness, provide energy availability for consumers, and increase employment.

As an example of the adoption of the strategy for the development of the electricity sector, an analysis will be given to Bosnia and Herzegovina (BiH), with a special focus on its one entity—the Republic of Srpska. The energy sector in the Republic of Srpska and BiH as a whole has significant development potential. Republic of Srpska and BiH currently in the region are rare examples of entities that have a positive electricity balance. Inadequate institutional and legal framework, political instability, undefined authorization procedures for the construction and selection of investors, along with the familiar problem of complicated and lengthy procedures for obtaining a large number of permits and approvals, represent an obstacle to significant investments in the energy sector in the Republic of Srpska and the Federation of BiH, and as such BiH as a whole. It should also be noted that BiH is significantly delayed in fulfilling its commitments undertaken by signing international treaties and agreements. The Energy Community Treaty provides for the creation of a legal framework for establishing a free energy market, promoting investment in the energy sector, and assisting the energy sector of countries in transition. The Stabilization and Association Agreement (SAA) also requires the adoption of European directives and standards related to energy. The development of MH Power Industry in the Republic of, with an emphasis on the future production portfolio of the company, should be traced, taking into account the stated goals of the EU and the legal settlement, and taking into account the initial and future technological, economic, legal and regulatory and social political situation in the Republic of Srpska as well framework positions related to BiH. It is particularly important to use our own energy resources and potentials available in the Republic of Srpska as a way to develop the economy, to increase employment and improve social opportunities. Initial analyzes made on the basis of previous observations indicate that for MH Power Industry in the Republic of Srpska, an optimal and realistic development scenario with the included mix of renewable energy sources (RES: hydro, wind, solar, biomass, geothermal energy, etc.) and replacement and modernized power plants on domestic coal. Such a mix implies optimized shares of individual RES and coal power plants in the function of the lowest costs of the electricity system, as well as in the function of other goals related to the ecological aspect (renewable growth, efficiency gains and emission reductions) and security of supply and energy independence. The current validated energy development strategy of the Republic of Srpska looks at three scenarios of development: S1—Higher GDP growth (the basic characteristic of this scenario is the rapid growth of gross domestic product (a desirable economic development scenario), the application of classic technologies without active government measures), S2—Of GDP with measures (the basic characteristic of this scenario is the rapid growth of gross domestic product with the application of energy efficiency measures and encouraging the use of renewable energy sources and S3—Lower GDP growth (the basic characteristic of this scenario is the slow growth of gross domestic product and the application of classical technologies without active government measures). Comparison of the three scenarios shows that the final energy consumption in the scenario with the measures and the high growth rate of GDP (S2) in the Republic of Srpska will be almost the same as the consumption of final energy in the scenario with a lower growth rate of GDP (S3), Fig. 4.

Energy efficiency has become one of the key issues of today. Energy savings as a result of more efficient use of energy sources have a significant impact on the economic and financial aspect of the company's operations, but also on the life of consumers. Increasing energy efficiency contributes to reducing greenhouse gas emissions (primarily CO_2), so more efficient production and rational energy consumption are actually key measures to combat global warming and climate change. Increasing energy efficiency should also be seen in the context of energy security, as well as the conservation of energy for future generations, or in the function of sustainable development. The EU emphasizes energy efficiency as the most economically efficient way to reduce emissions, improve energy stability and competitiveness, energy availability for consumers, and increase employment. The EU plans to incorporate energy efficiency into all relevant policies, including the implementation of education and training to change energy habits. The energy efficiency criterion will be imposed in all spheres, including the allocation of public funds. The current EU

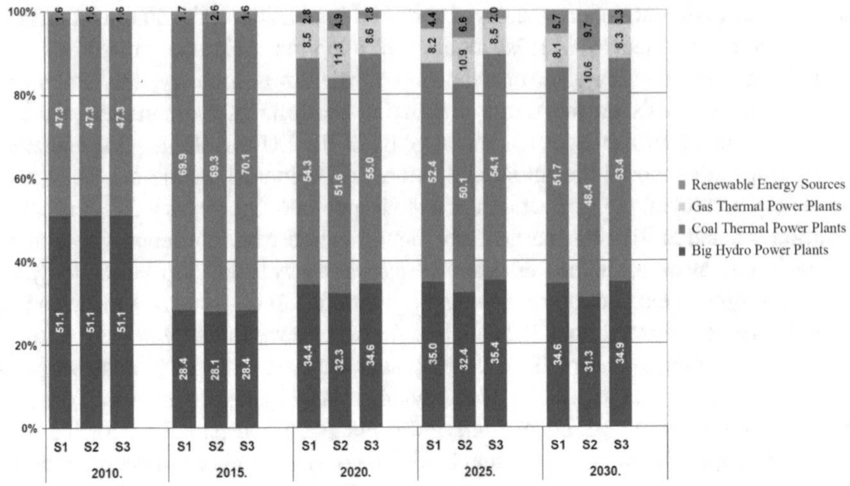

Fig. 4 Structure of electricity generation for three scenarios, with option B (prolongation of the lifespan of existing thermal power plants), [12]

energy policies and strategies will undoubtedly affect the development plans of MH Power Industry in the Republic of Srpska, when it comes to energy efficiency.

The EU 2020 Strategy (20-20-20) provides a solid framework for EU energy policy, based on five basic groups, where the first priority is to increase energy efficiency and energy savings. The EU 2020 Strategy sets ambitious targets for the energy sector, including the fight against climate change. In addition to reducing greenhouse gas emissions and increasing the share of renewable sources, the goal was also to increase energy efficiency by 20%. MH Electric power industry's policy of the Republic of Srpska is a continuous improvement of the aspect of energy efficiency in all its business processes, considering the complete chain of production—distribution—supply—energy consumption. Improving the aspect of energy efficiency should be at the very top of the strategic commitments of MH Power Industry in the Republic of Srpska. The activities needed to improve energy efficiency in existing production facilities are: expansion of cogeneration from TPP Ugljevik 1 and TPP Gacko 1 (heating of Gacka and Ugljevik), partial increase of energy efficiency of individual plants and processes in thermal power plants, reduction of own consumption, introduction of energy management, and the introduction of an energy efficiency management system. The new production facilities will have to meet the minimum level of efficiency according to the best available techniques (BAT) in relation to the type of plant and type of fuel (depending on the technology, this level for future blocks in Ugljevik and Gacko is 41–43%). Projections of the MH Electric power industry of the Republic of Srpska show that the average efficiency of the thermal power plant's production plant by 2030 could be increased from the current 30–33% to at least 35–37% and in the long-term over 40%. The Law on Electricity has ordered the opening of the electricity market 01 January 2015, which is a significant influence factor on the production activity of MH Power Industry in the Republic of Srpska. Customers should be able to choose a supplier from whom they will buy electricity in an unregulated open market, which is still not practiced in practice, and the responsibility lies solely with the Energy Regulatory Agency. Given that all activities are going, the truth at a somewhat slower pace, in this direction, the first determinant for the development of electricity generation activities relates to competitiveness. Considering the obligations of transposing EU regulations into domestic environmental legislation, renewable energy sources (RES) and energy efficiency, the second determinant that focuses on the development of production activity is ecological acceptability. Therefore, the long-term general goal of MH Power Industry in the Republic of Srpska in production activities can be formulated in the following way: to be competitive and to meet ecological standards. Competitiveness requires high productivity and cost reduction (the greatest impact of coal costs, operating costs due to employee redundancy and CO_2-related costs). This determines the priority directions of action, such as: energy efficiency increase, which reduces the necessary input raw materials and emissions, and hence costs, restructuring of the mines in order to obtain coal at the lowest price, realization of the activities from the aspect of environmental impact, i.e. realization of the first obligations to implement EU regulations on air emissions, energy efficiency and RES in final energy consumption (the above can be implemented through a strategy based on the use of domestic resources, both fossil fuels and renewable ones). Therefore, the activities and projects that are recommended through the long-term plan at the level of MH Power Industry in the Republic of Srpska refer to: planning and termination of activities related to the construction of new replacement thermo blocks that meet the criteria for border emission and minimum energy efficiency, continuation of reconstruction, modernization and revitalization of existing TPP units Gacko 1 and TPP Ugljevik 1, modernization of existing blocks from the aspect of fulfilling ecological norms, which means continuation of activities related to the construction of desulphurization and dentrification plants, especially the TPP Ugljevik 1 (ecological license requirement), construction of new power plants using renewable sources (hydroelectric power plants, wind farms, photovoltaic plants), the use of biomass and cogeneration (production of heat energy, including the construction of a hot water pipeline to expand the heat consumption). By optimizing these activities and individual projects, while respecting the framework long-term energy and financial projections at the level of MH Power Industry in the Republic of Srpska, it is possible to achieve concrete goals by 2030, as Table 4 is given for the activity of electricity generation. In order to achieve the set goals in the field of energy efficiency, MH Power Industry in the Republic of Srpska should adopt, follow and implement its own energy efficiency policy, as part of integral activities within the Energy Management System (EMS), which should be introduced into the company, all in accordance with EN 16001 and EN ISO 50001.

The introduction of EMS and the implementation of energy efficiency activities based on EMS is in line with the recommendations of the current long-term strategies and guidelines of the EU, as well as the new EU Directive 2012/27/EU on energy efficiency. Thus, MH Electric Power industry of the Republic of Srpska focuses its

Table 4 Presentation of concrete goals by 2030 for the activity of electricity production at the level of MH Power Industry in the Republic of Srpska, [1]

Production activity	Minimum necessary separation
Production of electricity and heat (and technological steam) If it uses renewable sources or waste or deals with combined heat and power production, it can obtain the status of a qualified manufacturer with a RERS solution	Accounting and management separation from other activities (completed process) Accounting separation between regulated (public service) and market production (completed process, established RERS Trebinje, FERK Mostar and DERK Tuzla)
Obligations of the manufacturer	
Compliance with the conditions of the license for the performance of the electric power industry Production of electricity that meets the quality requirements according to technical regulations, network rules and General conditions for the delivery of electricity Possession of measuring devices and equipment with which it is possible to measure energy and power which is delivered/taken over from the network Compliance with the prescribed rules of operation for the electricity market and the Network Code Satisfaction and compliance with the prescribed technical and operating conditions, as well as the environmental conditions from the environmental permit	
General goal	Key development areas
Be competitive and meet environmental standards using domestic resources	Energetic efficiency Reducing emissions of SO_2, NOx, PM and CO_2 Increased capacity on the basis of RES Cogeneration of electrical and thermal energy (tri generation and hybrid systems) Trigeneration (electric and thermal energy, cooling energy)

activities in the field of energy efficiency in the direction of the latest developments in this area, in the direction of the most developed countries. The process of mastering new technology or the development and application of a particular innovative solution must be based on upgrading existing knowledge and experience in a particular field, available to working personnel and their level of training and ethical awareness for the application of energy efficiency principles to energy installations within the Republic of Srpska. This implies that the technology of introducing the principles must be previously adopted, with the aim of integrating certain segments of energy efficiency into structured solutions. It requires a certain level of technical equipment for realization, as well as serviceability. Interest in achieving the project's earnings, in order to be able to maintain its business and its further development, besides engaging resources, require knowledge of the key factors for the successful management of energy efficiency projects.

These projects include the possession of certain competencies in the field of energy efficiency technology, as well as the existence of domestic demand for a certain energy efficiency technology, in order to achieve economies of scale and continue to further develop solutions and service improvements. The demand for energy-efficient technologies is under the significant influence of domestic regulatory authorities, which includes measures to raise standards and give financial incentives for implementation.

The realization of the project also requires the knowledge and skills of project management, which include project management methods, from project planning, risk management and change management to cost management and time management. Careful planning of the project and detailed definition of the offered solution results in the accuracy of the evaluation of the investment effects (calculated, taking into account not only the technical brochures of the producers, but also performed comparisons with similar realized systems, should provide an objective and reliable picture to the investor about the expected effects of the project). Before making an investment decision, an objective way is to reach these indicators through the making of a validity study, which will show the inventory and pricing of equipment and works, the costs of maintaining and using the system, using the sensitivity analysis method to determine the return indicators of investment in cases of deviations in relation to the planned, accessibility of financing conditions and access to financial institutions, as a prerequisite for successful implementation of energy efficiency projects.

Energy development planning is based on the security of consumer supply with minimal costs, with the accompanying rational use of domestic resources, which implies correct evaluation of imported energy forms, maximum prevention of monopolistic behavior (the only form of energy available) and achieving satisfactory environmental protection conditions. When planning the development of the electricity sector of the Republic of Srpska, the goal and criteria are uniquely determined through the settlement of the planned electricity consumption with minimum costs and assuming that certain financial, technical and environmental limitations on the availability of primary forms of energy are met. By adopting the Energy Development Strategy until 2030, as well as the Energy Efficiency Action Plan of the Republic of Srpska until 2018, prepared in accordance with the Law on Energy Efficiency (Official Gazette of the Republic of Srpska, No. 59/13), as well as the obligations under the Treaty establishing the Energy and on the basis of the requirements of the Directive 2006/32/EC of the European Parliament on energy efficiency in final energy consumption and energy services, the starting points for the introduction of the energy management system in companies, which became significant taxpayers as energy consumers of this Law, were created. On the other hand, projects for achieving energy savings are very important for thermal power plants within the Republic of Srpska EPS, as it enables, besides reducing costs and cleaner production, further improvement of the competitiveness of the company. Integrating these projects with existing business plans is in some way a further development of this company.

The proposed measures and their priority are given in accordance with the stated results, which does not mean that, in practice, depending on the current needs and priorities of the dependent company (subsidiaries) TPP Ugljevik and TPP Gacko within the MH Power Industry in the Republic of Srpska, Trebinje, different ranking of individual measures. When proposing energy efficiency measures, the current state of the equipment and financial capabilities of the CoP were taken into account, so that all proposed improvements in terms of reducing losses and increasing general energy efficiency are in realistic attainable limits. The new value of own consumption of the thermal power plant, which can be achieved by applying the proposed measures, is still above the reference value in comparison to similar systems in the world, which means that there is still a serious work on analyzing possible savings and reducing the losses as a whole. It should be noted that in the first period of implementation of the energy management system, several short-term measures will be identified, which do not require financial investments and are more of an organizational nature. After the introduction of the BAS ISO 50001 standard and the application of appropriate standardized procedures in practice, the number of short-term measures will be reduced, which also indicates an organized monitoring of the implementation of energy efficiency measures.

On the other hand, increasing energy efficiency and using renewable sources and environmental protection will be generated through the following activities:

- Establishing a mechanism for determining liability for environmental damage by the "polluter pays" principle;
- Drafting a strategy for encouraging investments in energy rehabilitation of existing residential and commercial buildings,
- Improving energy efficiency in public buildings,
- Adoption of the National Emission Reduction Plan for BiH (NERP) in accordance with the Report on the Plan for Reduction of Emissions of pollutants for the Electric Power Utility of the Republic of Srpska (EPU RS).

By adopting and applying the Law on Amendments to the Law on Waste Management, the basis will be created for the provisions of the proposed changes to achieve their compliance with the EU legislation, i.e. establish a mechanism for determining responsibility for environmental damage according to the "polluter pays" principle, i.e. that polluters be financially responsible for preventing damage to the environment. Amendments to the Law on Waste Management, whose enactment is planned in 2017, will create conditions for adopting regulations that will regulate the types of pollution, products that become special waste streams after use, criteria for calculating fees and bonds, the amount and method of calculation payment of fees, etc. When it comes to fiscal effects, there are no costs related to the activities of drafting and passing laws and by-laws. Regarding the impact on the business environment, budget, health and social status of citizens and the environment, this regulation has no effect on those areas. The proper application of the regulations and principles underlying EU legislation should not lead to an increase in the price of products since the financial instruments relate to imported products at which prices are already covered by the "polluter pays" principle.

According to the obligations arising from the decisions of the Ministerial Council of the Energy Community (MV EC), Bosnia and Herzegovina must implement the Energy Efficiency Directive, which, among other things, implies that the Republic of Srpska establish a long-term strategy for encouraging investment in the reconstruction of the housing and office buildings, public and private. This strategy should include a review of the real estate fund, based on statistical samples, the establishment of a cost-effective approach to renovations, depending on the type of buildings and climatic zones, policies and measures to encourage cost-effective large-scale renovation of buildings, including large-scale renovation procedures, a long-term perspective for directing decisions by individuals, the construction industry and financial institutions on investments, and assessing the expected energy savings and wider evidence-based benefits. According to the aforementioned decision of the MV EC, the Republic of Srpska is obliged to harmonize its legislative and regulatory framework for energy efficiency in the building sector by 15 October 2017, with the EU Directive, and adopt the aforementioned Strategy which, by 30 June 2017, should be an integral part of the NEEAP submitted to the Secretariat of the Energy Community. However, due to the lack of financial resources, it is realistic to expect that these deadlines will be extended for a minimum of 6–12 months, which will be harmonized with the international institutions in charge of technical assistance to fulfill these obligations. Implementation costs include the costs of developing the Strategy, which should be financed from public revenues of the Republic of Srpska and from grants from international organizations that provide technical assistance for the implementation of regulations related to energy efficiency. In addition to this, according to the preliminary estimates of the Energy Community in BiH, annually 15 buildings should be sanitized, and it takes about 7 million KM, of which the proportional part goes to the Republic of Srpska.

The basic risk for implementation is reflected in the lack of funds in the Environmental Protection Fund, which would, under favorable conditions and for the owners of buildings, be directed to energy rehabilitation and greater reconstruction of existing buildings, and the lack of interest of commercial banks to open favorable credit lines for lending to owners and to investors of existing facilities that need to perform major reconstruction to implement energy efficiency measures. This measure stems from the Republic of Srpska Energy Efficiency Action Plan, which is part of the Energy Efficiency Action Plan in BiH. This measure is in line with the Decision of the Ministerial Council on the implementation of the EU Energy Efficiency Directive.

By improving energy efficiency in public buildings, energy consumption in buildings will be reduced, which will directly affect the reduction of energy costs in the budget of the Republic and local self-governments, initiate economic activities in the construction sector and increase the number of jobs, which will positively affect GDP and public incomes. They are working on developing of financing mechanisms for improving energy efficiency, which will enable the involvement of the private sector in the financing of these projects. In 2017, the implementation of the World Bank "Energy Efficiency in BiH" project will continue, where investments of around 6 million KM are expected and the project "Green Economic Development", implemented by UNDP and the Environmental Protection and Energy Efficiency Fund of Republic of Srpska, where the planned investment value is 1–3 million BAM. It is planned that the reconstruction of public facilities will take place in the second and third quarter. During the period of 2018–2019 it is planned to implement new projects to be financed from credit and grants, as well as implementation of pilot projects to be financed through new models of energy efficiency financing. The realization of the mentioned activities will be based mainly on the engagement of domestic companies, which will enable the strengthening of their capacities in the realization of energy efficiency projects, especially in the areas of design, rehabilitation, energy audits, etc. Not providing additional credit and grant funds, and problems in the implementation of pilot projects given the fact that it would be a novelty for the market are potential risks for the implementation of the activities.

The Government of the Republic of Srpska has accepted the Information on the Draft of the National Emission Reduction Plan for BiH (NERP) in accordance with the Report on the Plan for Reduction of Emissions of Pollutants for Electric Power Industry of the Republic of Srpska, which was adopted by the Council of Ministers of BiH and addressed to the Energy Community Secretariat. The national plan has been prepared according to the guidelines of the Secretariat of the Energy Community. This plan refers to the reduction of emissions of sulfur dioxide, nitrogen oxides and solid particles from large combustion plants in the Republic of Srpska. The report on the plan for reducing pollutant emissions for the Republic of Srpska Electric Power Company will begin on January 1st, 2018 and will last until December 31st, 2027.

For the implementation of the adopted NERP, it is necessary to provide funds for the costs of financing the construction of a plant for the reduction of sulfur dioxide, nitrogen oxide and dust, which will be a significant burden for the Ugljevik TPP, which will result in an increase in total production costs, and thus the risk of jeopardizing the future market position of the power plant, and to a lesser extent the TPP Gacko, but will contribute to fulfilling obligations under the Energy Community Treaty and global reduction of air emissions from installations covered by the Plan. Because of this, ERS representatives have appealed to the relevant institutions to help find ways to fund the Plan.

The adopted National Emission Reduction Plan for BiH in accordance with the Report on the plan for reducing emissions of pollutants for the Republic of Srpska Electric Power Company has been prepared in accordance with the Policy Guidelines of the Secretariat of the Energy Community for the preparation of National Plans for Reduction of Air Pollution in Republic of Srpska and in accordance with the Order on Measures for Preventing and reducing air pollution.

It is important to emphasize that the Government of the Republic of Srpska in the area of energy efficiency will insist on the increased role of the Fund for Environmental Protection and Energy Efficiency of Republic of Srpska in providing financial support to economic activities related to the field of energy efficiency and environmental protection.

7 Conclusions

The methods used in planning the development of the electric power system differ with respect to: optimization technique (linear programming, nonlinear programming, etc.), type of approximation (linear, nonlinear) and economic valorization (with inflation, without inflation). None of the methods has proved to be absolutely acceptable for all problems so far, and a large number of methods have been developed for partial problems, i.e. or serve to optimize only the production facilities, or optimize only the transmission network, or optimize the overall energy, but with very simplified relations in the system. The basic assumption of applying optimization models is the high reliability of the parameters on which the calculation is based. Namely, if the parameters are not reliable enough, the question arises of the need for optimization. Sometimes, the sensitivity analysis, usually used in the final phase of each optimization model, cannot eliminate the effect of unreliable input data. Furthermore, it is never possible to include all the constraints within the model, so that only an approximately objective picture of the condition in the system is obtained. However, today, it is impossible to rely only on the intuition of planners in selecting the most favorable development option, so the application of the model is inevitable (multi-criteria evaluation methods).

Improving the efficiency of energy use also includes the need for quality energy management and the competence of staff to manage both equipment and energy in an adequate energy efficient manner. According to the recommendations of the international standard BAS ISO 50001, through the framework of energy management in organizations, it is necessary to establish energy saving programs that achieve energy efficiency and more competitive business, which should highlight the priority investments in energy efficiency projects, or to make lists of energy saving projects.

Reform of the BiH electricity sector through the implementation of strategic projects, aimed at securing high quality of universal service of energy supply and provision and protection of end customers or consumers, will create prerequisites for regional cooperation in the energy market in accordance with the South East European Energy Community Treaty. Although the current advantage of the Republic of Srpska is a surplus of electricity, the gradual and planned liberalization of the energy market will cancel the privileges of the monopoly position, and introduce a sharp competition in market competition that does not suffer from static and inflexible systems, which will have to be taken into account in the future.

The methods used in planning the development of the electric power system differ with respect to: optimization technique (linear programming, nonlinear programming, etc.), type of approximation (linear, nonlinear) and economic valorization (with inflation, without inflation). None of the methods has proved to be absolutely acceptable for all problems so far, and a large number of methods have been developed for partial problems, i.e., or serve to optimize only the production facilities, or optimize only the transmission network, or optimize the overall energy, but with very simplified relations in the system. The basic assumption of applying optimization models is the high reliability of the parameters on which the calculation is based. Namely, if the parameters are not reliable enough, the question arises of the need for optimization. Sometimes, even the sensitivity analysis, usually used in the final phase of each optimization model, cannot eliminate the effect of unreliable input data. Furthermore, it is never possible to include all the constraints within the model, so, only an approximately objective picture of the condition in the system is obtained. However, today, it is impossible to rely only on the intuition of planners in selecting the most favorable development option, so the application of the model is inevitable (multi-criteria evaluation methods).

References

1. Milovanović NZ, Papić Lj, Dumonjić-Milovanović RS, Milašinović NA, Knežević MD (2011) Sustainable energy planning: technologies and energy efficiency. DQM Monograph Library Quality and Reliability in Practice, Book 9, Prijevor, 779 p
2. Milovanović NZ (2011) Monographs: "Energy and process plants" Tom 1: Thermal power plants—Theoretical foundations. University of Banja Luka, Faculty of Mechanical Engineering Banja Luka, Banja Luka, 431 p
3. Milovanović NZ (2011) Monographs: "Energy and process plants" Tom 2: Thermal power plants—technological systems, design and construction, exploitation and maintenance. University of Banja Luka, Faculty of Mechanical Engineering Banja Luka, Banja Luka, 842 p
4. Milovanović NZ (2003) Optimization of power plant reliability. University of Banja Luka, Faculty of Mechanical Engineering Banja Luka, Banja Luka, 323 p
5. Papić RLj, Milovanović NZ (2007) Maintenance and reliability of technical systems. DQM monograph library quality and reliability in practice, Book 3, Prijevor, 501 p
6. Miličić D, Milovanović NZ (2010) Monograph of energy machines—steam turbines. University of Banja Luka, Faculty of Mechanical Engineering Banja Luka, Banja Luka, 897 p
7. Milovanović NZ, Miličić D (2012) Steam turbines for cogeneration energy production, library of monographs, Energy machines 3. University of Banja Luka, Faculty of Mechanical Engineering Banja Luka, Banja Luka, 520 p
8. Barbir F (2013) Optimal use of energy from the social aspect (Reflection on Ivan Illicha Energy and Equity essay). JAHR 4(8):673–680
9. Meadows DH, Meadows DL, Randers J, Behrens WW (2004) III: Limits to growth. Universe Books, New York, 1972; The book has been compiled by several more supplemented and updated editions: Meadows DH, Randers J, Meadows DL. Beyond the limits. Chelsea; Randers J, Meadows DL. Limits to growth: 30 years update. Chelsea Green Publishing Co., Post Mills, Vermont
10. Illich I (2010) Energy and equity. Calder and Boyers, London 1974. There is a Croatian translation: Illich I. Energy and righteousness, (Eugen Vukovic and Kaja Ocvirek Krusic, prev.), Discrepancy, vol 10, no 1 14/15
11. Heinberg R (2009) Searching for miracle, net energy limits and the fate of an industrial society. A joint project of the international forum on globalization and the post carbon Institute, (False Solution Series # 4)
12. Odum HT, Odum ECA (2001) Prosperous way down. University Press of Colorado, Boulder
13. Daly H (1996) Steady state economics, 2nd edn. Island Press, Washington, DC 1991; Daly H. Beyond growth, 1st edn. W. H Freeman, Beacon Press, Boston
14. Milovanović NZ (2009) The algorithm of activities for improving the competitiveness of power-process plant. Commun Dependability and Qual Manage 12(3):18–28
15. Milovanovic NZ (2009) Algorithm of measures to improve the competitiveness of thermal power plants part (in Russian). Prob Eng Autom 4:68–74
16. Milovanović NZ, Milanović D (2009) Current condition assessment and forecast of behavior features of complex power-process plants. Commun Dependability and Qual Manage 12(2):19–26
17. Granić G et al (2009) How to plan an energy after 2030. Oil 60(5):279–286 (Zagreb)
18. Milovanovic NZ (2000) Modified method for reliability evaluation of condensation thermal electric power plant. Ph.D. thesis, University of Banja Luka, Faculty of Mechanical Engineering Banja Luka

19. Milovanović NZ, Samardžić M (2015) Energy efficiency analysis of TPP Ugljevik for the period 2004–2014, Collection of works. In: Second scientific-expert symposium energy efficiency ENEF 2015, vol 1, pp 105–113
20. Milovanović NZ et al (2016) Study on energy efficiency of Ugljevik 1 Thermal power plant. IG Construction Institute, PC Trebinje
21. Law on Energy Efficiency (2013). Official Gazette of Republic of Srpska, No. 59/13, Banja Luka
22. Energy Efficiency Action Plan of Republic of Srpska until 2018 (2013). Government of Republic of Srpska, Banja Luka, Banja Luka
23. Energy Development Strategy of the Republic of Srpska until 2030 (2011). Energy Institute Hrvoje Požar Zagreb, Croatia and Economic Institute Banja Luka, RS-BiH, Banja Luka
24. Amendments to the Spatial Plan of the Republic of Srpska by 2025 (2015), Republic of Srpska, New Urban Planning Office of Republic of Srpska, Banja Luka
25. Milovanović NZ et al (2016) Introduction of energy management in thermal power plants of Republic of Srpska, Power Plants 2016, Collection of works, Zlatibor, 18 p
26. Milovanović NZ et al (2016) Models of energy sector development policy of Republic of Srpska from the aspect of energy efficiency, Power Plants 2016, Collection of works, Zlatibor, 19 p
27. Dumonjić-Milovanović RS et al (2016) Optimization of hybrid system based on energy of the sun and wind on Banjaluka region, Power Plants 2016, Collection of works, Zlatibor, 11 p
28. Babić V et al (2016) Optimization of pulverizing facility of steam boiler p-64 of 300 mW block of Ugljevik thermal power plant, Power Plants 2016, Collection of works, Zlatibor
29. Dumonjić-Milovanović RS et al (2017) Cleaner technologies for the production of energy and energy in the Republic of Srpska. Energy, Econ Ecol J Energy Union of RS
30. Milovanović NZ et al (2017) Analysis of solutions for the storage of electricity within the policy of the development of improved technologies based on renewable energy sources. Energy, Econ Ecol J Energy Union of RS
31. Dumonjić-Milovanović RS, Milovanović NZ (2017) The importance and role of distributed energy resources in the energy supply of residential and office buildings. Energy, Econ Ecol J Energy Union of RS
32. Milovanović NZ et al (2017) Analysis of the strategic plan for the development of the energy sector of the Republic of Srpska. Energy, Econ Ecol J Energy Union of RS
33. Milovanović NZ et al (2016) Analysis of the impact of the construction of a new block of TPP Gacko II of installed power of 1x350 MW on the environment, Power Plants 2016, Zlatibor, 19 p
34. Todorov M (2013) Stochastic approach to measuring the quality of electricity, Master work, University of Novi Sad, University Center for Applied Statistics—UCPS, Novi Sad, 101 p
35. Milovanović NZ et al (2009) Diagnostics of technical indicators for the maintenance of a thermal power plant, Part II—determination of reliability of the plant in the first approximation. Techn Diagn 8(3):3–8
36. Milovanović NZ et al (2016) Application of multi criteria decision-making for the selection of micro location of new thermal power plants, Thematic Collection of work, Analytical hierarchical process Application in energy, protection of work and environment and education, development center ALFA TEC ltd. Niš, Cent Complex Syst Res Niš 1:21–47
37. Papić RLj (2013) Concept of energy efficiency and international standard in the field of energy management ISO 50001. Research Center of the Center for Quality Management, Prijevor
38. Milovanović NZ et al (2016) Models of the development policy of the energy sector of the Republic of Srpska from the aspect of energy efficiency. Energy, Econ Ecol J Energy Union of Serbia, No. 1/2, Belgrade
39. Milovanović NZ, Samardžić M (2015) Analysis of energy efficiency of the operation of TPP Ugljevik for the period 2004–2014, Collection of works. In: Second scientific-expert symposium energy efficiency ENEF 2015, vol 1, pp 105–113
40. Milovanović NZ et al (2009) Diagnostics of technical indicators for the maintenance of a thermal power plant, Part III—determination of the physical causes of reliability decline. Tech Diagn 8(4):11–16

41. Milovanović NZ et al (2010) Extension of the life of equipment of thermal power plants in the function of the development of the power sector of the Republic of Srpska. Elektroprivreda 63(1):27–39
42. Vuković M (2014) Managing energy efficiency projects. Tech—Manage 64(5):855–860

The Integral Method of Hazard and Risk Assessment for the Production Facilities Operations

Alexander Bochkov

1 The Method of Integrated Safety Assessment at Hazardous Production Facilities[1]

Abstract Effective management of Hazardous Production Facilities (HPF), forecasting of deviations from nominal modes, prevention of failures, incidents and accidents is possible only on the basis of collecting and analysis of continuous flow of information on condition of HPF and also knowledge of processes complex happening at it (technological, organizational, behavioural and so forth). In the article problems of creation of integrated index within development of control methods of HPF industrial safety condition are designated and the problem of such objects management modelling because of precedents, based on classes of states is solved (there is an event/there is no event). The indicators describing conditions of industrial safety (in fact—risks factors) are offered to consider as signs of assessed situation. Assessments of analyzed states values of integrated index deviations are used. The offered approach to integration of methods of "data mining", conclusion based on precedents and adaptive management in consistent self-training advising system allows providing management of HPF safety and similar objects with weak formalizable behaviour. Based on a certain measure of proximity in capacity of which it is offered to use Hamming distance, one of similar precedents is chosen. The offered approach, using a method of support vectors, solves this problem according to precedents in

[1]Based on Bochkov, A., Lesnykh, V., Lukyanchikov, M. Problem of Creation of Integrated Index of Assessment of Production Safety Condition at Hazardous Production Facilities. Proceedings of the 29th European Safety and Reliability Conference. *Edited by* Michael Beer and Enrico Zio. Copyright ©2019 by ESREL2019 Organizers. *Published by* Research Publishing, Singapore. ISBN: 981-973-0000-00-0. https://doi.org/10.3850/981-973-0000-00-0 esrel2019-paper.

A. Bochkov (✉)
LLC "Gazprom Gaznadzor", Moscow 117418, Russia
e-mail: a.bochkov@gmail.com

M. Ram and H. Pham (eds.), *Advances in Reliability Analysis and its Applications*, Springer Series in Reliability Engineering,
https://doi.org/10.1007/978-3-030-31375-3_4

the past. A solver (auditor, qualifier, or recognizer) always keeps the found order until essential key signs on which partial order is formed are rejected. Examples of integrated index creation are presented.

Keywords Hazardous production facilities · Risk · Industrial safety · Integrated safety index · Precedent · Hamming distance · Method of support vectors · Risk focused approach · Trajectory analysis

1.1 Introduction

Hazardous Production Facilities (HPF) are complicated technical complexes where technological modes are supported in provided range of parameters according to project and standard documentation. Effective management of complicated technical system, forecasting of deviations from nominal modes, prevention of fails, incidents and accidents is possible only based on collecting and analysis of continuous information flow on its state and knowledge of set of processes proceeding in it. Over the last years, much attention is paid to questions of improvement of operating remote control (supervision) of industrial safety at HPF of oil and gas complex.

Enterprises of oil and gas sector actively discuss issues of organization of elements of remote monitoring and control system of industrial safety condition at HPF operated by them. Special acuteness to this problem is added with the circumstance that uniform understanding and methodical approach to its solving especially to the question of creation of integrated index of production safety condition is not created neither at the state level of management, nor at the level of enterprises of oil and gas complex.

A lot has been written about process safety indicators in General. However, there are few published empirical studies on this topic. Often in the literature, a distinction is made between so-called proactive and reactive indicators that provide an understanding of the level of system security [1]. The history of the development of the assumptions, models and theories of security described in the publications [2–4]. Consistent, epidemiological and system-dynamic assumptions, models and theories differ in this literature. Three models are often mentioned in the professional literature on process safety indicators: the Heinrich pyramid model [5–7], the classical Swiss cheese model [8–10], and the butterfly model [6, 7].

The large number of safety characteristics taken into account and the variety of their types [11–14] lead to the idea of the need to build a consolidated "comparative assessment", which, combining information on the values of individual characteristics and information on the significance of these characteristics, would allow to order all the situations under consideration by the degree of their overall priority [15].

In order to obtain such a consolidated assessment, the following questions must be addressed: "what qualities are to be considered in terms of the impact on the assessment of comparative advantage"; "how many of these qualities are individually measured"; "what method of grouping these numbers is accepted"; "what relative multipliers are attributed to the qualities to which preference is given". Answers to these questions determine the main stages of construction of summary (global, integral, generalized, General, synthetic, etc.) indicators, aggregating individual (local, differential, particular, marginal, analytical, etc.—respectively) indicators of the estimated object.

Really, possibilities of use of integrated indexes for decision-making in practice are strongly limited. This circumstance is noted by many authors [16, 17]. One of the reasons is integrated index, as a rule, does not allow to localize the cause of critical deviations, to mark field of activity or group of actions carrying out of which was insufficiently effective, in what process or production site threats and risks are most brightly shown. In other words, being effective at the stage of diagnostics of critical deviations, integrated indexes are useless for acceptance of "therapeutic" measures to the studied object. The second reason is a problem of weighing of private indexes of state which convolution directly affects value of integrated index. Weight are appointed randomly and with high degree of subjectivity, and, besides, the importance of this or that index can vary depending on object. The problem of identification of emergency situations at HPF is similar to classical task of management consisting in choice of scenario of behavior (prediction of possibility of negative event) in response to external challenges and threats.

Over the last years [18] "nonclassical" approach to management theory actively develops. This approach is connected with application of algorithms and methods of intellectual management of autonomous mobile objects based on fuzzy logic, neural networks and genetic algorithms. With the same approach, situational management based on hierarchical models with fuzzy predicates is connected; conclusion on precedents is method of decision-making in which knowledge of earlier arising situations or cases (precedents) is used. When considering new problem (the current case) similar precedent is found. Instead of every time to look for the decision from the very beginning, it is possible to try to use the decision made in similar situation, perhaps, having adapted it to the current case. Precedents in the case under consideration is accidents and incidents beginning in operating organization and their connection with undercontrol on one of indexes and/or actions directed to improvement of condition of industrial safety and dynamics of integrated index.

There is a problem of identification of object of management condition on its observed (known) parameters. For this purpose, it is necessary to be able to create based on prior statistical information generalized images—object conditions classes. It is possible to gain necessary knowledge from available data set by means of methods of data production—classification and clustering. In considered situations instead of mathematical model of object (HPF) of an exact type only prior information on it is available (assessment of requirements of the legislation to HPF, requirements to actions directed to improvement of industrial safety condition in this operating organization), operating impacts on it (discrepancies or deviations under these or those articles) and results of influences (actual data on occurred accidents or incidents).

Because of discrete nature of incidents and accidents, their rather small amount, use of data analysis device based on classical laws of large numbers for solution of described task is incorrect since convergence on probability in reality practically is never observed, except for statistics that is saved up in systems of mass service. It is also obvious that as true laws of distribution of analyzed accidental processes and, the main thing, factors defining them will be continuously adjusted (any hi-tech system changes quicker than adequate statistics collects), it is necessary to use criteria "free from distributions".

1.2 Goal Setting

In offered approach the goal of modeling of management of such objects on precedents based on classes of statuses is solved (there is an accident/there is no accident). In fact, it means that approach to integration of methods of data production, conclusion based on precedents and adaptive management in unified self-training advising system allowing managing objects with badly formalizable behavior is offered. Status of control object (HPF) is compared to precedents from in advance saved up database. On the basis of certain measure of proximity as which it is offered to use Hamming distance, one of similar precedents is selected. Hamming distance is number of positions in which corresponding characters of two words of identical length are different. In more general case Hamming distance is applied to lines of identical length and serves as distinction metrics (function defining distance in metric space) of objects of identical dimension. For example, descriptions of items of expenditure on improvement of industrial safety added with indexes of industrial and fire safety and labor protection and also indexes of observance by operating organizations of requirements of the legislation can serve as such objects.

In hierarchical system of natural and economic indexes and indicators used, for example, in PJSC Gazprom as generalized index LACE index (English, Lose of Average Capital Employed—losses of used capital)—dimensionless parameter (measured as a percentage) and determined as relation of amount of unplanned losses taking into account insurance protection to average used capital was offered [19]:

$$LACE = \frac{UPL \cdot (1 - M)}{ACE}, \tag{1.1}$$

where

UPL Unplanned Losses;
M Index of insurance safety of unplanned losses;
ACE Average Capital Employed.

As "base" for calculation of LACE average used capital—by analogy with definition of strategic target index (STI) of the first ROACE level (profitability of used capital) is chosen. By its nature, this index reflects result of efforts made by the company on providing at operated HPF levels of industrial and fire safety and labor protection demanded by the legislation.

This index acts as “switch” at classification of HPF that is subject to control. Positive dynamics of index values corresponds to decrease in current accident rate and total number of emergencies at HPF leading, respectively, to unplanned losses. Negative dynamics, on the contrary, speaks about growth of such situations and, respectively, about decrease in general level of production safety at HPF. In other words, positive (1) and negative (0) deviations of LACE from value at timepoint of t-1 “assess” corresponded to timepoint t level of production safety at HPF. As integrated index here acts not weighed convolution of basic indexes, but positive/negative deviation of index characterizing the level of unplanned losses and considering the level of insurance security of HPF of estimated organization and its average used capital. The set of indexes estimated by means of accounting of deviations of integrated index of LACE in this case also describes so-called precedents.

Reliable execution by the system of its functions is characterized by saving of some set characteristics (reflected in corresponding values of indexes) in set limits. In practice it is impossible completely to avoid deviations, however it is necessary to aim at minimization of deviations of current status from some set ideal—the purpose set, for example, in the form of values of criteria LACE values [20]. Measure of threat of negative deviations of LACE is considered in this case as variable value representing function of rather current status of the system: it increases at approach of assessed current situation to some admissible border after achievement of which the system cannot fulfill its obligations for improvement of values of integrated index.

Let the range of signs of the current situation of X (for example, the current values of indexes characterizing costs of ensuring production safety of HPF and also indexes of emergency and abnormal situations at them), the range of admissible realization of situations of Y (for example, the current value of integrated index of LACE is more (or less) than previous, etc.), to be set, and there is objective function of $y^*: X \rightarrow Y$, which values $y_i = y^*(x_i)$ are known only on final subset of objects $\{x_1, \ldots, x_l\} \subset X$ (for example, values of indexes describing condition of HPF corresponding to the current LACE value).

Pairs “object–answer” (x_i, y_i) we will call precedents. Set of pairs of $X_l = (x_i, y_i)_{i=1}^{l} = 1$ will make training selection. It is required to restore on selection of X_l dependence of y^*, that is to construct decision function of $A: X \rightarrow Y$ which would bring closer objective function of $y^*(x)$, and not only on objects of training selection, but also on all great number of X. As at the same time decision function of A has to allow effective computer realization, it is called also an algorithm. In such statement indexes (in fact—risks factors) act as signs of the current situation characterized by LACE value (risks indexes). Signs can be binary (1/0, red/green), nominal (set of values), order (set of ordered values) or quantitative type. In case the current value of integrated index of LACE in comparison with the previous value improved, the current situation is assessed positively (for example, it is coded as “1”), otherwise—is assessed negative (for example, it is coded as “0”).

1.3 Some Marks About Nonrandomness of Decision Rules

Object of management is HPF described by some feature set for which we take those items of expenditure on improvement of industrial safety on which financing was not carried out. Certainly, accident or an emergency are not caused directly by these signs. However, if to imagine the estimated company as "black box", these assessments are some kind of "sensors" brought to surface of this "box" and in unevident way reflecting its state. According to behavior of these "sensors" we do try to construct diagnosing system dividing situations into emergency and accident-free and allowing to predict their emergence in the future. Assessment models which are selected as a result of work entered the decisive rule are not those assessments which "provided" emergency at HPF, but they repeatedly "worked" in case of an emergency. If the number of such "operations" is big, it is impossible to ignore this fact, it means that in analyzed signs there is unevident information, there is hidden regularity characterizing condition of industrial safety of HPF.

Nonrandomness of such signs is shown in many works devoted to creation of complex diagnostic systems. For example, in [21] it is shown that the algorithm will generate nonrandom two-element piecewise linear rules if each of divided classes is presented by not less than 23 objects at $K = 2$ and not less than 56 at $K = 5$.

The precedent (within considered approach) is a description of problem or situation in total with detailed indication of actions taken (or not undertaken) in the considered situation or for solution of this problem. Conclusion on the basis of precedents is method of decision-making in which knowledge of previous situations or cases (precedents) is used. By consideration of a new problem (the current case) there is a similar precedent as an analog. It is possible to try to use its decision, perhaps, having adapted it to the current case instead of looking for the decision every time from the very beginning. After the current case is processed, it is brought into base of precedents together with decision for its possible subsequent use.

Offered model provides formation of generalized images of conditions of HPF of operating organizations on the basis of prior information; identification of condition of HPF of operating organizations in its output parameters—articles of programs and actions directed to improvement of condition of industrial safety; definition of influence of input parameters on transfer of HPF of operating organizations to various states.

At creation of such algorithms, one of the most important is the problem of choice of suitable precedent. After precedents are taken, it is necessary to choose "the most suitable" from them and to find "division border" of different groups of precedents. It is defined by comparison of signs of considered situation and in chosen precedents (i.e. events that earlier came to the end with accident or took place regularly).

1.4 Method of Solution

As a method on which finding of measure of similarity of precedents will be based the method of basic vectors is accepted [22]. Data of changes monitoring system in object of management (HPF) happening owing to these or those actions in the field of improvement of production safety which are carried out in operating organizations were taken for initial information. The authors, based on a certain measure of proximity between the situations under consideration, as which it is proposed to use the Hamming distance, propose the use of this method in the task of constructing an integral indicator of the safety of hazardous production facilities.

Statistical processing of forms of collecting this information allows marking those directions of works that are most relevant for effective and accident-free functioning and sustainable development of Society. Emergency events which were taking place in the past make knowledge base about earlier arising precedents, problems demanding the greatest concentration of efforts for their solving in the field of improvement of condition of production safety. In fact, it is possible to speak about questionnaires of quarterly assessment of situations at HPF as about element of early warning system and diagnostics of undesirable conditions of society, means of detection of its "diseases" at early stages.

Assessment of questionnaire of situation with costs of improvement of production safety for finding of steady combinations repeating at the majority of emergencies arising in the past at HPF is typical problem of classification of discrete objects,—task of images recognition theory. Offered approach, using method of basic vectors, solves this problem of precedents in the past. A solver (auditor, qualifier, and recognizer) always keeps found order until essential key signs on which partial order is under construction are rejected. Problem of creation of solver actually comes down to need to construct model of "grey box" testing entrances and outputs of some simulator of business processes of assessment of efficiency of investments to ensuring required level of industrial safety of operating organizations.

It is partially transparent model in which elements of decisions structure are visible and there is an opportunity to interfere with process of control of a solver. The idea of creation of piecewise linear decisive rule of general view is that hyperplane allowing to separate optimum in sense of some criterion (for example, maximum) a share of objects any one of the classes presented in the training selection is selected, then for the rest of objects new similar plane is looked for and so till full division of classes. Locally optimum planes used on each step of division of classes and creation of decisive rule of this kind in the procedure adopted by us are considered as orthogonal subspaces at first to initial individually considered signs, and then to new signs synthesized from these initial.

1.5 Case Study

For descriptive reasons we will illustrate the work of algorithm on illustrative example. Let some situation be characterized by nine criteria imposed on values of indexes reflecting condition of controlled HPF. Basic data of illustrative example are given in Table 1.

One of the reasons for execution of scaling of data or reduction to fixed range consists that the situation when, for example, one variable changes in the range from 10,000 to 100 000, and another in the range from 0.3 to 0.6, can have place. It is obvious that errors caused by influence of the first variable will influence assessment stronger than errors caused by the second one, changing in close limits. Having provided that each of variables changed within one range, we will provide equal influence of each of variables on change of weights in the course of assessment. Differentiable monotonic nonlinear increasing function which is often applied to "smoothing" of values of some value is called. Sigmoid function is often understood as logistic function.

On the first step it is necessary to execute analog-digital conversion (ADC) of data (actually their normalization) (Table 2). For sigmoidal curve it is possible to show the following converting table of rated data into probabilities (Table 3). The result of conversion is presented in Table 4. Scale of conversion into probabilities has

Table 1 Basic data of illustrative example

	+/-	1	2	3	4	5	6	7	8	9
1	1	1.00%	3000.00	2.00	5.00%	1.60	80.00%	200%	4.00	2.00
2	1	1.00%	3000.00	2.00	5.00%	1.60	80.00%	200%	4.00	4.00
3	1	1.00%	3000.00	2.00	5.00%	1.60	80.00%	200%	1.25	4.00
4	1	1.00%	3000.00	2.00	5.00%	1.60	80.00%	150%	1.25	4.00
5	1	1.00%	3000.00	2.00	5.00%	1.60	70.00%	150%	1.25	4.00
6	1	1.00%	3000.00	2.00	5.00%	1.00	70.00%	150%	1.25	4.00
7	1	1.00%	3000.00	2.00	10.00%	1.00	70.00%	150%	1.25	4.00
8	1	1.00%	3000.00	1.20	10.00%	1.00	70.00%	150%	1.25	4.00
9	1	1.00%	2500.00	1.20	10.00%	1.00	70.00%	150%	1.25	4.00
10	1	8.00%	2500.00	1.20	10.00%	1.00	70.00%	150%	1.25	4.00
11	0	8.00%	2500.00	1.20	10.00%	1.00	70.00%	150%	1.25	6.00
12	0	8.00%	2500.00	1.20	10.00%	1.00	70.00%	150%	0.50	6.00
13	0	8.00%	2500.00	1.20	10.00%	1.00	70.00%	100%	0.50	6.00
14	0	8.00%	2500.00	1.20	10.00%	1.00	60.00%	100%	0.50	6.00
15	0	8.00%	2500.00	1.20	10.00%	0.80	60.00%	100%	0.50	6.00
16	0	8.00%	2500.00	1.20	15.00%	0.80	60.00%	100%	0.50	6.00
17	0	8.00%	2500.00	1.00	15.00%	0.80	60.00%	100%	0.50	6.00
18	0	8.00%	2000.00	1.00	15.00%	0.80	60.00%	100%	0.50	6.00
19	0	6.00%	2000.00	1.00	15.00%	0.80	60.00%	100%	0.50	6.00
20	0	6.00%	2000.00	1.00	15.00%	0.80	60.00%	100%	0.50	6.00

Table 2 Result of basic data conversion

	+/-	1	2	3	4	5	6	7	8	9
1	1	0.00	1.00	1.00	1.00	1.00	1.00	1.00	1.00	1.00
2	1	0.00	1.00	1.00	1.00	1.00	1.00	1.00	1.00	0.37
3	1	0.00	1.00	1.00	1.00	1.00	1.00	1.00	0.44	0.37
4	1	0.00	1.00	1.00	1.00	1.00	1.00	0.58	0.44	0.37
5	1	0.00	1.00	1.00	1.00	1.00	0.54	0.58	0.44	0.37
6	1	0.00	1.00	1.00	1.00	0.32	0.54	0.58	0.44	0.37
7	1	0.00	1.00	1.00	0.37	0.32	0.54	0.58	0.44	0.37
8	1	0.00	1.00	0.26	0.37	0.32	0.54	0.58	0.44	0.37
9	1	0.00	0.55	0.26	0.37	0.32	0.54	0.58	0.44	0.37
10	1	1.00	0.55	0.26	0.37	0.32	0.54	0.58	0.44	0.37
11	0	1.00	0.55	0.26	0.37	0.32	0.54	0.58	0.44	0.00
12	0	1.00	0.55	0.26	0.37	0.32	0.54	0.58	0.00	0.00
13	0	1.00	0.55	0.26	0.37	0.32	0.54	0.00	0.00	0.00
14	0	1.00	0.55	0.26	0.37	0.32	0.00	0.00	0.00	0.00
15	0	1.00	0.55	0.26	0.37	0.00	0.00	0.00	0.00	0.00
16	0	1.00	0.55	0.26	0.00	0.00	0.00	0.00	0.00	0.00
17	0	1.00	0.55	0.00	0.00	0.00	0.00	0.00	0.00	0.00
18	0	1.00	0.00	0.00	0.00	0.00	0.00	0.00	0.00	0.00
19	0	0.86	0.00	0.00	0.00	0.00	0.00	0.00	0.00	0.00
20	0	0.86	0.00	0.00	0.00	0.00	0.00	0.00	0.00	0.00

Table 3 Converting table of probabilities

	1.0	0.9	0.8	0.7	0.6	0.5	0.4	0.3	0.2	0.1	0.0
1.0	0.00										
0.9	0.02	0.00									
0.8	0.08	0.06	0.00								
0.7	0.18	0.16	0.10	0.00							
0.6	0.32	0.30	0.24	**0.14**	0.00						
0.5	0.50	0.48	0.42	**0.32**	0.18	0.00					
0.4	0.68	0.66	0.60	**0.50**	0.36	0.18	0.00				
0.3	0.82	0.80	0.74	0.64	0.50	0.32	0.14	0.00			
0.2	0.92	0.90	0.84	0.74	0.60	0.42	0.24	0.10	0.00		
0.1	0.98	0.96	0.90	0.80	0.66	0.48	0.30	0.16	0.06	0.00	
0.0	1.00	0.98	0.92	0.82	0.68	0.50	0.32	0.18	0.08	0.02	0.00

to be nonlinear. From the theory of accidents and probability of deviation of values from the current value, this scale has to be proportional to squared distance from this index. The sigmoidal dependence allows converting basic data to new probability scale. Here we understand the probability of not accessory of considered situation as probability supernumerary (in fact, this probability is deviation measure from optimum value for considered index). On each of situations there will be indexes,

Table 4 Result of conversion of assessments into probabilities

	+/-	1	2	3	4	5	6	7	8	9
1	1	0.00	1.00	1.00	1.00	1.00	1.00	1.00	1.00	1.00
2	1	0.00	1.00	1.00	1.00	1.00	1.00	1.00	1.00	0.27
3	1	0.00	1.00	1.00	1.00	1.00	1.00	1.00	0.39	0.27
4	1	0.00	1.00	1.00	1.00	1.00	1.00	0.66	0.39	0.27
5	1	0.00	1.00	1.00	1.00	1.00	0.57	0.66	0.39	0.27
6	1	0.00	1.00	1.00	1.00	0.21	0.57	0.66	0.39	0.27
7	1	0.00	1.00	1.00	0.27	0.21	0.57	0.66	0.39	0.27
8	1	0.00	1.00	0.14	0.27	0.21	0.57	0.66	0.39	0.27
9	1	0.00	0.60	0.14	0.27	0.21	0.57	0.66	0.39	0.27
10	1	1.00	0.60	0.14	0.27	0.21	0.57	0.66	0.39	0.27
11	0	1.00	0.60	0.14	0.27	0.21	0.57	0.66	0.39	0.00
12	0	1.00	0.60	0.14	0.27	0.21	0.57	0.66	0.00	0.00
13	0	1.00	0.60	0.14	0.27	0.21	0.57	0.00	0.00	0.00
14	0	1.00	0.60	0.14	0.27	0.21	0.00	0.00	0.00	0.00
15	0	1.00	0.60	0.14	0.27	0.00	0.00	0.00	0.00	0.00
16	0	1.00	0.60	0.14	0.00	0.00	0.00	0.00	0.00	0.00
17	0	1.00	0.60	0.00	0.00	0.00	0.00	0.00	0.00	0.00
18	0	1.00	0.00	0.00	0.00	0.00	0.00	0.00	0.00	0.00
19	0	0.96	0.00	0.00	0.00	0.00	0.00	0.00	0.00	0.00
20	0	0.96	0.00	0.00	0.00	0.00	0.00	0.00	0.00	0.00

for example, can consist also of all 0. Some index from negative will have zero. After performance of stages 1, 2 we have in fact two matrixes of assessed values of probabilities of p(l, k) and q(n, k) for regular and emergency situations.

We look for contradictions according to Semi-Hamming. In further consideration of algorithm: values of $k = 1, \ldots, K$ is number of factor; values of $l = 1, .., L$ are lines of regular situations, and values of $n = 1, \ldots, N$ are lines of emergency situations, respectively.

Semi-Hamming matrix of $\Im_{sh}$ dimension (L, N) we calculate by the following formula:

$$\Im_{sh}(l, n) = \sum_{k} (p(l, k) - q(n, k)) \cdot b(p(l, k), q(n, k)) \tag{1.2}$$

Penalty function b(p(l, k), q(n, k)) equals to:

$$b(p(l, k), q(n, k)) = \begin{cases} 0, \text{ with } p(l, k) \le q(n, k); \\ 1, \text{ with } p(l, k) > q(n, k). \end{cases} \tag{1.3}$$

When all penalties ($k = 1, \ldots, K$) are equal to 0, they speak about complete dominance of "negative" example of n. There is an unremovable contradiction.

Results of calculation of Semi-Hamming distance are presented in Table 5. Analyzed area is marked in Table 5 with gray filling. For $l = 1, \ldots, L$ value of reserves equals to:

$$Z(l) = \min_n \Im_{sh}(l, n). \tag{1.4}$$

Table 5 Table of semi-hamming distance

	1	2	3	4	5	6	7	8	9	10	11	12	13	14	15	16	17	18	19	20	
1	0	0.7	1.3	1.7	2.1	2.9	3.6	4.5	4.9	4.9	5.2	5.6	6.2	6.8	7.0	7.3	7.4	8.0	8.0	8.0	**1**
2	0	0	0.6	1	1.4	2.2	2.9	3.8	4.2	4.2	4.4	4.8	5.5	6.1	6.3	6.5	6.7	7.3	7.3	7.3	**2**
3	0	0	0	0.3	0.8	1.6	2.3	3.2	3.6	3.6	3.8	4.2	4.9	5.4	5.7	5.9	6.1	6.7	6.7	6.7	**3**
4	0	0	0	0	0.4	1.2	2	2.8	3.2	3.2	3.5	3.9	4.5	5.1	5.3	5.6	5.7	6.3	6.3	6.3	**4**
5	0	0	0	0	0	0.8	1.5	2.4	2.8	2.8	3.1	3.4	4.1	4.7	4.9	5.2	5.3	5.9	5.9	5.9	**5**
6	0	0	0	0	0	0	0.7	1.6	2	2	2.3	2.7	3.3	3.9	4.1	4.4	4.5	5.1	5.1	5.1	**6**
7	0	0	0	0	0	0	0	0.9	1.3	1.3	1.5	1.9	2.6	3.2	3.4	3.6	3.8	4.4	4.4	4.4	**7**
8	0	0	0	0	0	0	0	0	0.4	0.4	0.7	1.1	1.7	2.3	2.5	2.8	2.9	3.5	3.5	3.5	**8**
9	0	0	0	0	0	0	0	0	0	0	0.3	0.7	1.3	1.9	2.1	2.4	2.5	3.1	3.1	3.1	**9**
10	1	1	1	1	1	1	1	1	1	0	0.3	0.7	1.3	1.9	2.1	2.4	2.5	3.1	3.1	3.1	**10**
11	1	1	1	1	1	1	1	1	1	0	0	0.4	1	1.6	1.8	2.1	2.2	2.8	2.9	2.9	**11**
12	1	1	1	1	1	1	1	1	1	0	0	0	0.7	1.2	1.4	1.7	1.8	2.4	2.5	2.5	**12**
13	1	1	1	1	1	1	1	1	1	0	0	0	0	0.6	0.8	1	1.2	1.8	1.8	1.8	**13**
14	1	1	1	1	1	1	1	1	1	0	0	0	0	0	0.2	0.5	0.6	1.2	1.3	1.3	**14**
15	1	1	1	1	1	1	1	1	1	0	0	0	0	0	0	0.3	0.4	1	1	1	**15**
16	1	1	1	1	1	1	1	1	1	0	0	0	0	0	0	0	0.1	0.7	0.8	0.8	**16**
17	1	1	1	1	1	1	1	1	1	0	0	0	0	0	0	0	0	0.6	0.6	0.6	**17**
18	1	1	1	1	1	1	1	1	1	0	0	0	0	0	0	0	0	0	0	0	**18**
19	0.96	1	1	1	1	1	1	1	1	0	0	0	0	0	0	0	0	0	0	0	**19**
20	0.96	1	1	1	1	1	1	1	1	0	0	0	0	0	0	0	0	0	0	0	**20**
	1	**2**	**3**	**4**	**5**	**6**	**7**	**8**	**9**	**10**	**11**	**12**	**13**	**14**	**15**	**16**	**17**	**18**	**19**	**20**	

Minimum value from all Z(l)

$$ZZ = \min_l Z(l) \tag{1.5}$$

defines "separability cutoff" of all of positive descriptions of situations from all negative.

In principle "separability cutoff" can be reached more than for one pair (l_1, n_1). In considered illustrative example separability cutoff is defined as $ZZ \approx 0.2724$. Respectively pairs of descriptions of situations composing separability cutoff 9 (regular)—11 (emergency), 10 (positive)—11 (emergency).

For assessment of probability of approachibility "goals" for regular situations, for all n_1 we will calculate $U(n_1) = \sum_k (1 - q(n_1, k))$. For these values there is semi-sum of distance $\Im_{sh}(l, n_1)$ therefore:

$$\begin{aligned} W(l, n_1) &= UR + (1 - UR) \cdot \left(\frac{\Im_{sh}(l, n_1)}{2 \cdot \left(U(n_1) - \frac{\Im_{sh}(l, n_1)}{2} \right)} \right) \\ &= UR + (1 - UR) \cdot \left(\frac{\Im_{sh}(l, n_1)}{2 \cdot (U(n_1) - \Im_{sh}(l, n_1))} \right). \end{aligned} \tag{1.6}$$

Concluding probability

$$W(l) = \min_{n_1} W(l, n_1). \tag{1.7}$$

Where UR—border level. In considered example border level we arrange equal to UR = 0.8.

Calculation of probability for emergency situations is identical. For all l_1 we will calculate $V(l_1) = \sum_k (1 - p(l_1, k))$. For these values there is semi-sum of distance $\Im_{sh}(l_1, n)$, therefore probability is calculated from the formula:

$$\begin{aligned} S(l_1, n) &= UR - UR\left(\frac{\Im_{sh}(l_1, n)}{2 \cdot \left(V(l_1) - \frac{\Im_{sh}(l_1, n)}{2}\right)}\right) \\ &= UR - UR \cdot \left(\frac{\Im_{sh}(l, n_1)}{2 \cdot (V(l_1) - \Im_{sh}(l_1, n))}\right). \end{aligned} \tag{1.8}$$

Concluding probability

$$S(n) = \max_{l_1} S(l_1, n). \tag{1.9}$$

i.e. in this case the maximum is taken.

Graphically carried out calculations can be illustrated with the scheme given on Fig. 1.

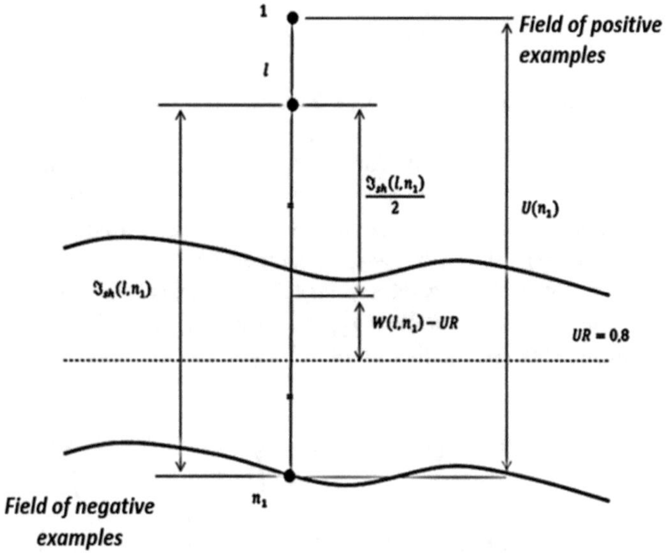

Fig. 1 Illustration of definition of situations transition probability from regular state into emergency

Concluding probabilities: $W(l) = \min_{n_1} W(l, n_1) = 0.805$ and, respectively, $S(n) = \max_{l_1} S(l_1, n) = 0.718$. These probabilities numerically characterize value of integrated index of under control risk "above" and "below", i.e. how controlled object can be reassessed or under assessed in terms of its inclusion in the plan of checks.

1.6 Conclusion

Development of offered method of analysis and management of situations in precedents can become basis of development of integrated index of condition of production safety at HPF and element of system of monitoring and early warning of breakdown and emergency situations within widely introduced in the present days so-called risk focused approach.

Generally, this approach covers both probabilistic methods of modeling of emergency processes and events, and deterministic methods. In wide sense, risk in the field of safety of objects of critical infrastructure should be considered as multicomponent vector parameters set of which can change.

Real assessment of safety level on the basis of use risk focused approach is impossible without rather informative base of rather quantitative and qualitative characteristics of risks factors and, on the other hand, data on objects state and technological processes at them which come under influence of these risk factors.

Risk assessment first of all aims at definition of its quantitative indexes that gives chance to use it not only for assessment of industrial safety condition, but also for proof of economic efficiency of actions, economic calculations of necessary compensation, or compensations of lost health to workers and environment when the question of "expense-profit" ratio appears.

As criteria of achievement of predictive goal, because of features of task specified in item 1 it is necessary to take not values of deviations of model and real data, but criteria used in methods of classification and recognition of images. For example, as forecast accuracy measurement it is possible to use values of errors of prediction of the first and the second kinds for various classes and types of situations, and if it is possible depending on classes of physical object and depending on value of parameters of forecast background. As criterion of decision-making it is possible to use statistical characteristic of assessed condition showing "distance" of current state, rated on unit, from "interface" of earlier observed and assessed (as regular or emergency) states—measure of threat (risk) of not achievement of preset values. Only based on correct primary analysis of long-term statistics it will be possible to draw the conclusion about the most preferable methods of integrated assessment of critical infrastructure objects safety.

Final goal of monitoring of HPF threats and risks safety is creation of such predictive model of dynamics of situations which will allow to reduce degree of uncertainty of probabilities of events and their scales by means of computing experiments and selection of acceptable parameters, that is to obtain forecasting information due to identification of insights which indicate either changes of object condition, or regularities of changes of external environment parameters significantly influencing its functioning [23]. Monitoring object control has to be organized so that it will be possible to realize management decisions in time if object condition comes close dangerous zone.

2 Some Methodological Aspects of Multicriteria Method of Decision-Making on the Sustainability and Security of Industrial Objects Exploitation

Abstract A real assessment of the level of safety based on the use of risk-based approach is impossible without a sufficiently informative base on the quantitative and qualitative characteristics of risk factors and, on the other hand, data on the state of objects and technological processes on them that are affected by these risk factors. Risk assessment always aims to determine its quantitative indicators, which makes it possible to use it not only to assess the state of industrial (and industrial) safety, but also to justify the economic efficiency of measures, economic calculations of the necessary compensation, or compensation for lost health workers and environmental damage, when there is a question of the optimal ratio of "cost-benefit". At the stage of solving the problem of risk assessment, it is necessary to establish links between the analyzed indicators and safety indicators with high-level indicators (for example, strategic targets of the Company) and the degree of their impact on the achievement of the target values of these indicators. It should be remembered that due to the discrete nature of incidents and accidents, their relatively small number, the use of data analysis apparatus based on the classical laws of large numbers to solve this problem is incorrect, because convergence in probability in reality is almost never observed, except for statistics accumulated in Queuing systems. It is also obvious that since the true laws of distribution of the analyzed random processes and, most importantly, the factors determining them will be continuously corrected (any high-tech system changes faster than adequate statistics accumulate), it is necessary to use the criteria of "free from distributions". In particular, for example, the criteria for achieving the prognostic goal should be taken not the values of deviations of model and real data, but the criteria used in the methods of classification and pattern recognition. For example, as a measure of prediction accuracy, you can use the values of prediction errors of the first and second genera for different classes and types of situations, and, if possible, depending on the classes of the physical object and depending on the value of the forecast background parameters. The theoretical basis and description of the developed, based on the use of mathematical theory of optimal control and the use of support vector machines, tools and algorithms of the so-called "trajectory group analysis" dynamics of key indicators characterizing the state of industrial safety of hazardous production facilities, with respect to a given target value.

Keywords Risk · Support vectors · Safety · Hazardous production facilities · Trajectory analysis

2.1 The Problem of Constructing an Integral Indicator Industrial Safety

Hazardous production facilities (HPF) are generally complexes integrated into a complex technical system. Effective management of such a system, prediction of deviations from nominal modes, prevention of production problems, failures, incidents and accidents is possible only on the basis of the collection and analysis of the continuous flow of information about its status and the processes occurring in it [24, 25]. There has always been a great deal of attention at the state level to improving oversight of the activities of organizations that operate CBOs. In recent years, in connection with the transition in the field of safety to the paradigm of risk-based approach, questions of the introduction of operational remote control (supervision) of industrial safety in the HPF oil and gas complex are widely discussed.

These issues have been repeatedly raised at the meetings of the Government Commission on the development of the fuel and energy sector, the reproduction of the mineral resource base and improving the energy efficiency of the economy. By the order of Rostechnadzor [29] the Federal regulations [30] were amended to ensure the operation of the system of remote control of technological processes at hazardous production facilities, providing exploration, development and operation of subsoil. Work is underway to make appropriate changes to the Federal law [31], which is fundamental for the organizations operating HPF.

Despite the fact that to date, the requirements regarding the organization of the remote control system on the HPF of other spheres of activity, except gas exploration and gas production, have not been formalized by Rostekhnadzor, the interested enterprises of the oil and gas sector are actively discussing the organization of elements of the system for remote monitoring of the state of industrial safety at the facilities they operate. The fact that there is no common understanding and well-developed methodological approach to its solution, especially to the issue of constructing an integral indicator of the state of industrial safety, is not formed either at the Federal level of management or at the level of oil and gas enterprises adds to the particular urgency of this problem. The very question of the possibility of obtaining an integrated assessment of the multifaceted activities related to the operation and safety of the HPF is questionable (and sometimes actively opposed).

Attempts, however, are being made. For example, the group of companies "Industrial safety" proposed[2] "Calculator of the integrated indicator of industrial safety". It is positioned that, quote: "… the Calculator is intended for inspectors of Rostekhnadzor and optionally can be used by experts in the field of industrial safety…"

[2]Electronic resource. Access mode: www.safety.ru/danger-analyse/#/.

implements algorithms "Methods of calculating the values of indicators used to assess the likelihood of potential negative consequences of non-compliance with the requirements of industrial safety", which was developed in the framework of the Federal law of 26.12.2008 №294-FL "On protection of the rights of legal entities and individual entrepreneurs in the implementation of state control (supervision) and municipal control". The technique, in turn, is based on the provisions of the Federal law of 21.07.1997 №116-FL "On industrial safety of hazardous production facilities", developing these provisions of regulatory legal documents of Rostekhnadzor in the field of industrial safety, registration of HPF, organization and implementation of industrial control, etc.; modern statistical data characterizing various subjects of the Russian Federation; expert assessments and actual data on the state of industrial safety of a particular HPF (statistics on accidents and injuries; information obtained as a result of control and Supervisory activities, the provision of public services, the implementation of industrial control; data contained in the register of expert opinions) …".[3]

The authors give numerous reservations of practical application of the proposed approach. For example, it is assumed that the specialists applying the Methodology and the Calculator have sufficient information about the evaluated object. At the same time, the criteria of "sufficiency" is not given. It is indicated that "the results obtained with the help of the calculator with incorrect or unqualified data entry can not be completely reliable, so they can not be used for management decisions…", but the criteria for checking the correctness of the data and the requirements for the qualification of experts are not given. Moreover, the applicability of this approach can only be discussed after the accumulation of representative statistics (5–10 years).

It should be said that the circumstances of the limited practical application of integrated indicators for decision-making are noted by many authors [32, 33].

One of the reasons is the integrated indicator, as a rule, does not allow to localize the cause of critical deviations, to identify the scope of activities or a group of activities, the conduct of which was not effective enough to determine: in which process or production site the most clearly manifested threats and risks. That is, being effective at the stage of diagnosis of critical deviations, integral indicators are useless for taking "therapeutic" measures to the studied object.

The second reason is the problem of weighing particular state indicators, the convolution of which directly affects the value of the integral indicator. Weights are assigned arbitrarily and with a high degree of subjectivity, and, in addition, the significance of an indicator may vary depending on the object. The distrust in expert assessments, which has been firmly rooted in recent years, makes the task of constructing an integral indicator insoluble.

At the same time, the problem of identification of accidents and incidents at HPF is largely analogous to the classical management problem, which consists in choosing a behavior scenario (predicting the possibility of a negative event) in response to external challenges and threats (for example, underfunding of programs and activities aimed at improving industrial safety) [24, 34].

---

[3]ibid.

2.2 The Main Provisions of the Group Analysis of Object Dynamics

Mathematical theory of optimal control (TOC) is a long-established branch of applied mathematics [35–37]. Various problems of optimization of systems of different levels of complexity in both deterministic and stochastic formulations are solved within the framework of the TOC. Almost all methods of solution in the theory of automatic control (TAC) are based on the calculation of control actions through feedbacks, which goes back to the fundamental work of N. Wiener is one of the founders of modern cybernetics [36].

Among the tasks of the TAC for solving the problems of assessing the state of safety and stability of the ESG operation, the closest analogy is the task of controlling the movement of the aircraft in the autopilot mode, which is formulated as the development of control actions aimed at returning to the standard dynamic trajectory, which in the case of an aircraft is the Central line of the air corridor permitted for flights. The analogy of the problem with the above task of the TAC is quite close in that the decision on the level of security of the HPF is not only a solution that affects the current state of the object, but also affects the subsequent state of the estimated object. In any case, from the point of view of the theory of systems, there is a specific case of power mode switching ("power control" type of surgery), rather than "parametric control" (type of homeopathic therapy) [38].

The effects are in portions—in the form of discrete gains, expressed in achieving the required levels of security. Figuratively speaking, in this case, the Corporation acts as a "refueling aircraft", which is one or another operating HPF enterprise in the form of financial injections into safety "refuels" (finances) with the assignment of responsibility for the implementation of the "route task" (for the performance of works provided for by state requirements and local regulations in the field of industrial safety of HPF).

The resulting finances give the operating organizations "forces" to perform their production functions, and if the forces are money, then the organizations selected for additional financing acquire "additional acceleration", inversely proportional to their size ("mass" of the object).

Thus, the same level of infusion (for example, in 1 million rubles) for a large operating enterprise (with a large number of employees and operated HPF) will have a weak "acceleration", and for medium and small operating organizations the same amount will be significant. At the same time, the amount of work will be the same in both cases, if the values of the "paths" passed by the organizations operating the HPF are equal.

Since any product is created by work A, and the work, respectively, is equal to the product of (M) mass (number of employees) per (a) (acceleration—specific infusions of resources) and per path (S) (productivity of the unit of allocated funds per employee engaged in ensuring the security of the HPF).

This approach inevitably raises natural questions:

- whether the work will be done (passed the way with the proper level of quality of work performed);

– how the proposed works are interfaced with other works and obligations of the operating HPF of the organization, whether there will be an unreasonable overload of "obligations" from the Corporation received together with additional finances;
– how to ensure the planned load in such a way that in the event of the need for urgent work, there was a possibility of additional involvement of the organization to work without compromising the safety and quality of the production process.

There is a "fundamental task" for the management of the Corporation: to make all applicants for additional funding "moderately hungry". Moderately underutilized HPF operating organizations are prone to innovation, search for hidden reserves of productivity and safety: their present is not so good, so they "care about the future", that is, they are more responsive to strategic planning issues.

And, finally-subsidizing the security of HPF is really discrete-and it is difficult to be decomposed into parts, due to the fact that solutions such as "one and a half-lumberjack" are not always even technically feasible—there will always be "ahead-going", which were financed in 2 of 1.5 situations, and lagging behind, which received 1 volume of 1.5 necessary funding. In this case, the initial analogy with the aircraft works well: you can not make 1.5 flights. Rather, you can on average-two pilots [(1 + 2)/2 = 1.5]. And this, just means that the average value (optimum) is not achievable by the analyzed objects (in one calculation period), but is achievable by one pilot in two calculation periods: on the first day—1 plus on the second day—2, total: 1.5 on average per day. This in principle changes the relation to dynamic charts (Figs. 2 and 3).

The cyclic drop from 2 to 1 in the load of each of the objects (Fig. 2) should not be considered as their "failure"—in fact, the Corporation can give a total load of only 3 units. The situation of "discrimination", when two equal objects" repeat " their successes (Fig. 3)—not the best. The first object in the calculation periods will have chronic underfunding, and the second will have an obvious "overload".

The consequence of the above reasoning is the conclusion that in security matters it is necessary to abandon the principle of keeping on any one "optimal" trajectory and take as a basis the principle of keeping the trajectory within the optimal corridor. This corridor has a "Central line", for example, equal to 1.5, there is a half—width-0.8. All objects in the range of 1.6 ± 0.8 are equally good, with the nuance that the one

Fig. 2 Illustration of the case of the cyclic load

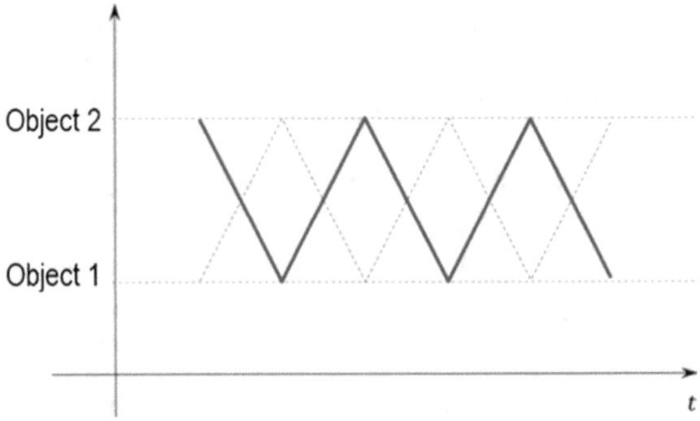

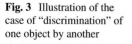

Fig. 3 Illustration of the case of "discrimination" of one object by another

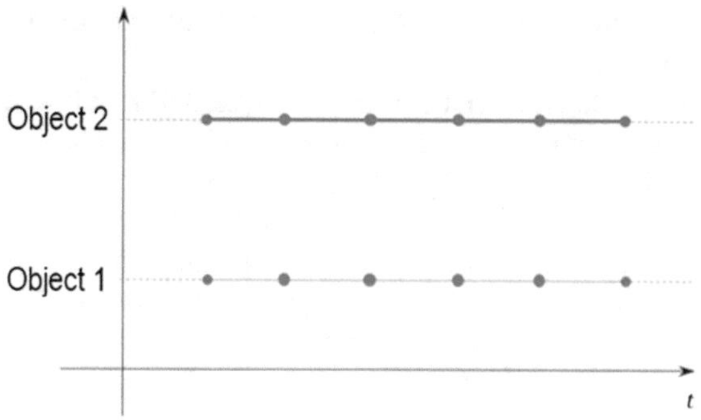

who shows indicators equal to two, is above (or ahead) the "center of mass" of other objects, and the second with indicators equal to one, is closer to the bottom. Note that 1.6 here is not a typo, it is some value greater than 1.5 should be the optimum. Both "leaders" and "laggards" should have stimulating conditions for their growth. That is, in the "optimum" dynamic behavior—objects with an average of 1.5—is, in fact, "underutilized" objects with a reserve of 0.1.

At the present time to fight for "performance" inflation causes. Even someone who has a constant value of "2 "when the center of the corridor is shifted (1.6 → 1.7 → … → 2.1 → …), it will cease to be a "leader" over time, if it calms down, and will not increase the volume of its obligations. In addition, the width of the corridor may vary. At times when it is "narrower" than the average, participants in the security process are more limited in the choice of work and optimize their desire to receive additional funding with their ability to meet their commitments.

On the contrary, if the width of corridor allowed for more freedom for "overfulfillment" and "underperformance", and therefore allowed the risky search of work with cost, there is inadequate production (overpayment in the hope of success).

Key points of the above reasoning:

- the dynamics of safety indicators of objects should be evaluated according to the work performed by the operating organizations to ensure safety in terms of the trajectory (overcoming some path);
- evaluation of the object should be normalized to its "mass" (for example, with respect to contractors participating in the procurement, for which this method was used in practice, this means that we are interested in the performance indicators for 1 person working in the company, as an indicator of productivity);
- "average position" is unattainable due to various circumstances (the average trajectory is replaced by a set of trajectories passing through some corridors; the section of the corridor can change, and this section is similar to some "target", while being constantly in the middle of a constant target (stabilization) is not the best solution in terms of mobility and workload of the group);
- the dynamics of indicators is subject to General laws for the group of participants, determined by the amount of funding (which may not depend on the quality of participants), inflation, the strategic line of development of the Corporation or the

requirements of the state in the field of security (this is the so-called trend acting at all);

- the General trend, however, acts on different operating organizations in different ways, not only because of their inertia, but also because of the folding of certain circumstances, expanding and vice versa narrowing both their ability to obtain the necessary amount of funding and their ability to fulfill their commitments (therefore, there is a phenomenon that in the theory of non—equilibrium systems called diffusion—operating organizations "run" by activities in which they are more successful-specialize);
- the size of the corridors, and not only their centres, may change over time (this should be taken into account in the design of the monitoring system, with appropriate adjustments).

2.3 *Diffusion Approximation of Markov Processes. Fokker-Planck Equation*

Typically, the group behavior of elements in the system is described by the probability distribution function $P(x_1, \ldots, x_n)$, which is associated with the field of forces defined by the potential function $V(x_1, \ldots, x_n; c_1, \ldots, c_k)$ by the Fokker-Planck equation [39, 40]:

$$\frac{\partial P}{\partial t} = \nabla \cdot (P \cdot \nabla V) + \nabla^2 (D \cdot P). \tag{2.1}$$

While the control parameters $\langle c_1, \ldots, c_k \rangle$ depend on time, the potential $V(x_1, \ldots, x_n; c_1, \ldots, c_k)$ does not explicitly depend on it. The function $P(x_1, \ldots, x_n)$ depends on $\langle x_1, \ldots, x_n \rangle$ and $\langle c_1, \ldots, c_k \rangle$, and $\langle c_1, \ldots, c_k \rangle$ depends on the time t.

The right part of Eq. (2.1) has two terms. The first term $\nabla \cdot (P \cdot \nabla V)$ is the drift component. It reflects the General trend in the movement of the "collective of particles": they "mainly" move from areas with high potential to areas with low potential of the force field (Fig. 4).

And the stronger the gradient-derivative ∇V, the faster the movement of all particles on average to the local minimum-equilibrium, if it exists.

The second term of the Eq. (2.1) is $\nabla^2 (D \cdot P)$—the diffusion component. The main role of this term is that it describes the scope of the distribution function—the size of the corridor within which objects are observed, concentrating near the local minima—the Central lines of the corridors.

The diffusion term of Eq. (2.1) is responsible for smoothing "inhomogeneities" even when there is no trend. Diffusion is an indicator of "relaxation" (see Figs. 5 and 6).

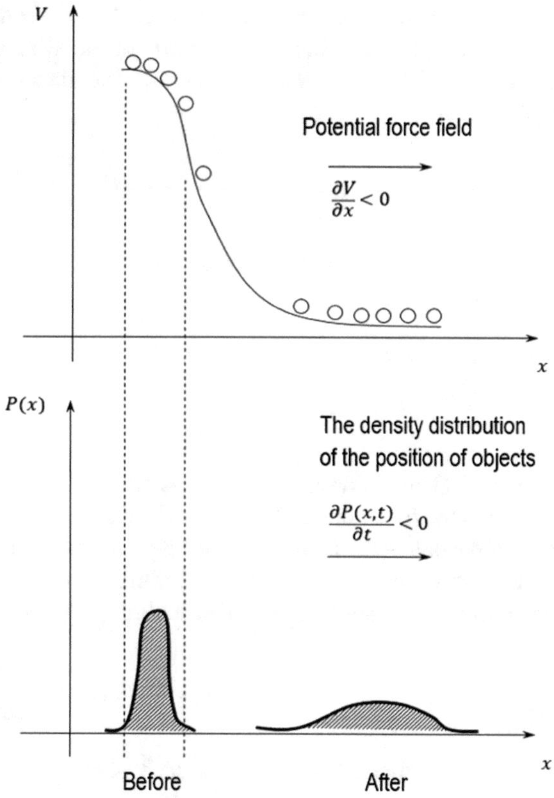

Fig. 4 Movement of particles in the field of potential forces

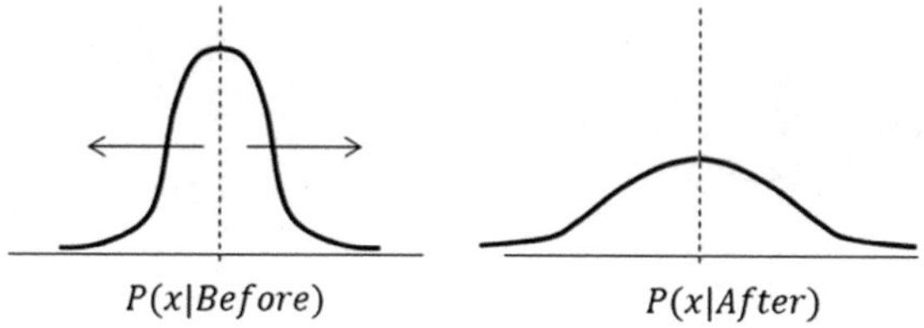

Fig. 5 Illustration of the smoothing of heterogeneities (case 1)

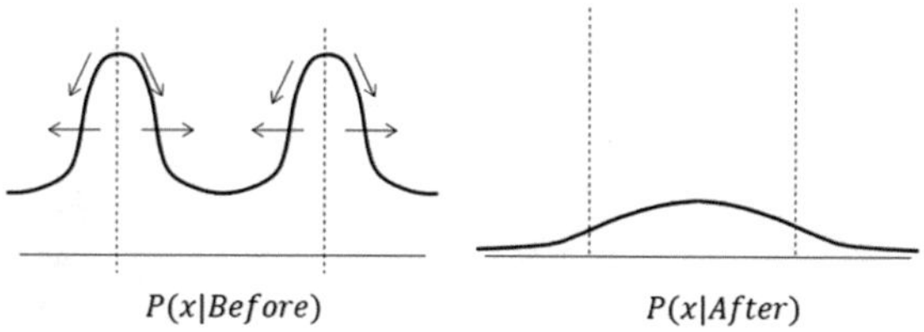

Fig. 6 Illustration of smoothing inhomogeneities (case 2)

Let us illustrate this with an example. Let the control parameters are fixed and $V(x_1, \ldots, x_n; c_1, \ldots, c_k)$ are not time dependent. Then Eq. (2.1) in asymptotics has a stationary form $\frac{\partial P}{\partial t} = 0$ at $t \to \infty$ and the corresponding asymptotic solution as a distribution:

$$P(x_1, \ldots, x_n; c_1, \ldots, c_k) = N_1 \times \exp\left\{-\frac{V(x_1, \ldots, x_n; c_1, \ldots, c_k)}{D}\right\} \tag{2.2}$$

if the diffusion index D is a constant in the phase space $\langle x_1, \ldots, x_n \rangle$, and

$$P(x_1, \ldots, x_n; c_1, \ldots, c_k) = N_2 \times D^{-1}(\vec{x}) \times \exp\left\{-\int^{x} D^{-1}(\vec{x}') \cdot \nabla V(\vec{x}'; \vec{c}')dx'\right\} \tag{2.3}$$

in a more General form, where N_1, N_2 are normalizing coefficients.

Consider, for example, an unimodal relaxation with $V(x; k) = \frac{1}{2}kx^2$—the equation of the potential energy of the spring. The solution of the Fokker-Planck equation is as follows: near the center of equilibrium, $x = 0$, the particles are distributed according to the law of normal distribution (Fig. 7):

$$P(x; k) = N_1 \times \exp\left\{-\frac{kx^2}{2D}\right\}. \tag{2.4}$$

Next, we give a "premium" to each point of size w. "Perturbed" distribution $P(x; k) = N_1 \times \exp\left\{-\frac{k(x-w)^2}{2D}\right\}$ with center w (Fig. 8) over time, t will shift to

Fig. 7 Illustration of the unperturbed state

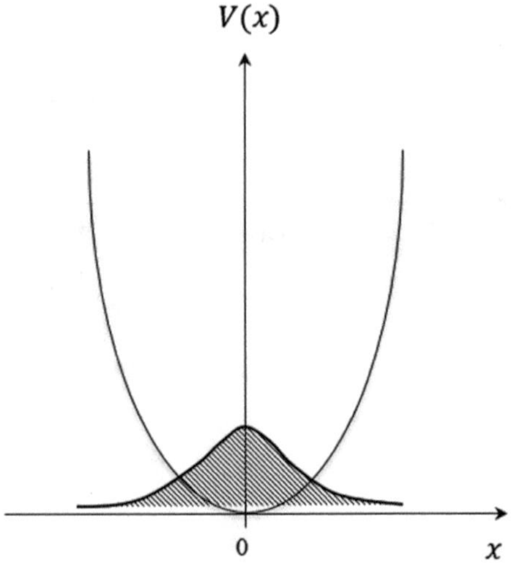

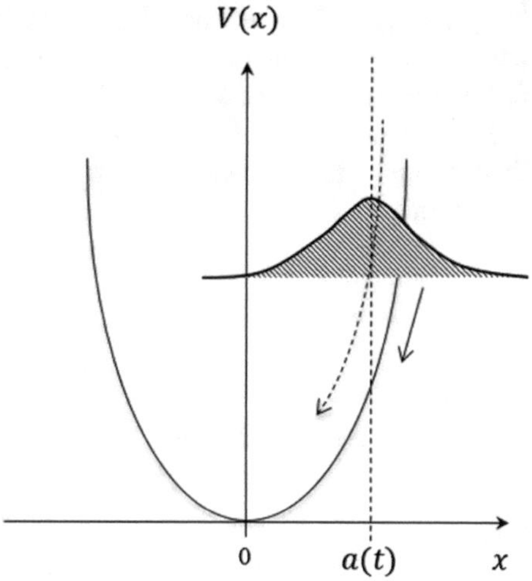

Fig. 8 Illustration of the disturbed condition

(2.4) according to the distribution laws $P(x;k) = N_1 \times \exp\left\{-\frac{k(x-w(t))^2}{2D}\right\}$, where $w(t) = w \cdot \exp\{-kt\} = w \cdot \exp\left\{-\frac{kt}{T_{relax}}\right\}$.

Relaxation time $T_{relax} = \frac{1}{k} = \left(\frac{d^2V}{dx^2}\right)^{-1}$. That is, the stiffer the spring, the faster the points will return to the unperturbed position. In the multidimensional case, when a small perturbation w, the magnitude of the relaxation time is estimated as $T_{relax} = \max\left\{\frac{1}{\lambda_i}\right\}$, $i = 1, \ldots, n$, where λ_i—the eigenvalues of the matrix stability V_{ij}.

By definition, $V_{ij} \underset{=}{\mathrm{def}} \left\| \frac{\partial^2 V}{\partial x_i \partial x_j} \right\|$ is the Hesse matrix (the Hessian of a dynamical system at the local minimum point of a potential function).

2.4 Estimation of Average Time of Passing «Way» for the Fulfilment of Obligations

The Fokker-Planck equations considered above have a close connection with the probability theory [41]. This connection is especially evident in the theory of Markov random processes. Consider, for example, the random path described by the Fokker-Planck equation, which contains a potential function understood in the analogy given earlier as energy (financing) giving forces to the operating HPF organizations to perform work on some "path" (to ensure the implementation of the required safety measures).

For simplicity, let's limit ourselves to the one-dimensional case. Let the one-dimensional function $V(x)$ be given and there is a constant diffusion coefficient D. For concreteness, let the path begin at some initial point x from the segment

$[x_{start}, x_{final}]$. Since any path starting at the point x must end after some time T at one of the ends of the segment $[x_{start}, x_{final}]$, try to estimate the average time T_{fp} that a point (object) needs to "stop".

If the path of a point begins at point x with probability $P_0(x)$, then the average time to "stop" is determined by the mathematical expectation, which is calculated by the formula:

$$T_{fp} = \int_a^b P_0(x) \cdot T(x)dx. \tag{2.5}$$

Put $P_0(x)$ equal to the Dirac Delta function:

$$P_0(x) = \delta(y - x), \tag{2.6}$$

that is, we assume that the object at the time $t = 0$ has a known position x. Substitute (2.4) in (2.5) and, after the transformations, we obtain:

$$-1 = -\frac{dV}{dx} \cdot \frac{dT}{dx} + D\frac{d^2T}{dx^2}. \tag{2.7}$$

Since the path starting at the boundaries $[x = x_{start}]$ (work not started) and $[x = x_{final}]$ (work completed) is zero, this allows you to write zero boundary conditions:

$$T(x_{start}) = T(x_{final}) = 0. \tag{2.8}$$

Before giving the solution (2.7) and (2.7) in General, we consider two special cases: $V = const$. Hence $\frac{dV}{dx} = 0$ and

$$T(x) = \frac{(x_{final} - x) \cdot (x - x_{start})}{2D} \tag{2.9}$$

We observe the operating HPF organization, giving it funds only to compensate the inflationary component and look when the safety indicators of the operated HPF "deteriorate" (the organization will move to $[x = x_{start}]$) or "improve" (the organization, respectively, will move to $[x = x_{final}]$).

The maximum value of $T(x)$ is observed in the middle of the segment $[x_{start}, x_{final}]$:

$$T(x) = \frac{(x_{final} - x_{start})^2}{8D}. \tag{2.10}$$

The result is interesting because the larger the size of the observation window, the object will be longer "in sight" in proportion to the square of the size of the observed area (on average). In addition, the more freedom ($D \to \infty$), the faster or "better" or "worse" the values of the safety indicator of the observed HPF.

In order to avoid abrupt changes in situations, and so they do not occur "often", there are laws that limit economic behavior. The tendency of D to infinity, for example, is limited by laws that prohibit so-called "financial pyramids."

But there is inflation, which "eats" money, that is, the trend is "work on obligations" which allows you to earn no less than inflation eats.

We believe that

$$V(x) = const - \left(\alpha_{econ} - \alpha_{inf}\right) \cdot x \tag{2.11}$$

$\alpha = \left(\alpha_{econ} - \alpha_{inf}\right) > 0$ sufficient funding (average);
$\alpha = \left(\alpha_{econ} - \alpha_{inf}\right) < 0$ "insufficient" funding (average);
$\alpha = \left(\alpha_{econ} - \alpha_{inf}\right) = 0$ balance.

Substituting $V(x)$ from (2.11) into Eq. (2.7), we obtain:

$$+\alpha \cdot \frac{dT}{dx} + D\frac{d^2T}{dx^2} = -1 \tag{2.12}$$

with boundary conditions

$$\begin{cases} T(x_{start}) = 0, \\ T\left(x_{final}\right) = 0. \end{cases} \tag{2.13}$$

The homogeneous differential equation corresponding to (2.12) has two eigenvalues $\left\{\lambda_1; \lambda_2 = -\frac{\alpha}{D}\right\}$, so the General form of the solution can be written as

$$T(x) = \left(x_{final} - x\right) \cdot c_1 + c_0 + c_2 \cdot \exp\left\{-\left[\frac{\alpha}{D}\right] \cdot \left(x_{final} - x\right)\right\}. \tag{2.14}$$

By supplying a General form $T(x)$ to (2.13) and (2.12), we obtain a system of three linear equations with three unknowns:

$$\begin{cases} c_0 + c_2 = 0; \\ \left(x_{final} - x_{start}\right) \cdot c_1 + c_0 + c_2 \cdot \exp\left\{-\left[\frac{\alpha}{D}\right] \cdot \left(x_{final} - x_{start}\right)\right\} = 0; \\ \alpha \cdot (-c_1) = -1. \end{cases} \tag{2.15}$$

Solving (2.15), we obtain

$$\left\{ \begin{array}{c} c_2 = \frac{\left(x_{final} - x_{start}\right)}{\alpha \times \left(1 - \exp\left\{-\left[\frac{\alpha}{D}\right] \cdot \left(x_{final} - x_{start}\right)\right\}\right)}; \\ c_0 = -c_2; \quad c_1 = \alpha^{-1}. \end{array} \right\}. \tag{2.16}$$

Substitute (2.16) in (2.14), we get a General view of the solution:

$$T(x) = \frac{(x_{final} - x)}{\alpha} - \frac{(x_{final} - x_{start})}{\alpha} \times \frac{1 - \exp\{-[\frac{\alpha}{D}] \cdot (x_{final} - x)\}}{1 - \exp\{-[\frac{\alpha}{D}] \cdot (x_{final} - x_{start})\}} \tag{2.17}$$

Analysis of formula (2.17) suggests that the presence of diffusion in General accelerates the achievement of the right boundary, and between the drift α(management of hard financing) and diffusion D (economic freedom of "part-time" market participants "on the side") there is competition (competitiveness).

Consider the asymptotic cases.

Case 1. ($\alpha \to 0$; $D - constant$). We introduce the parameter $z = \frac{\alpha}{D} \to 0$ and decompose the solution (2.17) by the small parameter z. Get rid of the members of the smallness z^2 and above, the following approximation:

$$T(x) \cong \frac{1}{2D} \times \frac{(x_{final} - x)(x - x_{start})}{1 - 2z \cdot (x_{final} - x_{start})}, \tag{2.18}$$

which at $z \to 0$ approaches the formula (2.9) linearly in z.

Thus, with a weak dependence on the source of additional funding, the General condition of the object, which we "on average" find, is determined by "random walks" (diffusion D).

Case 2. On the contrary, let ($D \to 0$; $\alpha - constant$). Then

$$T(x) \cong \frac{(x_{final} - x) - (x_{final} - x_{start}) \times \left[exp\left\{-\frac{\alpha \times (x - x_{start})}{D}\right\}\right]}{\alpha}. \tag{2.19}$$

The factor in square brackets decreases rapidly with the growth of x, that is, the effect of diffusion is significant only in a small neighborhood of the point x_{start}. The result depends only on the competence of the subsidiary and on the source of additional funding.

If $(x_{final} - x)$ is considered as a path, parameter α corresponds to the speed of fulfillment of obligations on works related to safety. Note that this option is very tough, limiting freedom to the maximum.

Note also that diffusion is often useful, because it allows you to "snatch on the side of "the point of growth of technologies" that have not yet taken root in competitors, but the implementation of such a policy is risky. Models describing the size of the fee for increased efficiency by increasing the risk of "failures" require separate studies.

A few additional comments on the Fokker-Planck equation. The external field, giving the finances in order to exist and fulfil their obligations, the company operating the HPF, creating partially or completely the Corporation for its orders and its policies in the field of security. Diffusion occurs when the work is financed, but the allocated funds are not optimal (balanced)—surpluses can be spent by operating organizations on "plugging holes" on other projects. And, on the contrary, that, the truth is rare, the "contractor" is looking for additional resources to meet the requirements of the Corporation and the state, as it is important to a stable customer (for example, in the "image" plan). Therefore, externally in the group portrait (until the "trend" reason is proved), it looks like diffusion. Diffusion is needed, including, when funded by the innovative direction of work—qualification, when a new sector of activities no one of the candidates authentically verified story that could convince you that failures in the execution of works will not.

It is necessary to alternate the phases of freedom of choice with the phases of "self-reliance" in order not to encourage business structures that parasitize on excessive outsourcing. In these cases, diffusion (risk) develops through the delegation of expert functions to "narrow specialists" who transform the work into "their highly specialized understanding".

2.5 Diffusion Model in Economic Behavior. Approaches to Developing a Methodology for Analyzing the Risk of Default of the Operating Organization

The simplest model of economic behavior States that "in the course of the business cycle, money through the production of goods and services grows in proportion to the capital invested at the beginning of the business cycle."

$$\frac{dx(t)}{dt} \cong \varepsilon(t) \times x(t), \tag{2.20}$$

where $\varepsilon(t)$ is the specific capital gain at time t.

The "average" value of ε as a whole of the random variable $\varepsilon(t)$ and the variance $D_\varepsilon = \sigma_\varepsilon^2$ are indicators of the efficiency and riskiness of the entire business cycle. The trend potential function is linear. The quasi-stationary solution-logarithmically normal distribution—is well studied—and the asymptotics of the median of this distribution indicate that sooner or later all 100% of participants will leave any business [42].

The model (2.20) is known as the Brownian motion model for large amounts of capital. In General, it is inadequate if the cycles are carried out by discrete impulses and portions that make up a significant share of fixed capital [43]. However, to describe the trends in the economic dynamics of the team with a relatively large number of participants (their small shares in the total capital), the model (2.20) is quite acceptable. Diffusion occurs due to accelerating (slowing down) circumstances in the implementation of business processes. Another reason for the inadequacy of the model under discussion is that it is a model of a closed system, it does not take into account neither the inflow nor the outflow of capital. In this model there is no diversification by "industries" and stages of work (not all of them are equally profitable, but all are equally necessary).

The second component of the business is that past achievements and results are "forgotten" after a series of failures, or "get used to them"—previously achieved "innovative breakthroughs" become the property of if not all, then many, and this is a decrease in integral profitability. That is, in an adequate model there should be a negative component that reacts to the rate of change in the volume of capital over time. In principle, this component should be nonlinear, exceeding the quadratic dependence.

The fact is that the classical model of the growth regulator is the Ferhulst-Wiener model [36, 37]

$$\frac{dx(t)}{dt} \cong \alpha(t) \times x(t) - \gamma(t) \times x^2(t) \tag{2.21}$$

it has a solution in the form of a logistic curve reflecting only the phenomenon of "growth limit".

When degree when variable x is greater than 2 there is a possibility of degeneracy of the Hessian of the system, and as a result, the presence of points of bifurcations, in which the "selection" potentially promising trajectories and the rejection of the trajectories starting ruinous business cycles.

It is expected that excessive growth will lead to a decrease in the "quality of work" and the inhibitory components will take effect. Models of formation of waves of economic activity in developing systems due to the natural alternation of acceleration and deceleration are presented, in particular, in the review [39], the description of models of evolution can be found, for example, in the monograph [43].

For monitoring purposes, but not for tasks of strategic planning and work areas, sufficient diffusion model in the view "roller coaster" (Figs. 9, 10, 11 and 12).

Figure 9 illustrates the ideal option-a uniform impact of Finance close without "deviations" to the target at a constant speed α ("as scheduled"):

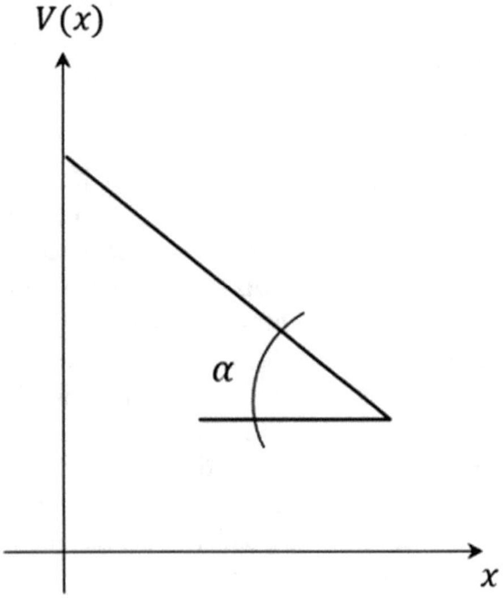

Fig. 9 Illustration of the ideal case of performance

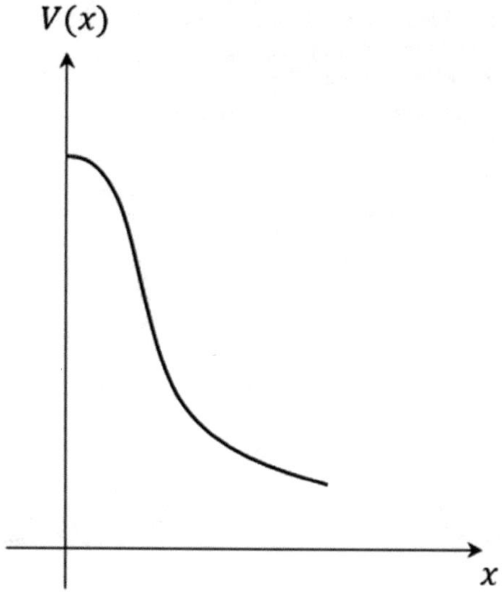

Fig. 10 Illustration of too high a rate of commitment

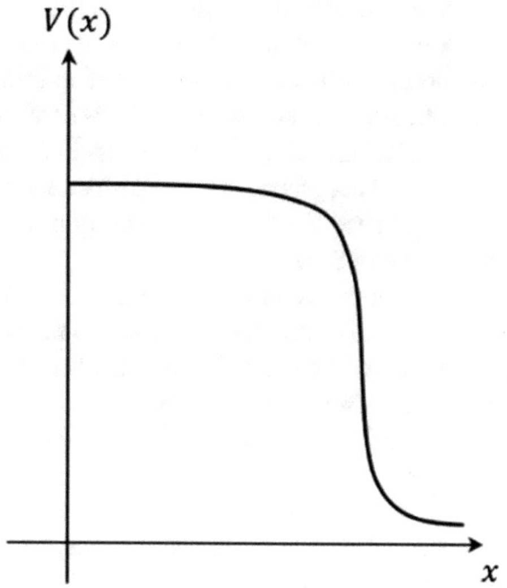

Fig. 11 Illustration of "seasonality" in the performance of obligations

$$\alpha = \frac{\left(x_{final} - x(t)\right)}{\left(T_{final} - t\right)}. \quad (2.22)$$

Figure 10 illustrates a variant, when a subsidiary of the society soon begins to fulfill its obligations. But, since not all of them are equally time-consuming, evenly painted resources are spent on work that is "easier to perform on their own." Then, when the "fit" results of "outsourcing", it turns out that something was done "in vain", and the results of the aggregate does not fit. Subsequent approvals and corrections are made

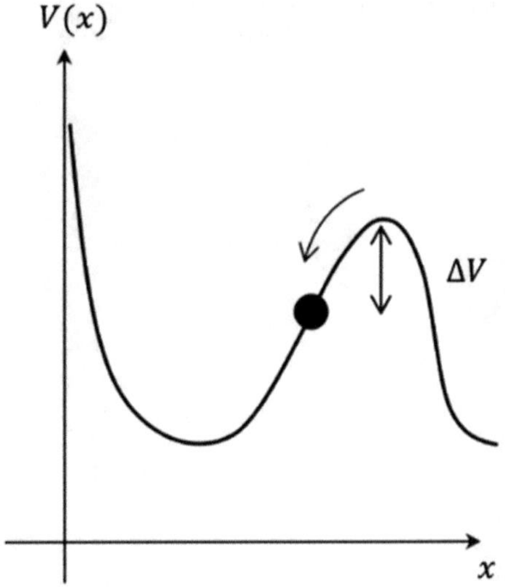

Fig. 12 Illustration of the occurrence of force majeure in the performance of obligations

"at their own expense". Sometimes the work is "seasonal" in nature, then the funds are really spent at the initial stage of preparation for the "field season" (Fig. 11). In the summer, with the involvement of additional employees, the main stage of work on the collection of research material is carried out, and the end of the business cycle—data analysis, packaging the results into the overall system picture—again "paper" relatively quiet phase. Figure 12 illustrates the situation known as "force majeure"—the worst case: no funds for alteration, and the "potential barrier"—another "hill" can not be overcome.

As a result, we have at each time a General distribution of the "visible and controlled" state of Affairs in relation to the volume of work performed (including, for example, one employee) (Fig. 13). Some firms and work will be "ahead", but this may be the result of the organization of work on the option reflected in Fig. 11, others are "behind", but it may just be "not their season".

Having worked out the "limit" cases, it can be argued that there are some boundaries of the corridor in which there should be a situation with the state of industrial safety of HPF operating organizations, depending on the resources allocated to them and those characteristics that can be learned from their background.

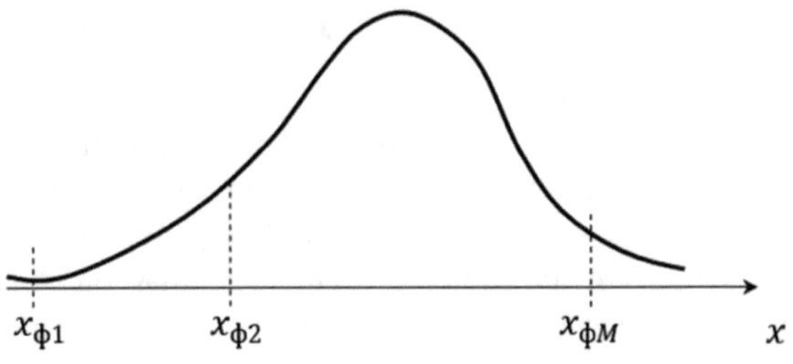

Fig. 13 View for general distribution

The boundaries of the corridors are not symmetrical with respect to their "centers of gravity", since equal losses (the consequences of emergency and crisis situations) and acquisitions are estimated differently. But for "homogeneous" organizations dealing with similar types of HPF in approximately the same industry segments, the differences should be minor.

Jumps in differences can occur locally for two reasons:

(a) implementation of robotic systems (and the model considers the output per employee);
(b) "local" conditions: bad, for example, rainy season (the equipment gets stuck in dirt) and good-assistance of local authorities, existence of branches in the region of work, etc.

So, according to the "roller coaster" model, we have to allocate one corridor for resources (energy invested in the operating organization), and the second corridor according to the estimates of their "provisions" in relation to the "planned" indicators. It is clear that for different values of drift α and dispersion D for different types of distribution works (Fig. 13) will vary, and the Union (mixture of distributions) will represent some figure with a complex profile (Fig. 14).

At the same time, it is not necessary that the points of "control measurements" are in the center of this figure, since they are taken from the data of different "slides".

So, the "suspicious point" (operating organization F1) can "fall behind" when it allocated little money. And it can be reloaded with works.

The other "points" (the operating organization F7) can actually stay ahead of the competition. Here it is necessary to conduct an additional analysis: whether the operating organization itself performed the work, or it is only a "mediator".

Variants of interpretation of the lot—the essence is the same: to monitor the desired model of the dynamic corridor in which investments are treated as investments in the acceleration of the implementation of commitments.

In the next section, we will give an example of building a dynamic corridor model, built in the analysis of data from a limited sample of counterparties that have passed (not passed) prequalification on a certain trading platform for a certain type

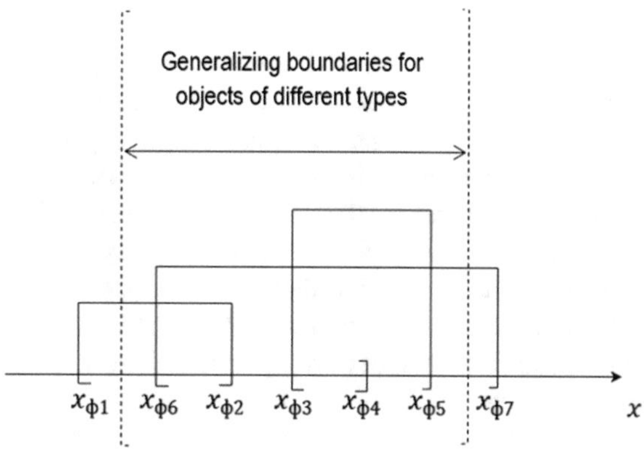

Fig. 14 Mixture of distributions

of activity. Note that the correct choice of contractors to carry out works related to the safe operation of critical infrastructure facilities, is decisive in the task of managing the probability of emergency, crisis and emergency situations, i.e. reducing the associated risks.

2.6 Example of Building a Dynamic Corridor Model

Using the trajectory approach described earlier, we present the output intensity per person by the time function $y(t)$. Its analogue in physics is instantaneous speed.

Then for any period of time $[t_1, t_2]$ the volume of output per person will be the sum (integral)—$\int_{t_1}^{t_2} y(t)dt$.

At the initial stage, in the absence of a full—format monitoring system, three dimensions can be taken as the basis—the values of integrals:

- I_0 output (expected, declared, supported by contracts, but not actual) per person for the current year;
- I_{-1} volume of output per person for the previous year (actual, documented by acts of acceptance of works); and
- I_{-2} output per person two years ago (from current).

Formally, we can write these three integrals as:

$$\int_{0}^{1} y(t)dt = I_0; \int_{-1}^{0} y(t)dt = I_{-1}; \int_{-2}^{-1} y(t)dt = I_{-2}. \tag{2.23}$$

Next, for each operating organization, a trajectory of change in its "performance" $y(t)$ is constructed. In the simplest case, the law of variation of the values of $y(t)$ has the form of uniformly accelerating (uniformly decelerating) motion, i.e.

$$y(t) = y_0 + v \times t + a \times \frac{t^2}{2}. \tag{2.24}$$

Here: y_0—estimation of the intensity of production on the "first day" of the current year; v—average rate of increase $y(t)$; a—average acceleration (rate of increase V).

If $v > 0$, this means that (according to I_0, I_{-1}, I_{-2}) the specific output in the operating organization increases; if, on the contrary, $v < 0$, it falls; if $v = 0$, the output is maintained at the same level on average for three years.

Similarly, is interpreted an indicator of acceleration a: the value of $a > 0$ means that the operating organization is on the rise - the rate at the end of the analyzed three-year period, growing faster than the average; the value of $a < 0$ may indicate that although the rate of development and growing, but the magnitude of the growth rate is decreasing.

Formally, we can obtain a correspondence between the triple of values y_0, v, a and the triple of values I_0, I_{-1}, I_{-2}. Substituting (2.24) into (2.23) we obtain:

$$\begin{aligned} I_0 &= y_0 + \tfrac{v}{2} + \tfrac{a}{6}; \\ I_0 &= y_0 - \tfrac{v}{2} + \tfrac{a}{6}; \\ I_0 &= y_0 - \tfrac{3v}{2} + \tfrac{7a}{6}. \end{aligned} \tag{2.25}$$

Solving three Eq. (2.25) with three unknowns, we determine the values y_0, v, a:

$$\begin{cases} y_0 = \frac{2I_0 + 5I_{-1} - I_{-2}}{6}; \\ v = I_0 - I_{-1}; \\ a = I_0 - 2I_{-1} + I_{-2}. \end{cases} \tag{2.26}$$

Values of y_0, v, a vary from object to object and make sense of "averages" for different objects of estimates of indicators of their trajectory for 3 years. Since these indicators smooth out "seasonal" and other diffusion factors, it is desirable to normalize them. When displaying General characteristics, group indicators are primarily of interest, and then the relative position of the values of each individual object in relation to the group indicator. We introduce two similar indicators.

The first is the relative acceleration factor (q_a), calculated as the ratio of the sum of workings in the current (I_0) and pre-issuing years (I_{-2}) to the development in the previous year (I_{-1}). The output for the previous year (I_{-1}) was chosen because it is the closest of the estimates, confirmed by the facts of the work.

$$q_a = \frac{I_0}{I_{-1}} + \frac{I_{-2}}{I_{-1}} = \frac{a}{I_{-1}} + 2. \tag{2.27}$$

From (2.27) it can be seen that q_a is the "failure" meter in the development in the previous $(I_{-1})(q_a \gg 2)$ or current $(I_0)(q_a \ll 2)$ years.

That is indirectly a factor q_a - factor in the evaluation of instability of the trend (in case of large deviations from two) and, on the contrary, the resilience of the trend for values q_a close to 2. A priori, we can assume that the instability of q_a and the downtrend are "suspicious", since they are associated with the reorganization of firms and it is necessary to "find out": whether the firm in its new state is the successor of the qualification and other capabilities of its predecessor (the name may not change, but the functional readiness—due to the influx-outflow of personnel—Yes). In addition, a steady "fall" ($q_a \cong 2$) in a negative trend is not exactly what you need for development, it is important to know the value of the second factor.

The second factor (q_v)—shows the relative growth rate of performance of the operating organization over the past two years:

$$q_v = \frac{I_0}{I_{-1}} = \frac{v}{I_{-1}} + 1. \tag{2.28}$$

This indicator is naturally correlated with the first indicator, being a part of it. The adequacy of this indicator is somewhat weaker, since the value of (I_0) at the time of allocation of funding (especially if it occurs at the beginning of the year) has "normal" error ranges for some operating organizations (with steadily passing works), while for others the estimate (I_0) is only the declared amount of obligations that will require additional conditions (outsourcing, etc.) to be met.

To visualize the changes in the possible value of factors depending on the financing process, it is useful to move from the indicators q_a and q_v through affine transformations to a pair of related normalized indicators—F_a and F_{va}, which give us a geometric image, which we will call "target".

The first factor F_a is simply the normalized factor q_a

$$F_a = \frac{q_a - c_a}{L_a}, \quad (-1 \le F_a \le 1). \tag{2.29}$$

Here: c_a is the center of the target along the abscissa axis, and L_a is the size of half the side of the cross-section square.

Normalized factor F_{va}—linear combination of acceleration factors q_a and speed q_v:

$$F_{va} = \frac{q_a - c_a - \mu \times (q_v - c_v)}{L_{va}}, \quad (-1 \le F_{va} \le 1). \tag{2.30}$$

The effect of increasing the velocity with increasing acceleration is taken into account by subtracting this "additive" from the coefficient $\mu > 0$, which is determined as the constants c_a, c_v, L_a, L_{va}, based on the analysis of group data. Geometrically, the position of a square target ($F_a \times F_{va}$) of size $[-1, +1] \times [-1, +1]$ on the plane of factors q_a (abscissa) and q_v (ordinate) will be displayed as a parallelogram ABCD (Fig. 15).

Point A corresponds to point $\{F_a = -1;\ F_{va} = +1\}$. Point B corresponds to point $\{F_a = -1;\ F_{va} = -1\}$. Point C corresponds to point $\{F_a = +1;\ F_{va} = -1\}$. Point D corresponds to point $\{F_a = +1;\ F_{va} = +1\}$. Due to the fact that conversion is affine, it is obvious that the center of the target with coordinates (c_a; c_v)—the point of intersection of the diagonals corresponds to the center of a square target coordinates $\{F_a = 0;\ F_{va} = 0\}$. For Fig. 15 it is shown that the coefficient $\mu > 0$ is equal to the tangent of the angle of inclination of the side (DA) to the ordinate axis.

According to the group analysis, the size and slope of the target were determined using a modified support vector machine. In principle, the target does not have to be a parallelogram, it can be an arbitrary polygon according to the support vector machine.

Note that the relations are fair

$$\begin{cases} F_a = \dfrac{\frac{I_0}{I_{-1}} + \frac{I_{-2}}{I_{-1}} - c_a}{L_a}; \\ F_{va} = \dfrac{F_a \times L_a - \mu \times \left(\frac{I_0}{I_{-1}} - c_a\right)}{L_{va}}. \end{cases} \tag{2.31}$$

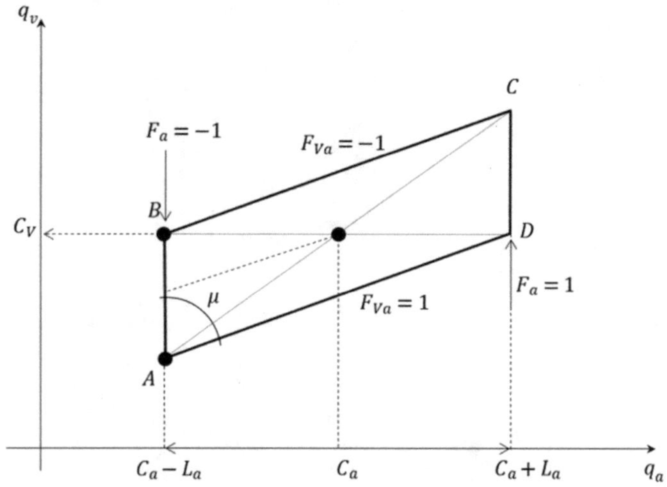

Fig. 15 The appearance and settings of the target

At $\mu > 1$, the minimum distance of the trajectory $\langle F_a(q_a); F_{va}(F(q_a))\rangle$ is reached at the point that in the normalized coordinates touches the angle of the concentric square with the center (0; 0) regardless of whether the trajectory crosses the target "above" the center (point $P1$) or "below" the center (point $P2$), or does not cross the target at all (points $P3$ and $P4$, respectively)—Fig. 16.

At these points, the validity of the equation $F_a^* = F_{va}^*$, whence it follows that

$$\begin{aligned}
F_a^* &= F_{va}^* = \frac{B_1}{A} + \frac{B_2}{A} \times \frac{I_{-2}}{I_{-1}};\\
A &= 1 - \frac{(1-\mu)L_a}{L_{va}} > 0;\\
B_2 &= \frac{\mu}{L_{va}} > 0;\\
B_1 &= B_2 \times (c_v - c_a) < 0.
\end{aligned} \tag{2.32}$$

If $|F_a^*| \leq 1$—the trajectory crosses the target, when $|F_a^*| > 1$—passes it at a distance $|F_a^*| - 1$ Since F_a^* depends on I_0, it is natural to determine the optimum I_0^*.

Its value is calculated from (2.32) as a linear combination of previous results:

$$I_0^* = K_{-1}I_{-1} + K_{-2}I_{-2}, \tag{2.33}$$

where

$$K_{-1} = \frac{B_1 L_a}{A},\ K_{-2} = \frac{B_2 L_a}{A} - 1. \tag{2.34}$$

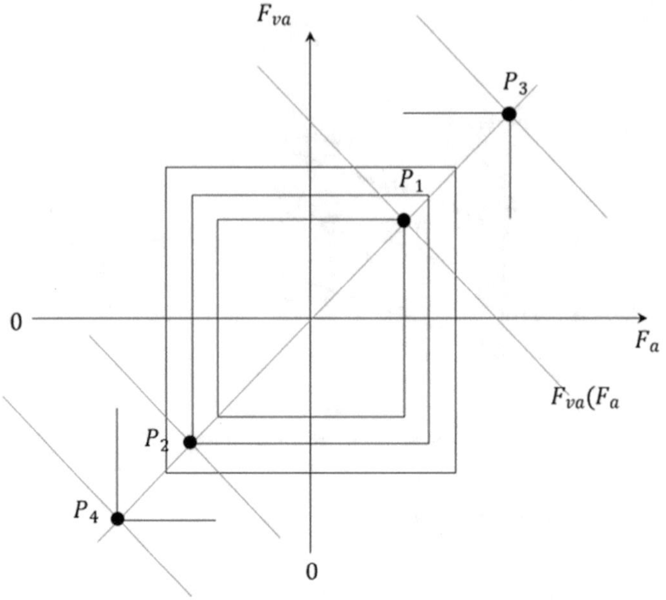

Fig. 16 Illustration of determining the minimum distance of the trajectory

The above reasoning (and the study of the calculated values obtained for the case of application of the described model for the needs of prequalification of counterparties of PJSC Gazprom) mean that there is a simple rule—in the current year "on average" optimally load the average employee by 86% from the previous year and add to this load another 31% of the load from the pre-issuing year.

For the resulting difference Eq. (2.33) the characteristic equation is:

$$\lambda^2 - K_{-1}\lambda - K_{-2} = 0. \tag{2.35}$$

Where the designated

$$\lambda_{1,2} = \frac{K_{-1}}{2} \pm \sqrt{\frac{K_{-1}^2}{2} + K_{-2}}. \tag{2.36}$$

For the mentioned example of prequalification of counterparties, the first eigenvalue $\lambda_1 = 1.1315 > 1$ corresponds to the growth of the system in the Brownian regime (the achieved successes initiate greater confidence in trading), but $\lambda_2 = -0.2723 < 0$—shows the rate of "forgetting merits", if they are not confirmed by the result. In this case, due to the negativity. λ_2—component of the solution describes the damped oscillations. A General view of the phenomenological solution is the sum of two geometric progressions (as opposed to Brownian motion).

$$I_t = c_{-1}(\lambda_1)^t + c_{-2}(\lambda_2)^t, \tag{2.37}$$

where the constants c_{-1} and c_{-2} are defined from the initial conditions

$$\begin{cases} I_{-1} = c_{-1}(\lambda_1)^{-1} + c_{-2}(\lambda_2)^{-1}; \\ I_{-2} = c_{-1}(\lambda_1)^{-2} + c_{-2}(\lambda_2)^{-2}. \end{cases} \tag{2.38}$$

Further from formulas (2.32) it is visible that F_a^* changes the sign when the trajectory passes through the center of a target. When $F_a^* = 0$, the optimal ratio of the volume of work performed in the previous year is Θ^*.

$$\Theta^* = \frac{I_{-2}}{I_{-1}} = \frac{|B_1|}{|B_2|} = |c_{va} - c_a|. \tag{2.39}$$

In the case of prequalification of contractors $\Theta^* = 0.9079$, that is, it is optimal that the increase in output (on average) should be—10%: $[\Theta^*]^{-1} = 1.1014$. Note that $[\Theta^*]^{-1} < \lambda_1$, which indirectly confirms the "correctness" of the chosen approach—it would be optimal 10%, but 13% is better, as part of the "merits" will inevitably be forgotten.

So, knowing the "reference" trajectory passing through the center of the target, it is possible to estimate in percentage terms the level of "marginal suboptimality" of the trajectories of operating organizations ψ in the form of overload or underloading to the standard, if they were given the opportunity to implement the "optimal loading scenario"

$$\psi = \frac{I_0 - I_0^*}{I_0^*}. \tag{2.40}$$

Closer than this relative value (in percent) the firm will not approach the optimum. Calculating the deviation of the trajectory of the company from the optimum in the current year δ_0 and evaluating optimal "expected" deviation δ_0^+ in the coming year, according to the formulas (2.41) (on the assumption that you will not need to "shift" the target).

$$\begin{aligned} \delta_0 &= \tfrac{I_{-2}}{I_{-1}} - \Theta^*; \\ \delta_0^+ &= \tfrac{I_{-1}}{I_0} - \Theta^* \end{aligned} \tag{2.41}$$

it is possible to develop an indicator of the "progress" of the operating organization ξ in the form of the difference of the above estimates normalized through logistic curves. The type of logistic curves is selected from the considerations that to remain a "leader" is also a progress, and a slight improvement in the "outsider"—it is better not to consider a significant progress.

It is quite obvious that the described target model considers the trajectories of the objects without taking into account the "diffusion" components. Therefore, it is an interesting and basic element for building up aggregates and indicators for a future branched monitoring system.

It is quite expected that the assessments of the current state, and especially the assessments of "progress" on different indicators, will partially contradict each other. In this case, you will need to "close" the cycle of work—to treat indicators and indicators as elements of the logic of decision-making, to carry out analog-to-digital conversion of indicators into logical variables, to build a "new" logic of removing contradictions. The development of such a methodology will require additional research and is not the purpose of this work.

2.7 *Target Construction Based on the Support Vector Machine*

In two-dimensional problems, the classical support vector machine and its modifications have a visual interpretation.

Let the plane show two sets of points specifying positive (+) (green zone) and negative (−) (red zone) examples. For each of them there is the nearest neighbor (nearest neighbors) of the opposite sign. In the middle of the segments connecting the points with the nearest neighbors, the reference points (support vectors) are placed, having the property that they are equidistant from the green and red zones—(Ø). A priori support vectors are in the "orange zone": neither good nor bad. Through the support vectors are perpendicular to the corresponding segments.

Polylines consisting of points equidistant from all "positive" and "negative" examples are formed. The fractured lines (•) come from the fact that the composition of a pair of "positive neighbor of negative neighbor" at these points is changed (Fig. 17). Belonging of the point under study to the "inner" or "outer" parts of the plane,

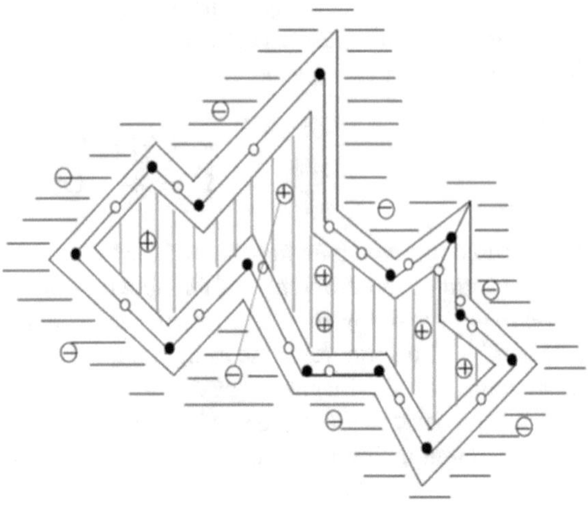

Fig. 17 Illustration of the provisions of the support vectors and the kinks, defining the geometry of the solutions

delimited by broken lines, indicates whether it is a "positive" or "negative" example. Otherwise, we have to point out that it is in the orange zone—the area of the support vectors. The most "unpleasant" configurations for analysis are constructions of the "ring" type, when the monotony of the topology of the regions is not observed (Fig. 18).

In cases where there are logical restrictions on the number of segments that define a polyline boundary (often the correct one is not the most "accurate", but "acceptably accurate", but a simpler solution), the search for "generalizing" support vectors is carried out.

For example, often searches for the smallest number of segments of the polyline, that is, minimizing the number of "intersecting" lines (in the multidimensional cases, hyperplanes) which are part of these segments of the polyline. The idea of such an improvement is simple (Fig. 19)—it is required to find lines L such that the minimum

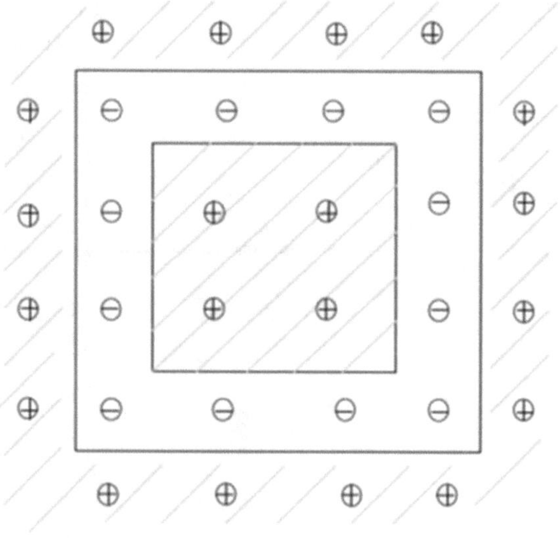

Fig. 18 Case of nesting of decision-making areas

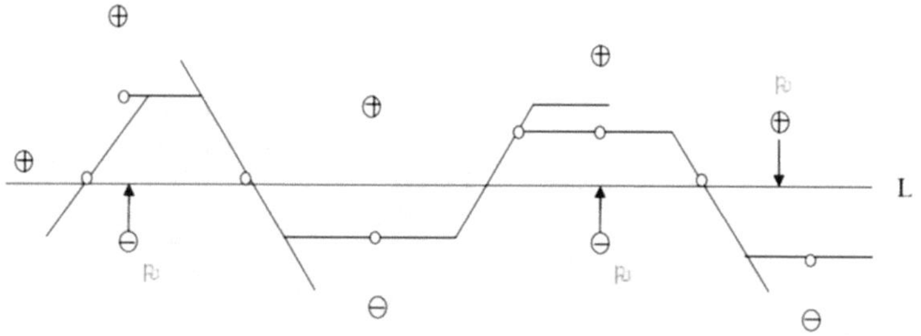

Fig. 19 Illustration of the reference vector, specifies the position of the section hyperplane

heights lowered on these lines from positive examples, and the minimum heights raised on these lines from negative examples were the maximum possible; equal and opposite in sign.

Configurations of L lines are found by rather complex numerical optimization methods [44, 45]. In multidimensional cases, "jump-like" solution procedures are obtained. In General, as in neuromathematics, the solution is ambiguous and depends on the basic strategy of its construction. For Fig. 20 shows the classical symmetric solution-the boundary is constructed as a Union of contours of separating lines.

For Fig. 21 the strategy of step-by-step separation of "positive" examples from all neighboring "negative" examples is implemented, and in Fig. 22, on the contrary, the "green area" is constructed by cutting off the "excess".

The author proposes an approximate method including the following steps:

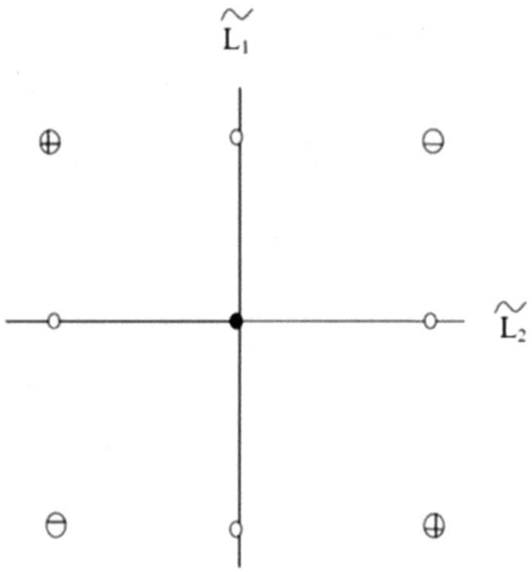

Fig. 20 Symmetric case. Classical support vector machine

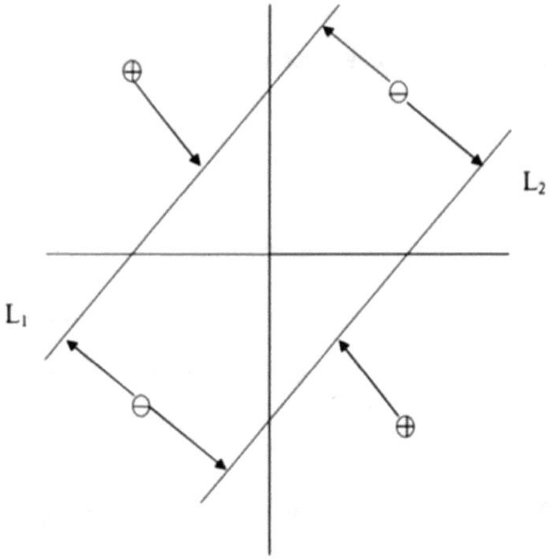

Fig. 21 Selection "positive" examples

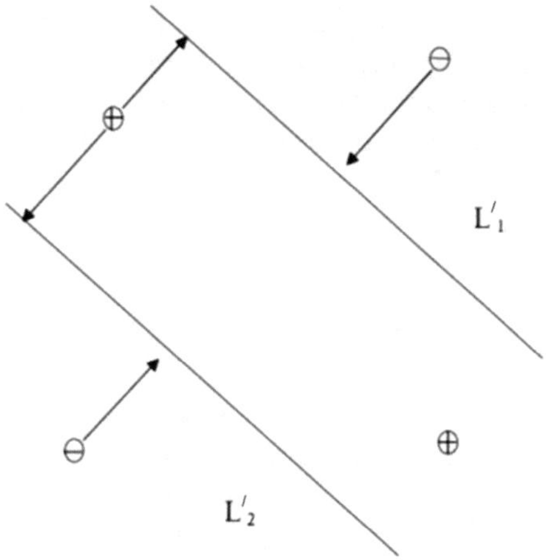

Fig. 22 Clipping "negative" examples

- delineation of positive examples visually hit the target and drawing through the points of the external lines of these examples parallelogram-polygon of the desired type;
- according to the rules of geometric similarity—homothety—definition of homothety center (point of intersection of the diagonals—the center of the target);
- "inflating the contour line" evenly over all the rays emanating from the center of the target (which is the essence of homotetics—"projective similarity"—the shadow is similar to the object it casts) until the condition of equality of distances (Fig. 23).

The expansion coefficient defining the boundary (K_H^*) is calculated by the expansion coefficient at the "touch" of the inflated parallelogram of the nearest "negative" example K_H^- by the formula: $K_H^* = 0.5 \times K_H^* + 0.5$.

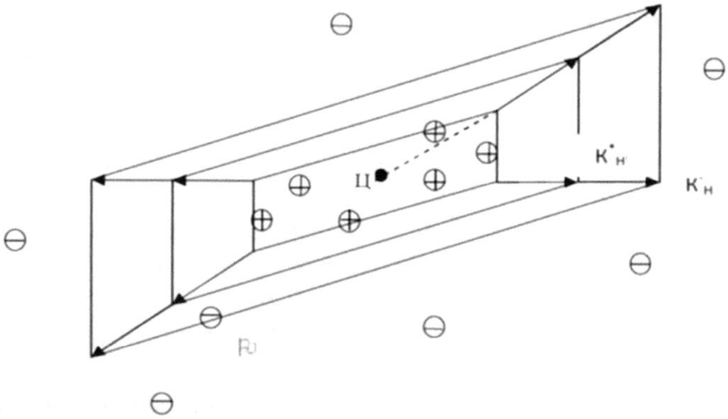

Fig. 23 Illustration of the inflating of the contour line of the target

If it is more important not to make mistakes of the second kind—not to evaluate positively "negative questionnaires", the 2nd stage can be omitted ($K_H^* \equiv 1$). There is an additional margin of "rigidity" of the selection rules in the number of "positive" examples. If, on the contrary, it is important not to make mistakes of the first kind—"inflate" the contour line should be "all the way" ($K_H^* \equiv K_H^-$).

2.8 Case Study

Let us illustrate the above reasoning by the example of constructing a zone of acceptable values of performance indicators of contractors involved in procurement procedures. As a test sample, the results of competitions for the design of LF in the amount of 40 points were considered (see Table 6).

Based on data from Table 6 by formulas:

$$X = \frac{\text{Output for the current year} + \text{Output for the year before}}{\text{Output for the last year}}$$

$$Y = \frac{\text{Output for the current year}}{\text{Output for the year before}}$$

the X and Y values for each firm were obtained.

The points on the x, Y plane were distributed as follows the small sample size Attracts attention, which complicates the statistical approach to classification. It is also worth noting that the very possibility of a successful classification on a complex basis is not previously a criterion in the selection requires additional justification.

The configuration of the positive data concentration region shows that the linear separation of positive and negative solutions is impossible, since the negative solutions are inside the positive region, even if one strongly spaced point of the positive solution is neglected. As one of the solutions it was supposed to make a separation using the support vector machine, where the straightening space is given by means of a Gaussian kernel. The result of this automatic object classification is shown in Fig. 24.

Figure 24 shows that the area of positive solutions has a ring topology and is poorly suited for SVM clustering. Thus, the problem of finding the boundaries of the area of positive solutions was reduced to the problem of minimizing errors of the first kind (negative solutions are mistakenly recognized as positive) with a fixed value of errors of the second kind. The type of the area of positive solutions (parallelograms) was given. Taking into account this a priori information, the problem was divided into parts: first there were left and right borders of the parallelogram, then its upper and lower borders.

To find the right boundary, based on the conditions of the problem, the function was formed:

Table 6 Source data

Participant Id	Output for the current year	Output for the last year	Output for the year before	Result
2	2366.99	2671.19	2746.8	1
3	68	68	42	0
4	121	131	128	0
9	1027.88	866.1	1185.67	0
11	548.5740183	1001.93	1058.6	1
20	196	189	93	0
21	400	2034	1225	1
22	73.6	70.1	50.6	0
27	1575.15	1165.2	2111.9	0
29	4.209	2.659	3.808	0
31	2626.78	2686.69	2551.67	1
35	3431	3312	3161	1
36	2331	2256	2334	1
39	3094	3103	3212	1
62	707.8	675.6	842.88	0
90	2243	3100	2664	1
101	1363.787746	1795.5	2444.85	0
103	1319	2073	2346	0
108	408.45	1037.8	967.77	1
112	5268.6	30729	6303	1
119	2794	4058	3851	0
121	2756.23	2694.85	1542.33	0
122	2765	2739	5823	0
125	1161.42	4838	3629	0
133	2158	1640	1589	0
137	1176.53	1634.16	1563.2	0
143	141.25	145.63	391.96	0
147	117.2	89.8	123.2	0
148	1303	991	1102	0
149	1836	2640	2508	0
151	1355.301397	1153.66092	874.9508	0
161	1350.92	1474.3	1284.5	0
163	998	1408	670	0
164	278.08	1095.63	1040.25	0
178	648.81	2747.5	52	0

(continued)

Table 6 (continued)

Participant Id	Output for the current year	Output for the last year	Output for the year before	Result
182	1798.49	1012.477	855.472	0
184	86248.5	219168.37	221001.72	0
186	1290.175	1290.175	975	0

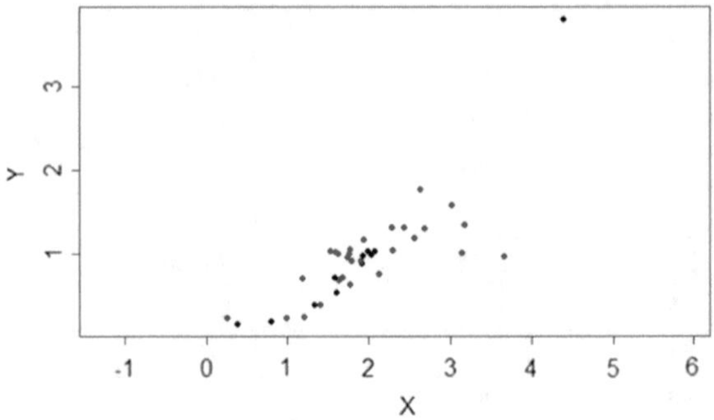

Fig. 24 Results of the selection competitions for the design of objects in the Champions League. Red shows negative results, black-positive

$$f(x) = \begin{cases} \frac{\sum_i I_{\{\text{neg}, \ x<x_i\}}}{N} & \text{if } x > \max(\forall x^{\text{pos}}) \\ 0 & \text{if } x < \max(\forall x^{\text{pos}}) \end{cases} \tag{2.42}$$

where $I_{\{\text{neg}, \ x<x_i\}}$ is an indicative function that takes a value of 1 or 0 depending on whether the conditions in braces are true or not, N is the total number of negative examples. A similar (up to sign) function was formed to find the left border. The maximum of the function (2.42) was the solution for the right (left) boundary of the region (Figs. 25, 26 and 27).

The function of the form (2.42) has a fairly simple form and depends on only one variable, which makes it convenient for further work. Using transformations of the coordinate system, the problem of finding the upper and lower boundaries of the

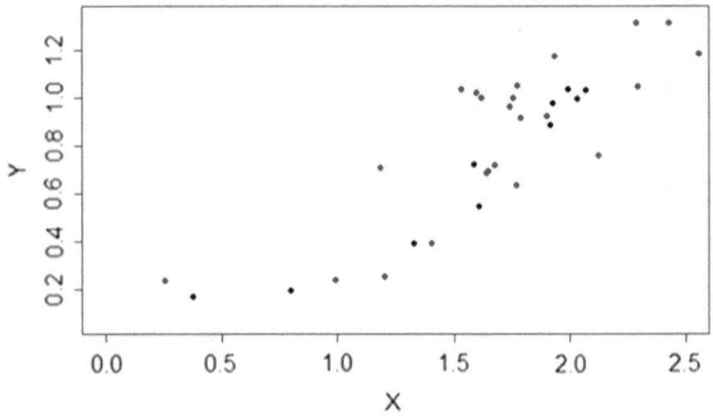

Fig. 25 Area of maximum concentration of positive solutions

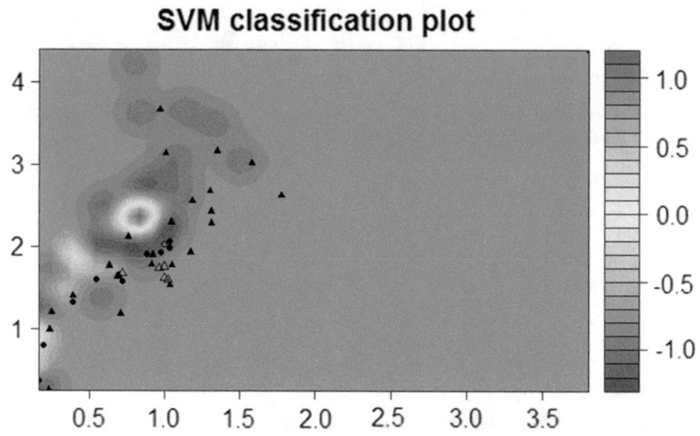

Fig. 26 Classification method SVM, where rectifiable space specified with a Gaussian kernel

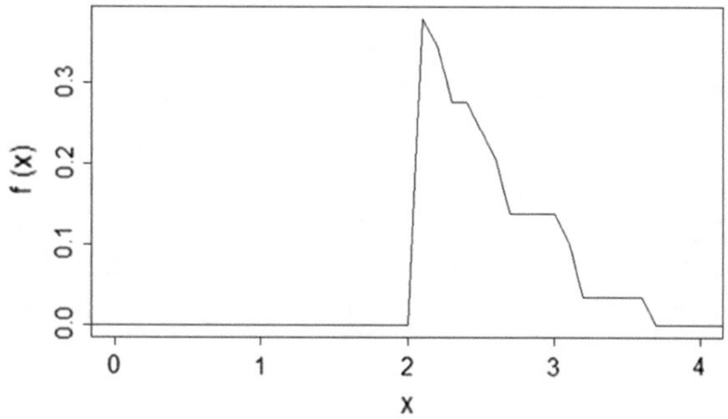

Fig. 27 Graph of the function (2.42) for the right border areas of positive values

area of positive solutions can be reduced to the problem of finding the left and right boundaries in the space transformed as a turn around the center of the coordinate system:

$$\begin{bmatrix} x' \\ y' \end{bmatrix} = \begin{bmatrix} \cos i & \mp \sin i \\ \pm \sin i & \cos i \end{bmatrix} \begin{bmatrix} x \\ y \end{bmatrix} \tag{2.43}$$

where i is the angle of rotation.

Thus, using the rotation of the coordinate system in the range from 0° to 90° (see Fig. 28), the optimal angle of inclination can be found, at which the number of rejected negative solutions of the upper and lower boundaries of the positive solutions reaches a maximum.

To do this, we introduce a function of the form:

$$f(x_n, i) = \begin{matrix} \frac{\Sigma_i I_{\{\text{neg},\ x_n' <> x'_i\}}}{N} & \text{if } x_n' <> \sup(\forall x'^{\text{pos}}) \\ 0 & \text{if } x_n' <> \sup(\forall x'^{\text{pos}}) \end{matrix} \tag{2.44}$$

Fig. 28 The area of positive values when the coordinate system is rotated. The blue lines show the optimal left borders, the green lines show the right borders

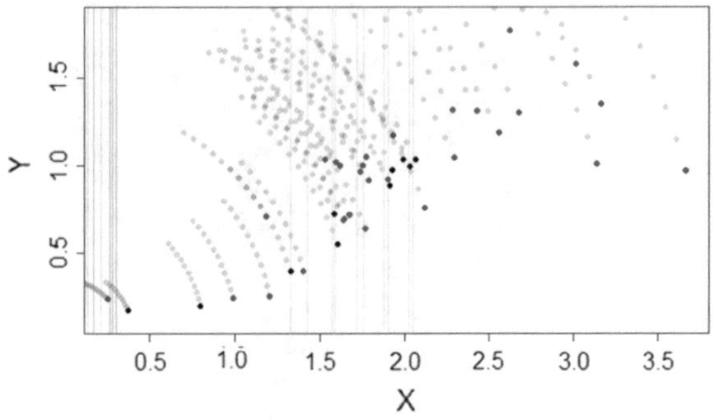

Here x'—the value of x in the converted with the help of (2.43) space, x_n is a vector of the form $\begin{bmatrix} x_{lev} \\ x_{priv} \end{bmatrix}$. From the formula (2.44) it can be seen that at a given angle of rotation, this function reaches a maximum for both the upper and lower bounds.

Figure 29 shows a graph for the function (2.44) applied only to the upper bound. Such degeneration allowed us to represent a function of three variables on the graph (recall that x_n is a vector). Figure 29 shows that the function is not only smooth, but not continuous, which is consistent with the discrete nature of the problem and the type of function (2.44). Thus, the problem of finding the upper and lower boundaries of the region of positive solutions is reduced to finding the maximum of the function (2.44) when rotating the coordinate system. This problem belongs to the class of

Fig. 29 Plot of function (2.44) in the case of finding only the upper limit

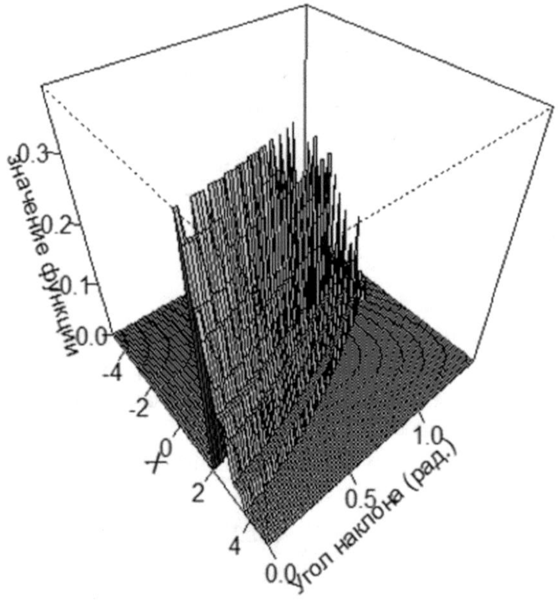

multiparameter nonlinear extremal problems with constraints. A genetic algorithm was used as a solution method. Genetic algorithms (GA) are a subclass of optimization algorithms based on the concepts of natural selection and genetics, such as inheritance and mutations.

The advantages of GA are their ability to simultaneously manipulate many parameters, resistance to the non-unimodality of the optimized function, and applicability to problems with a changing environment.

The classical scheme of functioning of the genetic algorithm is as follows:

(1) To initiate the initial time $t = 0$. Randomly generate an initial population consisting of k individuals $B_0 = \{A_1, A_2, \ldots, A_k\}$.
(2) Calculate the fitness of each individual ($F_{A_i} = fit(A_i), i = 1, \ldots, k$) and the population as a whole $F_t = fit(B_t)$ (function F_t is also sometimes called the term "fitness"‘). The value of this function determines how well the individual described by this chromosome is suitable for solving the problem.
(3) Select individual a_c from population: $A_c = get(B_t)$.
(4) Select the second individual $a_{c1} = get(B_t)$ from the population and make the crossing operator ('crossing over'): $A_c = Crossing(A_c, A_{c1})$. Randomly determined point inside the chromosome, in which both chromosomes are divided into two parts and exchange them.
(5) Execute mutation operator: $A_c = mutation(A_c)$. The randomly selected bit in the chromosome is inverted. The mutation operator is necessary to remove the population from the local extremum and helps to protect the algorithm from premature convergence.
(6) Place the resulting chromosome in a new population: $insert(B_{t+1}, A_c)$.
(7) To execute the operation k times, starting from point 3.
(8) Increase the number of the current epoch: $t = t + 1$.

Verification: if the condition is met the algorithm stops, then to complete the work, otherwise move to step 2. Usually, as criteria for stopping, restrictions on the maximum number of epochs of the algorithm operation are applied. Table 7 shows the parameters used for the operation of the genetic algorithm.

Figure 30 shows the result of the genetic algorithm to find the optimal values of the vector x_n. Thus, we have defined all the steps to find the area of positive values as a parallelogram for a given error of the second kind. In R language the code was implemented, the result of which is presented in Fig. 31.

The GA package was used to find the maximum of the function (2.44).

Table 7 Values of parameters of the genetic algorithm

Population size	50
Maximum number of epochs	100
The crossover probability	0.8
Probability of mutation	0.1
Elitism	0.05

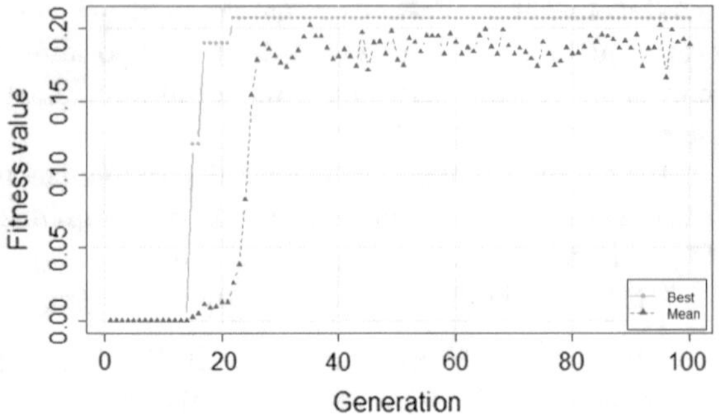

Fig. 30 The result of the genetic algorithm

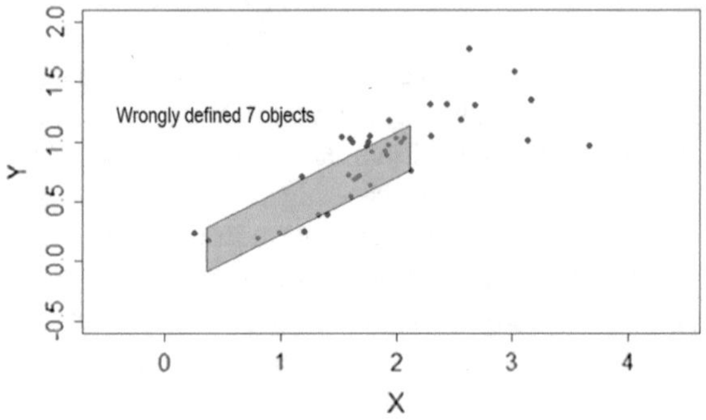

Fig. 31 Found the positive values

To count the number of negative results, within the area has been point.in.polygon of SP.

2.9 Conclusion

The development of the proposed method of analysis and management of emergency situations by precedents (taking into account the calculation of the lags of the impact of the relevant articles of financing of industrial safety measures on accidents and injuries in the operating organization) can become the basis for the development of an integral indicator of the state of industrial safety and an element of monitoring and early warning of emergency and emergency situations in the framework of the widely implemented currently in the industry so-called risk-based approach.

In General, the proposed approach covers both probabilistic methods for modeling emergency processes and events, and deterministic methods. The use of probabilistic and deterministic estimates has taken a significant place in studies to improve safety and operational procedures. However, the experience of using, for example, in the nuclear industry purely probabilistic safety analysis (in fact-a single-criteria tool), showed that this approach does not cover all the necessary aspects of safety.

The security risk of HPF, in particular and, more generally, of critical infrastructure facilities, should be considered as a multicomponent vector, the set of parameters of which may vary.

The control monitoring unit (HPF) in the implementation of a system for remote control (supervision) should be organized in such a way as to be in time to make management decisions, if the state of the object approaching the danger zone. This task is divided into a number of subtasks, because vertically integrated companies usually have several decision-making centers at different levels of management.

Methods for assessing the reliability of achieving targets and methods of group analysis can be promising in solving this problem (the latter are more preferable, since they allow to build trajectories of indicators without taking into account the "diffusion" components and, as a result, can serve as a basic element for building up aggregates, indicators and indicators in a future branched monitoring system).

References

1. Allford L (2009) Process safety indicators. Saf Sci 47(4):466
2. Swuste P, C Gulijk, W Zwaard (2010) Safety metaphors and theories a review of the occupational safety literature of the US UK and the Netherlands, till the first part of the 20th century. Saf Sci 48:1000–1018
3. Swuste P, van Gulijk C, Zwaard W, Oostendorp Y (2014) Occupational safety theories, models and metaphors in the three decades since WO II, in the United States, Britain and the Netherlands: a literature review. Saf Sci 62:16–27
4. van Gulijk C, Swuste P, Zwaard W (2009) Developments in safety science during the interbellum period, and Heinrichs contribution. J Appl Occup Sci 22(3):80–95
5. ANSI/API (2010) Process Safety performance indicators for the refining and petro-chemical Industries, first ed. ANSI/API RP 754
6. CCPS (2010) Guidance for process safety metrics. AIChE, New Jersey
7. OGP (2011) Process safety, recommended practice on key performance indicators. Report nr 456, November, London
8. HSE (2006) Process safety indicators, a step-by-step guide for the chemical and major hazards industries, HSG 254. The Office of Public Sector Information, Information Policy Team, Kew, Richmond, Surrey
9. Hopkins A (2007) Thinking about process safety indicators. In: Oil and gas industry conference paper 53. Manchester, November
10. UK Oil and Gas Industry (2012) Step change in safety, leading performance indicators, guidance for effective use (Aberdeen)
11. Dombi J (1982) Basic concepts for a theory of evaluation: the aggregative operator. Eur J Oper Res 10(3):282–293
12. Walker W (1981) Rankings and ranking functions. Can J Math 23:395–399
13. Wittmuss A (1985) Scalarizing multiobjective optimization problems. Math Res 27:255–258
14. Smith A (1976) An inquiry into the nature and causes of the wealth of nations, vol I. Oxford, pp. 44–46
15. Swuste P, Theunissen J, Schmitz P, Reniers G, Blokland P (2016) Process safety indicators, a review of literature. J Loss Prev Process Ind 40:162–173. https://doi.org/10.1016/j.jlp.2015.12.020
16. Gumerov R (2016) Methodological problems of measurement and assessment of condition of national production security. Economist 4:33–41

17. Landrini G (2009) Integrated levels of safety according to IEC standards 61508 and 61511 and analysis of their connection with technologic operation [Electronic resource]. Modern technologies of automation, no 1, pp 72–78
18. Karpov L, Yudin V (2007) Adaptive management on precedents based on classification of conditions of operated objects [Electronic resource]. Works of System Programming Institute of RAS (electronic magazine), vol 13, no 2, pp 37–58
19. Barsukov A, Bykov A, Lesnykh V (2008) Creation of a System of Emergency and Crisis Situation Indicators and Indexes at UGSS's facilities. Promyshlennaia i ekologicheskaia bezopasnost obieektov gazovoi promyshlennosti [Industrial and environmental safety of gas industry facilities]. Collection of Research Papers. Moscow, VNIIGAZ LLC, pp 76–86 (In Russian)
20. Barsukov A, Bochkov A, Lesnykh V (2015) Situation centers. Monitoring, forecasting and management of crisis events in the gas industry. Part 1: monitoring and forecasting. Moscow, NIIgazekonomika, SAM Poligrafist LLC, 596 p (In Russian)
21. Pereverzev-Orlov V (1990) Specialist's guide. Experience of development of partner system. M. Science (Nauka), 133 pp
22. Bochkov A, Zhigirev N (2017) About some applied problems of safety and situational management of the unified gas supply system on the basis of assessed data flow analysis with support vector machine. gas science guide: scientific and technical collection/Gazprom VNIIGAZ LLC, no 1 (29). Increase in reliability and safety of gas industry objects, 278 pp, pp 129–141
23. Bochkov A, Ponomarenko D (2017) Scientific and methodical bases of monitoring and forecasting of production safety condition of PJSC «Gazprom». Gas Ind 3(749):20–30
24. Bochkov AV, Lesnykh VV, Ponomarenko DV (2018) Modern approaches to monitoring the state of industrial safety of hazardous production facilities. Problems Risk Anal 15(1):6–17
25. Bochkov AV (2016) Methodological issues of security of critical infrastructure of structurally complex systems. In: Proceedings 7th DQM international conference life circle engineering and management ICDQM-2016, 29–30 June 2016, Prijevor, Serbia; Čačak. XIII, 543, pp 43–71
26. Bochkov AV (2016) Development of an integrated approach to risk management of critical infrastructure. In: Proceedings of the international scientific School of IADB (P.-Pb., 25–28.10.2016). GUAP, St. Petersburg, pp 87–96
27. Bochkov AV, Zhigirev NN, Lesnykh VV (2014) AHP modification for decision making under uncertainty. In: International symposium of the analytical hierarchy process, Washington DC, USA
28. Bochkov AV, Gigiri NN (2015) The development of methods of comparison of alternatives when making decisions in conditions of incomplete definiteness of the expert's preferences. In: Proceedings of international scientific school IADB-2015 (S.-Pb., 17–19.11.2015)/SPb. GUAP 2015, 304, pp 224–232
29. The order of Rostekhnadzor of 12.01.2015 No. 1 «About modification of Federal regulations and rules in the field of industrial safety "Safety Rules in the Oil and Gas Industry"»
30. Federal norms and rules in the field of industrial safety "Safety Rules in the Oil and Gas Industry"
31. Federal law "On industrial safety of hazardous production facilities" dated 21.07.1997, № 116-FZ
32. Gumerov R (2016) Methodological problems of measurement and assessment of the state of national industrial safety [Electronic resource]. Economist 4:33–41. https://elibrary.ru/item.asp?id=18853513. Date accessed 22.02.2018
33. Landrini G (2009) Integrated safety levels in accordance with IEC 61508 and 61511 standards and analysis of their connection with technological maintenance [Electronic resource]. Modern automation technologies, № 1, pp 72–78. https://www.cta.ru/cms/f/382073.pdf. Date accessed 22.02.2018
34. Bochkov AV, Demidova NS, Bashkin VN (2012) Mathematical modeling of transport processes in life support and human protection systems in extreme environmental conditions. A series of "gas to conduct science." Monograph—Gazprom VNIIGAZ, Moscow, 282 p

35. Pontryagin LS, Boltyansky VG, Gamkrelidze RV, Mishchenko EF (1961) Mathematical theory of optimal processes. Fizmatgiz, Moscow, 392 p
36. Wiener N (1983) Cybernetics, or control and communication in animal and machine. Under the editorship of G. N. Povarova, 2nd ed. Science, Moscow, 344 p
37. Popov EP (1989) Theory of linear systems of automatic regulation and control. In: Proceeding textbook for technical colleges, 2nd ed, pererab. I DOP. M.: Science. GL. ed. Fiz. Mat. lit., 304 p
38. Romanovsky YM, Stepanova N, Chernavsky DS (1984) Mathematical biophysics. M., 304 p
39. Wang MC, Uhlenbeck GE (1945) On the theory of Brownian motion II. Rev Mod Phys 17:323–342
40. Bogolyubov NN, Krylov NM (1939) On Fokker-Planck equations, which are derived in the perturbation theory by the method based on the spectral properties of the perturbed Hamiltonian. Notes of the Department of mathematical physics Of the Institute of nonlinear mechanics of the USSR Academy of Sciences, vol 4, pp 5–80
41. Arnold VI (1983) Theory of catastrophes. 2nd ed Izd., DOP. M.: Moscow state University Publ., 80 p (New ideas in natural science)
42. Wiener N (2002) Cybernetics and society. M.: Tidex Ko, p 184
43. Remizov AN, Maksina AG. et al (2003) Medical and biological physics. 4th ed, DOP. And Pererab. M.: bustard, 560 p. ISBN: 5-7107-9844-4
44. Leonov AA, Leonova MA, Fedoseev YN (2000) The synthesis of a neural network to solve the problem of identifying the state of the object. In: Reports of the all-Russian conference "Neuroinformatics-99", Part 1. M.: MEPhI, pp 100–109
45. Segal IH, Ivanova AP (2003) Introduction to applied discrete programming. In: Models and computational algorithms. M.: Fizmatlit, 204 p

Multi-level Hierarchical Reliability Model of Technical Systems: Theory and Application

Igor Bolvashenkov, Jörg Kammermann, Ilia Frenkel and Hans-Georg Herzog

Abstract This chapter describes an assessment methodology for various sustainability indicators of technical systems, such as reliability, availability, fault tolerance, and reliability associated cost of technical safety-critical systems, based on Multi-Level Hierarchical Reliability Model (MLHRM). As an application case of the proposed methodology, the various sustainability indicators of electric vehicle propulsion systems are considered and evaluated on the different levels of the hierarchical model. Taking into account that vehicle traction drive systems are safety-critical systems, the strict requirements on reliability indices are imposed to each of their components. The practical application of the proposed technique for reliability oriented development of electric propulsion system for the search-and-rescue helicopter and icebreaker LNG tanker and the results of computation are presented. The opportunities of improvement regarding reliability and fault tolerance of such technical systems are investigated. The results of the study, allowing creating highly reliable technical systems for the specified operating conditions and choosing the most appropriate system design, are discussed in detail.

Keywords Multi-level hierarchical reliability model · Technical system · Reliability oriented design · Fault tolerance · Electric propulsion system · Markov model · Reliability associated cost

I. Bolvashenkov (✉) · J. Kammermann · H.-G. Herzog
Institute of Energy Conversion Technology, Technical University of Munich (TUM), Arcisstrasse 21, 80333 Munich, Germany
e-mail: igor.bolvashenkov@tum.de

I. Frenkel
Center for Reliability and Risk Management, SCE—Shamoon College of Engineering, 84100 Ashdod, Beer Sheva, Israel
e-mail: iliaf@frenkel-online.com

M. Ram and H. Pham (eds.), *Advances in Reliability Analysis and its Applications*, Springer Series in Reliability Engineering,
https://doi.org/10.1007/978-3-030-31375-3_5

1 Introduction

The rapid modern development of new technical systems in various areas of the industry is directly related to a significant increase in their complexity. In addition, the levels of integration of subsystems, units and components and, accordingly, their mutual effect largely increase as well. This, in turn, has a very strong impact on the reliability, fault tolerance, and maintainability of the designed technical systems. Reliability concepts can be applied to virtually any engineered system. In its broadest sense, reliability is a measure of performance.

All of the above fully applies to the traction drive of electric vehicles, the creation of which is a major challenge in the modern way to electrification of the different types of vehicles: ships, planes, trains, helicopters, buses, and cars. For transport facilities that are safety-critical systems, the issues of assessing and optimizing reliability indicators are of particular importance.

As can be seen from Fig. 1, the magnitude of the level of technical excellence of an electric traction drive is determined by three comprehensive criteria: sustainable functioning, efficient functioning, and environmental level. It follows from Fig. 1 that the maximum number of factors affects the amount of sustainable functioning criterion of the traction drive. Accordingly, the above criterion has the maximum potential to increase the value of the level of excellence of the traction electric drive and an electric vehicle as a whole. In addition, the most stringent requirements are imposed on reliability, fault tolerance, and survivability of electric vehicles, which are safety-critical systems.

In this way, reliability-oriented design of the vehicle electric propulsion system and, accordingly, all its subsystems, units, and components is a very urgent and complex task while considering their interactions. In recent years, a multilevel approach

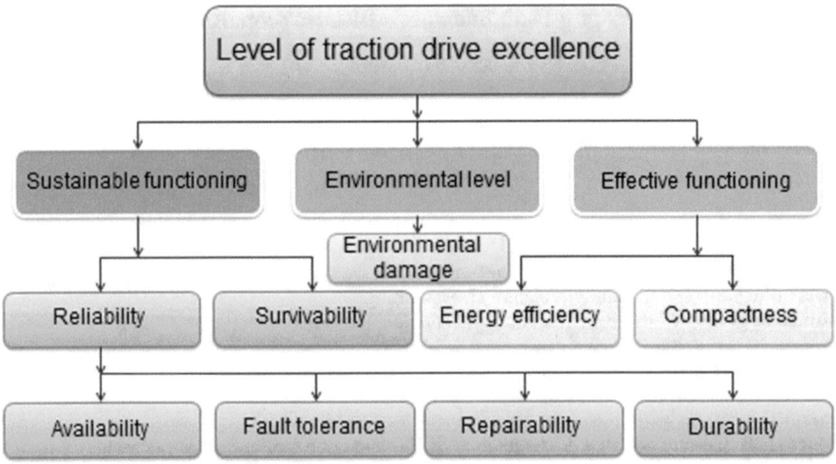

Fig. 1 Level of excellence

in the development, design, and optimization of various technical systems and their particular parameters has become quite widespread. In addition, when using a multilevel approach in most cases, the various levels are interconnected hierarchically. Depending on the complexity of the system being developed, the multilevel hierarchical reliability model may consist of a different number of levels. In the simplest case, it can consist of three levels.

Attempts to develop the methods for solving such a problem were undertaken by various research groups. The first group of scientists, whose works are presented in [1–4], uses the method of hierarchical decomposition of the technical system, better known as analytic hierarchy process (AHP). It was developed by Thomas L. Saaty in the 1970s and represents a structured technique to organize and analyze complex decisions, described in detail in [1]. This approach has significant advantages when important components of the decision are difficult to quantify or to compare, or when communication between team members is made difficult by their different specializations, terminology, or perspectives. Due to the relatively simple mathematical formula, as well as the easy data collection, AHP has been widely applied by many researchers.

The integral shortcoming of the AHP is the fact that the criteria are assumed to be completely independent, even though in real world problems, the criteria are often dependent. In [2] the AHP approach was applied in the four-level hierarchical tree to identify the main attributes and criteria that affect the level of accuracy of the models used in probabilistic risk assessment. The main disadvantage of AHP approach is the inability to consider the uncertainties of the process. In order to overcome this limitation the application of different hybrid combination of fuzzy theory and AHP, so-called called Fuzzy AHP, and analytic network process (ANP) method have been used in [3] for inter-criteria dependencies definition and in [4] for the vehicle safety analysis. It should be noted that in real life, most of the decision problems are represented by a network and not only structured as a hierarchy.

Various hierarchical stochastic models have proven to be a powerful tool for analyzing the reliability of complex technical systems for different application. The authors in [5] described a method, called the hierarchical Markov modeling (HMM), which allows to perform the predictive reliability assessment of distribution electrical system. This method can be used not only to assess the reliability of existing distribution systems, but also to estimate the reliability impact of several design improvement features.

HMM creates a primary model based on the system topology, secondary models based on integrated protection systems, and tertiary models based upon individual protection devices. Once the tertiary models have been solved, the secondary models can be solved. In turn, solving the secondary models allows the primary model to be solved and all of the customer interruption information to be computed.

An interesting approach to solving the complex problem of performance, availability, and power consumption analysis of infrastructure as a service (IaaS) clouds, based hierarchical stochastic reward nets (SRN) is presented in [6]. In order to use the resources of an IaaS cloud efficiently, several important factors such as performance, availability, and power consumption need to be considered and evaluated carefully. The estimation of these indicators is significant for cost-benefit prediction and quantification of different strategies, which can be applied to cloud management.

Possible techniques and ways to solve the problem of a multistage reliability-based design optimization (MSRBDO) based on Monte-Carlo method, and its application to aircraft conceptual design, which is described in detail in [7] and with subsequent corrections and development in [8]. In recent years, a multilevel (tiered) systematic approach has become increasingly widespread for analyzing and optimizing the various characteristics of technical systems, the theoretical foundations of which are described in detail in [9–12].

In the work [9] the four-level (system, subsystem, assembly, and device-component) representation of variable-speed drive systems is proposed for analysis of reliability, availability, and maintainability. The calculations were performed analytically and step by step. Paper [10] describes the rules and properties of multilevel hierarchical representation of the vehicles propulsion systems life cycles and the optimal types of stochastic methods and models for use at each individual level.

A new look at solving the problem of assessing various system resilience, based on the three-level (tiered) approach is proposed in [11]. Reference [12] presents a systematic four-level approach to develop the reliability design of the mechanical system—the refrigerator, which is similar to the target of this chapter, but it does not present any analytical optimization.

A significant amount of research works is related to the assessment of the reliability of particular units or component at one of the local levels of the multilevel model and the development of appropriate methods and models [13–16]. In references [13, 14], several options for assessing reliability at the component level are presented. In the first case [13], it is proposed to do this using failure mode and effects analysis (FMEA) with weighted risk priority number (RPN), and in the second case [14] it is proposed to do this based on a multi-state Markov model, which allows to consider random environmental conditions.

The hierarchical model for lithium-ion battery degradation prediction, discussed in [15], represents reliability assessment technique at the unit level of multilevel model. The three-level (system, subsystem, and component) aircraft engine model's hierarchical architecture is described in [16]. This paper concludes that in a large system, such as an aircraft engine, failure prognostics can be performed at various levels, i.e. component level, subsystem level, and system level. A similar approach for estimation of remaining useful life (RUL) for the multiple-component systems—when using the prognostics and health monitoring (PHM) technologies in modern aircraft—is proposed in papers [17, 18]. This methodology combines particular components RUL estimations into a single system level RUL estimation. This characteristic becomes more relevant when the number of components within the system increases.

2 Methodology of Multilevel Hierarchical Reliability Model

In order to solve the problem of implementing the reliability-oriented design for electric propulsion system, the authors, based on previous own research and research of other scientists, developed the methodology for creating and using the multilevel hierarchical reliability model (MLHRM) of electric vehicles' functioning. The main features, techniques, and potentials of the model are presented below.

The proposed method of reliability oriented design of vehicle electric propulsion system based on the MLHRM, allows to solve a complete set of tasks related to the full range of indicators of comprehensive reliability for the safety-critical electric traction systems, such as failure-free operation probability, fault tolerance, availability, maintainability, durability, reliability associated cost, etc.

2.1 *Structure of MLHRM*

Figure 2 shows the general view of the MLHRM structure. The number of levels of the model can vary depending on the complexity of the technical system and the tasks to be solved. The model presented in Fig. 2 has six levels, which correspond to the task of analyzing and optimizing the reliability characteristics of electric vehicles, taking into account their interaction in random environment.

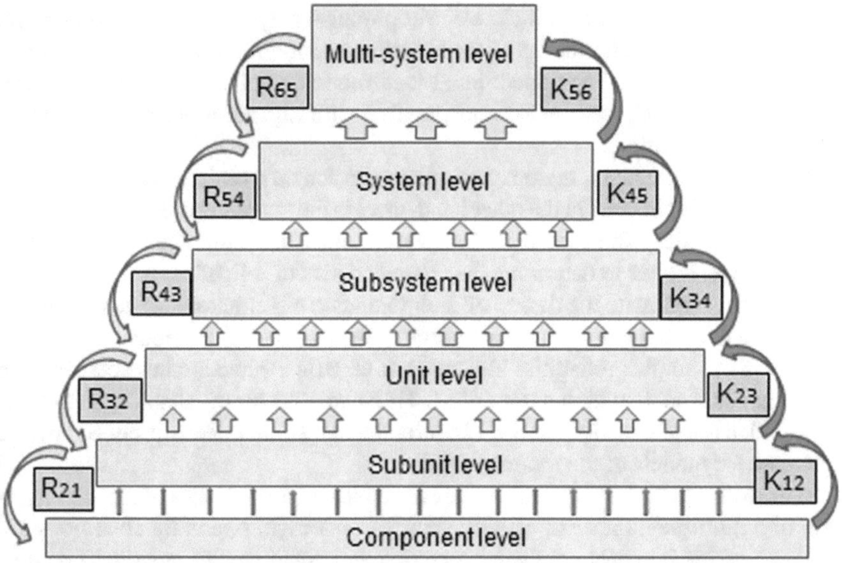

Fig. 2 General structure of MLHRM

The coefficients K12–K56 determine the magnitude of the influence of the reliability of the lower level of the model on the neighboring upper level. The coefficients R21–R65 determine the ratio of the required values of the performance of the upper level of the model relative to the neighboring lower level.

As noted above, the MLHRM shown in Fig. 2 includes six levels, namely, component level (CL), subunit level (SUL), unit level (UL), subsystem level (SSL), system level (SL), and multi-system level (MSL).

At the CL, based on statistical reliability data, analytical calculations, or using Markov models for binary-state components, reliability characteristics of element of the next level (SUL) are determined. In operational mode, component failures can lead to the degradation of the whole system performance.

Respectively, the performance rate of any component can range from fully functioning up to complete failure. The failures that lead to a decrease in the element performance are called partial failures. After partial failure, elements continue to operate at reduced performance rates, and after complete failure the elements are totally unable to perform their missions.

At the SUL the initial parameters for the analysis of reliability indicators of the red level are determined. As subunits, the independent functional parts of the next level (UL) can be considered. In turn, at the UL, an analysis and evaluation of independent functional units, which are integral parts of the next level, SSL, are carried out.

The reliability indicators calculated at the UL are the input data for the models used within the next level—the SSL. In the case of electric vehicles simulation, the SSL corresponds to the level where the assessment of the reliability characteristics of the entire electric traction drive takes place.

The basic model of the vehicle electric propulsion system at this level can be represented as stochastic model of multi-state system with the change of discrete operating load modes. Each operational load mode complies with specific power characteristics, which have to be implemented with highest probability for safety operation of the vehicle.

Thus, on the one hand, there are requirements for safe vehicle operation, which form a model of demand. On the other hand, there is the guaranteed generated electric power, which values form the model of performance. The combined performance-demand model allows to determine the characteristics of reliability, based on which it is possible to estimate the degree of fault tolerance of vehicles electric propulsion system and to optimize its values according to the project requirements.

At the SL, complex reliability indicators of electric vehicle are investigated. The input data for modeling at this level of MLHRM are the output reliability characteristics, which are obtained at the SSL. In turn, the output characteristics of SL are the input data for models of the top-level MSL.

At the MSL, the reliability associated economical characteristics of the joint operation of a multiple number of electric vehicles under real operating conditions are estimated taking into account their interaction and random environment. The problems solved at this level were not the purpose of present study and, therefore, are not considered in this chapter.

Based on the presented MLHRM, an algorithm was developed for accelerated estimation of the compliance of the propulsion system reliability indicators with the project requirements, which is shown in Fig. 3.

In accordance with the above algorithm, the main task of a simplified rapid assessment of reliability indicators is to determine the critical important components of

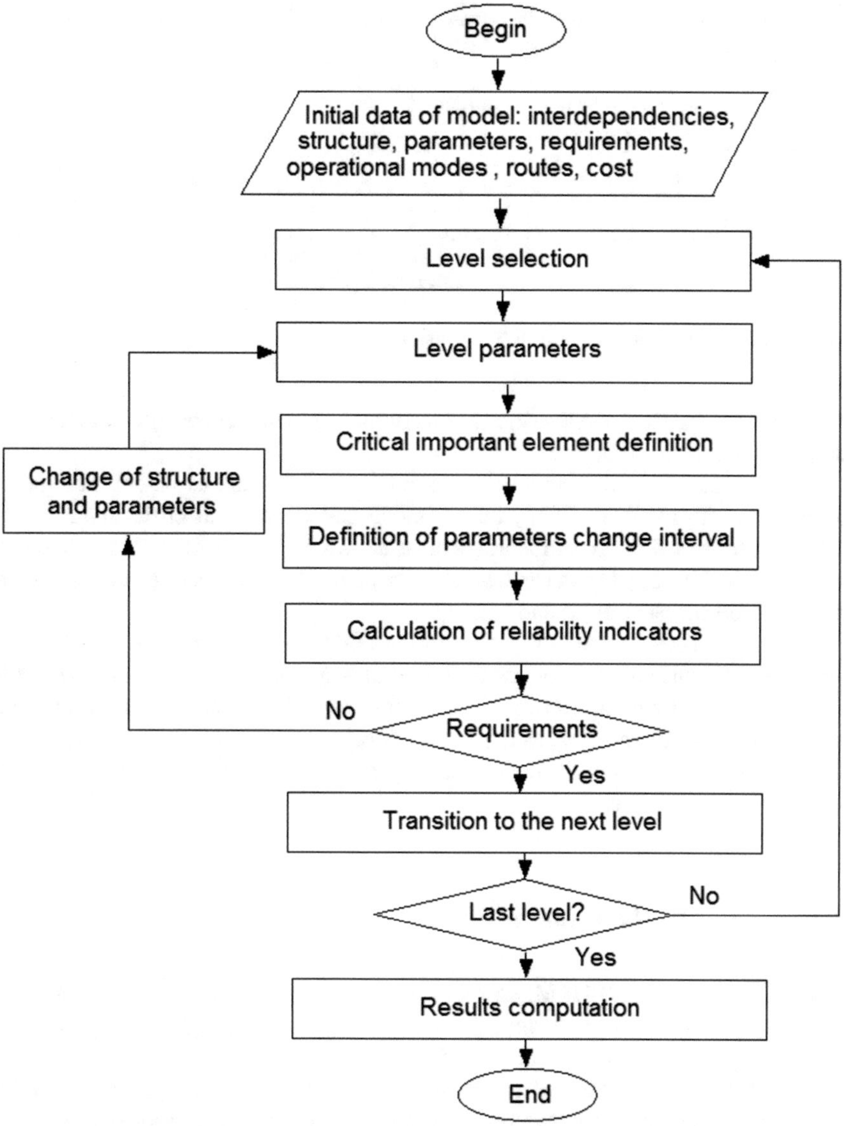

Fig. 3 Algorithm for rapid analysis of the reliability characteristics of a technical system

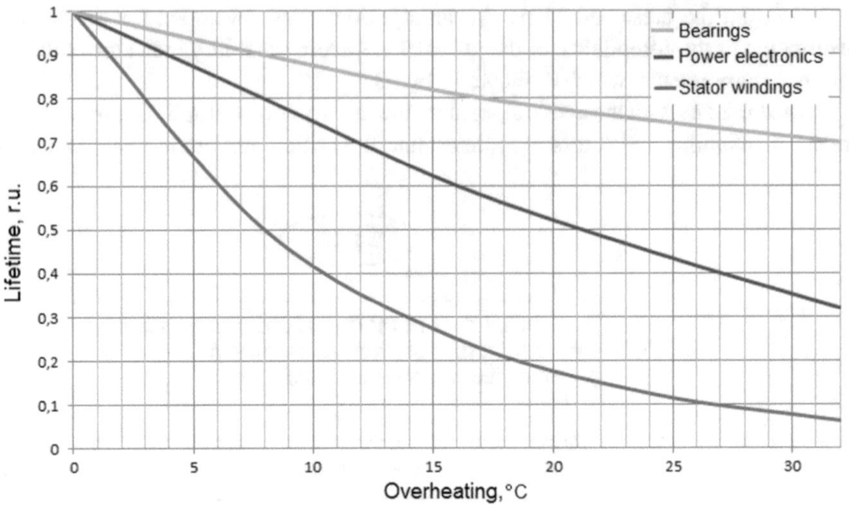

Fig. 4 Critical importance analysis of the subunits

each level of MLHRM and the degree of its influence on the reliability characteristics of the neighboring upper level.

In this case, the critical important parts of each level can be determined based on risk priority number (RPN), failure mode and effects and criticality analysis (FMECA) or based on experimental data, as shown in Fig. 4, which was previously presented in [19–21] for the main subunits of the traction electric motor: stator windings, power electronics and bearings.

From the results shown in Fig. 4, it follows that the most sensitive parts to thermal effects in various operating conditions and in terms of reliability, are the stator windings of the traction electric motor. In this case, for further investigations, the stator windings are accepted as a critical important subunit for the unit—the traction electric motor. Similarly, the critical important parts for the remaining levels of MLHRM can be defined.

2.2 *Goals, Methods and Models*

At each level of the MLHRM, specific models are used to solve specific tasks in order to achieve the corresponding goals at each level. Figure 5 graphically presents the problems associated with the reliability characteristics of electrical propulsion systems that can be solved by means of MLHRM. In addition, Fig. 5 presents the methods and models recommended in order to assess the reliability indicators of different MLHRM levels.

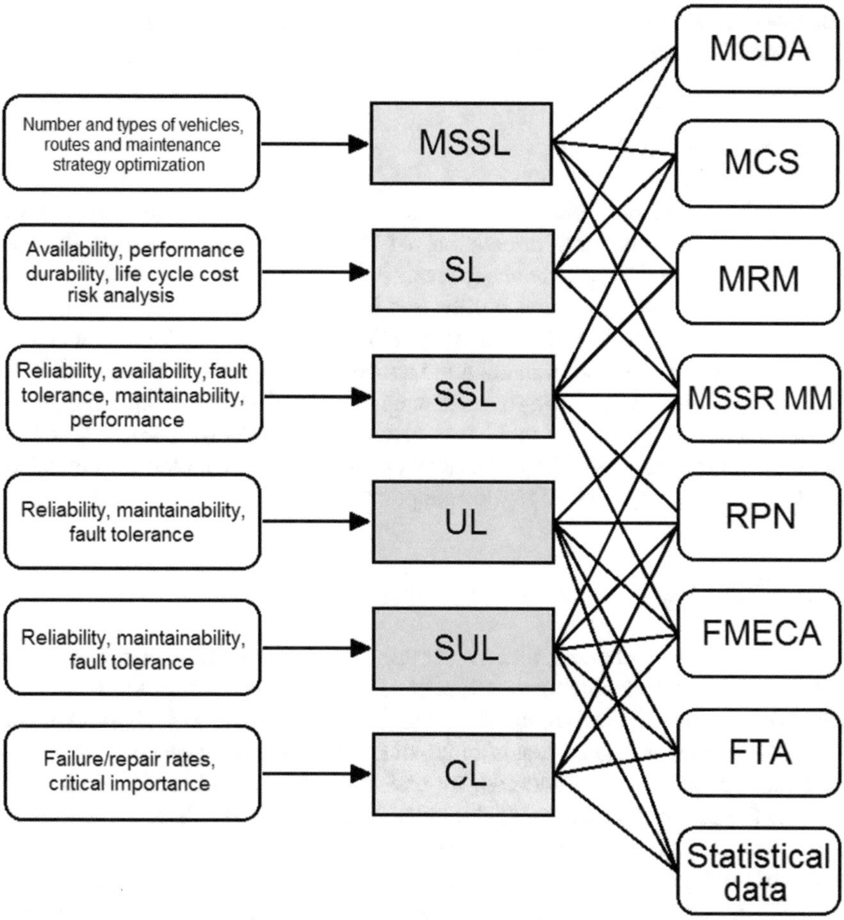

Fig. 5 Tasks and methods of their solutions for different MLHRM levels

Below, a detailed description of the tasks and methods for their solution, applied to each level of MLHRM is given.

2.2.1 Component Level

The main tasks that are solved at the CL are the collection, analysis, and structuring of statistical data on the reliability of all components that affect the reliability of the neighboring top level of the MLHRM. It also identifies the critical important components and their degree of influence on the reliability features of the next level—the SUL. Possible methods for achieving these goals are fault tree analysis (FTA), failure mode and effects analysis (FMEA), FMECA, and RPN. Several examples of the reliability characteristics analysis of electric propulsion systems at CL of MLHRM are described in [21–23].

2.2.2 Subunit Level

As subunits, this chapter examines individual, relatively independent parts of units having a specific functional orientation. At the subunit level, based on the data obtained in the previous component level, it is advisable to determine the characteristics of reliability, maintainability, and fault tolerance of the subunit groups, forming the corresponding elements of the next level—the UL. The recommended methods for analyzing and evaluating the above reliability characteristics are FTA, FMEA, FMECA, and RPN using experimental failure and repair statistics. If there are blocks that are not binary, but multi-state elements (elements with degraded states), the multi-state system reliability Markov models (MSSR MM), described in details in [20, 23, 24], can be applied for the computation.

2.2.3 Unit Level

At the UL, the tasks of computation and optimization of reliability, maintainability, and fault tolerance of autonomous functional parts (units), within the propulsion system of electric vehicles, are solved. Taking into account that the units are elements with several degraded states, that is, multi-state systems, it is advisable to use MSSR MM for their research. In addition, by means of MSSR MM, one can take into account the actual load modes of the units, regarding overloads capacity and the aging processes. The transition probabilities for MSSR MM can be calculated by means of the degree of fault tolerance (DOFT) [24] using statistical operational data or can be determined at the design stage based on the requirements to the safety and sustainable vehicle operations. In order to determine the critical important elements of the UL for further optimization, RPN, FMECA, FTA, and experimental test methods can be used.

2.2.4 Subsystem Level

At the SSL the problems of determining and optimizing the reliability characteristics of operational availability, maintainability, fault tolerance, redundancy (functional and structural), and performance of entire electric propulsion system should be solved. In order to build the corresponding combined stochastic model of the electric vehicle propulsion system including electric energy source, the concept of balanced relationship between demand (required power) and performance (available power) have been applied.

Hence, the model of the electric propulsion system operation can be represented as a MSSR MM with the change of discrete operating modes: start (takeoff), acceleration (climb), constant speed (cruise), deceleration (reduction of altitude), and stop (landing). Along with MSSR MM, Markov reward models (MRM) and Monte-Carlo simulation (MCS) can be widely apply.

2.2.5 System Level

At this level, the most preferred are the various stochastic models of the electric vehicle's lifecycle, which allow to assess the reliability indices of repairable systems by optimizing maintenance strategies according to intensity of the scheduled and unscheduled repairs, the use of functional systems of monitoring, forecasting reliability, and diagnostics. These may be MSSR MM, MRM, MCS and multi-criteria decision analysis (MCDA).

A definition of current and forecasted values of reliability indices, performed considering the external and internal operation conditions of the vehicle, as well as taking into account the availability (or non-availability) of structural or functional redundancy. Thus, the study and optimization task of the so-called reliability associated costs (RAC) estimation, based on MRM, is most interesting and promised [20].

In order to build such a model, the process of the vehicles operations can be represented by a chain of the lifecycles: operational, non-operational, working, standing, etc. The data on duration of each cycle are obtained based on the analysis of statistical operational data of a particular type of vehicle on certain routes and areas.

3 Application Cases

As application examples of the proposed MLHRM methodology for assessing and optimizing the reliability characteristics of electric traction drives, the electric propulsion systems of a search-and-rescue (SAR) helicopter and an icebreaker liquefied natural gas (LNG) Arctic tanker are considered.

The selected objects of investigation differ significantly in almost all operational vehicle indicators, such as operational conditions, the values of nominal performance, the possibility of repairs during operation, etc. The main purpose of the selection of such objects is to show the universality of the proposed model and methodology for various types of electric vehicles.

3.1 Electrical Helicopter

Functionally, the MLHRM of the electrical helicopter is presented in Fig. 6. As a basic conventional prototype for the design and development of the full electric propulsion system, the Airbus helicopter EC135 currently being in operation has been considered. The traction system of the EC135 has two gas turbine engines. Accordingly the turbines Turbomeca Arrius-2B2 or Pratt and Whitney PW206B2 are installed as gas turbine engines on the EC135.

In consideration of statistics from the German automobile club (ADAC), every SAR helicopter in Germany is operated by a daily average of 8–10 h, i.e., the average ratio of operational time in one year is 0.33–0.42. Thus, in the further simulation an annual flight of the helicopter assumed to be equal to 3000 h. Table 1 shows the weight and dimensions of a traditional traction drive of the EC135 with two turbines.

Generally, the propulsion system of the electrical helicopter consists of a various units, such as electric energy source (EES), power electronics (PE), control unit (CU), and traction electric motor (EM), as is shown in Fig. 7. In addition to the units indicated in Fig. 7, the electrical propulsion system of the helicopter includes switchboard (SWB), sensors (SENS) and other blocks, that affect its reliability.

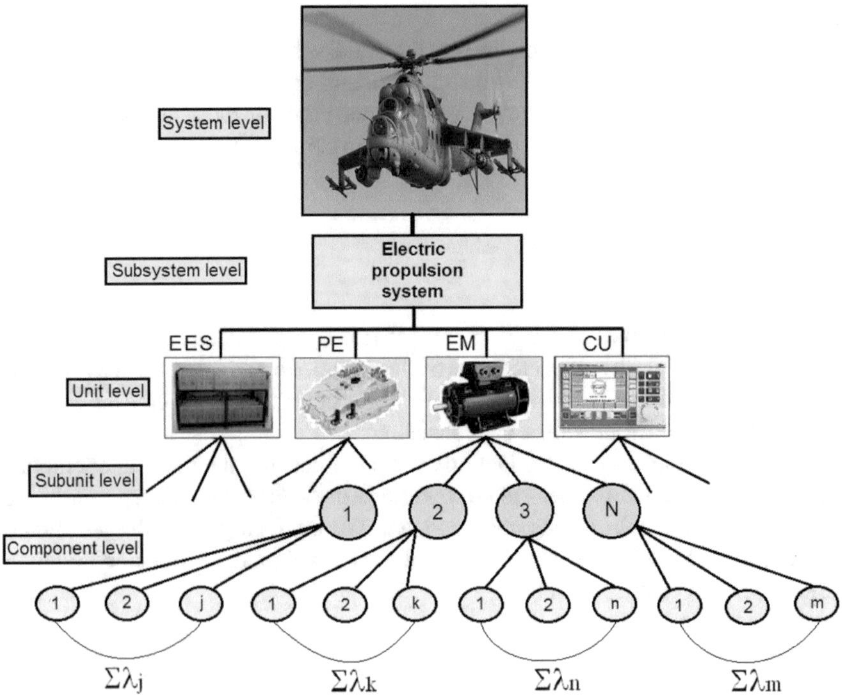

Fig. 6 MLHRM structure of helicopter with electric propulsion

Table 1 Technical data of EC135 traction drive

Component	Weight (kg)	Volume (l)
Two turbine engines	228	330
Fuel tank	650	737
Total	878	1067

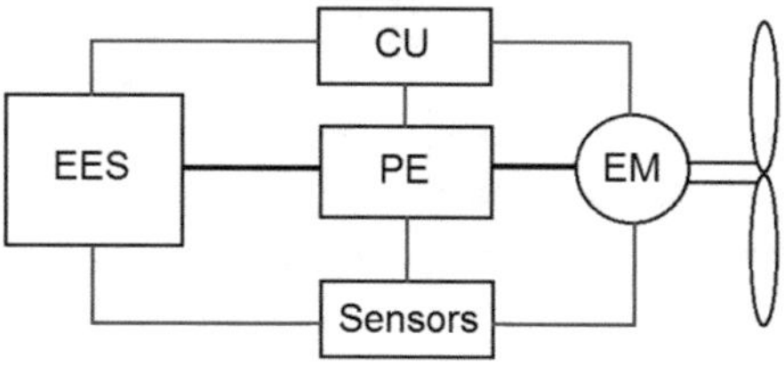

Fig. 7 Structure of the helicopter's electric traction drive

3.2 Icebreaker LNG Tanker with Electric Propulsion

In general, the MLHRM of the icebreaker tanker is presented in Fig. 8. The new Arctic LNG tanker "Christophe de Margerie", built in 2017 by Daewoo Shipbuilding &

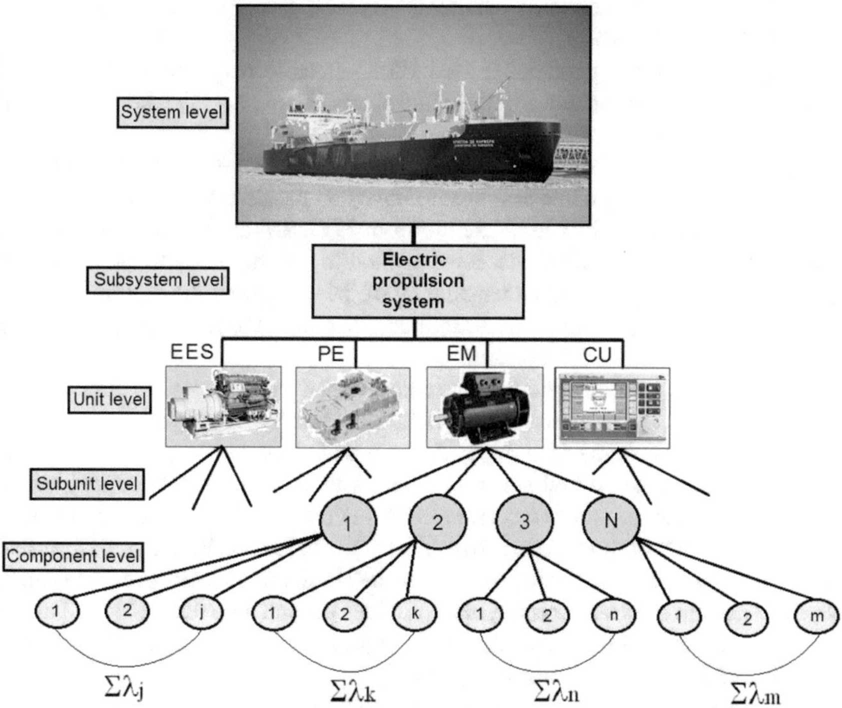

Fig. 8 MLHRM structure of icebreaking cargo ship with electric propulsion

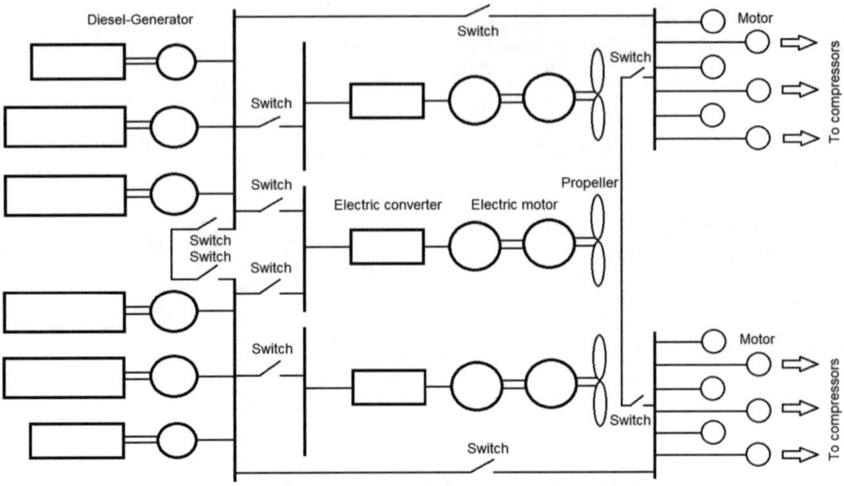

Fig. 9 Structure of the whole power system of LNG tanker

Marine Engineering in South Korea, was selected as the research object to investigate reliability features of the overall electric propulsion system. The characteristics of the LNG tanker "Christophe de Margerie", as well as its propulsion system, are described in detail in [25].

In Fig. 8, the following notation is used: EES—electric energy source, PE—power electronics, EM—electric motor, CU—control unit, λ_j, λ_k, λ_n, λ_m—failures rates of various components.

The main goal of the ship's propulsion system is to ensure the safe and efficient transportation of cargo and/or passengers. Based on the stated main goal, the functions that should be performed at each level of MLHRM are analyzed. Below is a detailed description of each model level applied to the ship's electrical propulsion system. For a more complete understanding of the essence of the multilevel structure of MLHRM, Fig. 9 shows the simplified diagram of the fully integrated power system of the icebreaker LNG tanker.

The entire ship's power system can be conventionally represented as three subsystems: the electric energy source system (EES), the ship's electric propulsion system (EPS), and the subsystem of the ship's consumers of electric energy (EEC). The first subsystem includes six diesel-generators with a total power of 62,000 kW, which supply electric energy to a two-section main switchboard. The electric propulsion subsystem consists of three electric traction drives, including electric converters and three two-section electric traction motors, located in steering gondolas of the Azipod system. The ship's consumer subsystem provides general ship needs, as well as the critical important consumer, namely the gas liquefaction and storage system (LSS), consisting of 12 powerful motor-compressors.

When transporting LNG, specifically stringent requirements are imposed on the whole power system of the tanker in terms of safe and sustainable operation. On the one hand, in the heavy ice conditions of the Arctic, it is necessary to ensure the maximum possible power on all three propellers of the vessel, and on the other hand, in the same time, it is necessary to ensure uninterrupted functioning of the LSS for the safety and keeping the cargo—liquefied gas. This feature should be unconditionally observed during the simulation on SL and MSL. It should be noted that this requirement extends over 50% of the operating time of LNG tanker.

3.3 Component Level and Subunit Level

At the component level, based on available failure statistics [21–23] and the above methods of analytical reliability calculation (FTA, FMEA, RPN, etc.) the total failure rates of all components, of which the subunits are composed, can be analyzed and estimated. For EM, as the part of UL, the subunits are a stator with windings, a rotor with magnets, a bearing, and others, as shown in Fig. 10.

Considering the above data of Fig. 10, generally reliability of electric motor λ_{EM} can be determined by the formula:

$$\lambda_{EM}(t) = \Sigma\lambda_{Si}(t) + \Sigma\lambda_{Rj}(t) + \Sigma\lambda_{Bk}(t), \tag{1}$$

where λ_{Si}, λ_{Rj} and λ_{Bk} are the failure rates of parts of the all parts of electrical machine, respectively of stator, rotor and bearing.

For EC, as the part of UL, the subunits are the semiconductors, printed circuit boards (PCB), capacitors, and others, as shown in Fig. 11.

Based on the above data of Fig. 11, generally the failure rate of an electric power converter λ_{EI} can be estimated considering the reliability values of its components by the equation:

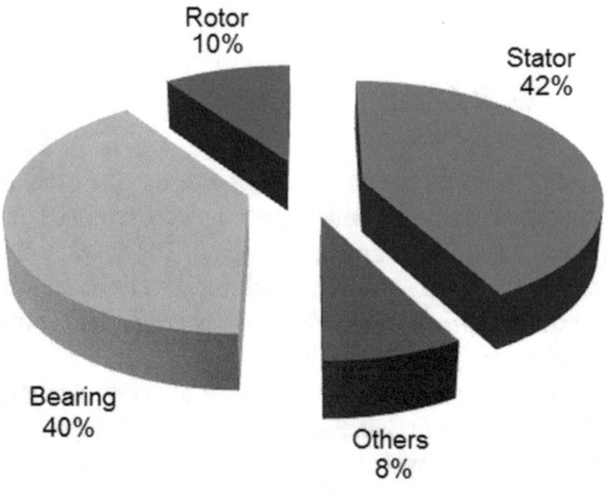

Fig. 10 Failures statistics of traction electric motor

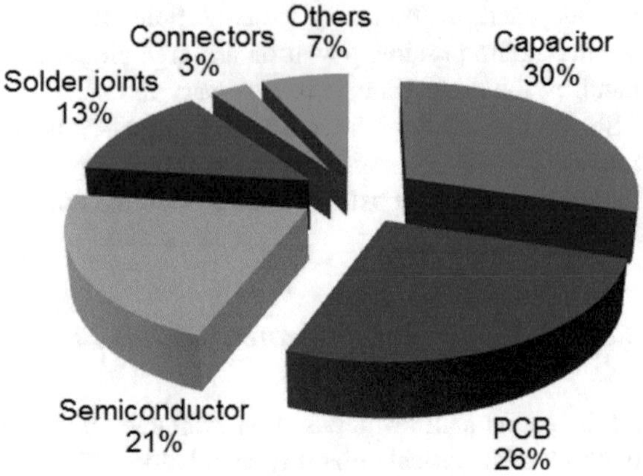

Fig. 11 Failures statistics of electric converter

$$\lambda_{EC}(t) = \Sigma\lambda_{Ti}(t) + \Sigma\lambda_{Dj}(t) + \Sigma\lambda_{Ck}(t) + \Sigma\lambda_{Bn}(t), \quad (2)$$

where λ_{Ti}, λ_{Dj}, λ_{Ck} and λ_{Bn} are the failure rates of the all components of electric inverter, respectively of transistor, diode, capacitor, and printed circuit board.

The similar calculations are performed for all other subunits of the SUL, which are taken into consideration. Based on the results of the calculation, the sensitivity of changing the values of the reliability indicators at the subunit relatively to the change of the components' failure rates is determined. The obtained results are used further in the models at UL and SSL.

Increased reliability features on the CL can be obtained while using components and materials with higher reliability values and by various methods of critical components redundancy. In order to achieve the required performance characteristics of the SUL, as shown in [21], it is necessary to optimize the type of stator windings, permanent magnets, bearings, semiconductors, etc. In addition, redundancy of critical important parts of subunits can be used.

3.4 Unit Level

At this level of the MLHRM, the tasks of providing reliable performance of all functional elements, which form the subsystem of the electrical propulsion systems presented in Figs. 6 and 8, are solved. The detailed descriptions of the use of various techniques to improve the reliability and fault tolerance of electric energy sources, traction electric motors, electric converters, and control units at this MLHRM level are given in [19, 20, 23, 26].

The correct choice of the type of electric machine, the methodology of which is presented in [21], has a significant impact on the reliability indicators of an electric propulsion system. Based on the completed studies, it was proposed to use a synchronous motor with permanent magnets as the most promising one in terms of reliability and fault tolerance.

One of the most effective methods to improve the reliability and fault tolerance of traction electric motors is the use of a multi-phase motor topology with concentrated windings and galvanically uncoupled phases, described in [19, 26]. A significant influence on the characteristics of fault tolerance and overload capacity of the traction electric motor is provided by the parameters and the location of the permanent magnets on the rotor. In the work [21], it is shown that the most preferable design is the permanent magnet synchronous motor with internal v-shaped arrangement of permanent magnets on the rotor.

3.4.1 The Choice of the Phase Number of Traction Electric Motors

In order to select the suitable number of phases of the traction electric motor in accordance with the requirements on reliability and fault tolerance, it is advisable to represent the multiphase traction electric motor in the form illustrated in Fig. 12.

A multiphase electric motor means a motor with more than 3 phases. A critical electric motor failure in this chapter means the loss of one or several phases of an electric motor with a corresponding decrease of its performance—the shaft power.

Each electric motor, which has more than three phases, has a certain level of fault tolerance, i.e. is able to function in degraded states after one or more phase failures. The state space diagram of MSSR MM for the fault tolerance estimation is presented in Fig. 13.

The results of calculations, presented in Fig. 14, showed that the 9-phase electrical machine meets the requirements of the project on the fault tolerance for the propulsion system, which is equal to one FIT. In this regard, a further increase in the number of motor phases for the considered application case is inexpedient.

3.4.2 The Choice of Power Electronics

As a converter of electrical energy for the study, a multilevel inverter was chosen which has important advantages from the point of view of fault tolerance in comparison with the conventional one. At the same time, for the given parameters of the electric propulsion system, a 17-level cascaded H-bridge inverter (CHB) is defined as the most promising topology of a multilevel inverter. One submodule of CHB is presented in Fig. 15.

Results of calculations, presented in Fig. 16, shown that to the requirements of the project on the fault tolerance for electric propulsion system satisfy the 7- and 9-phase topologies.

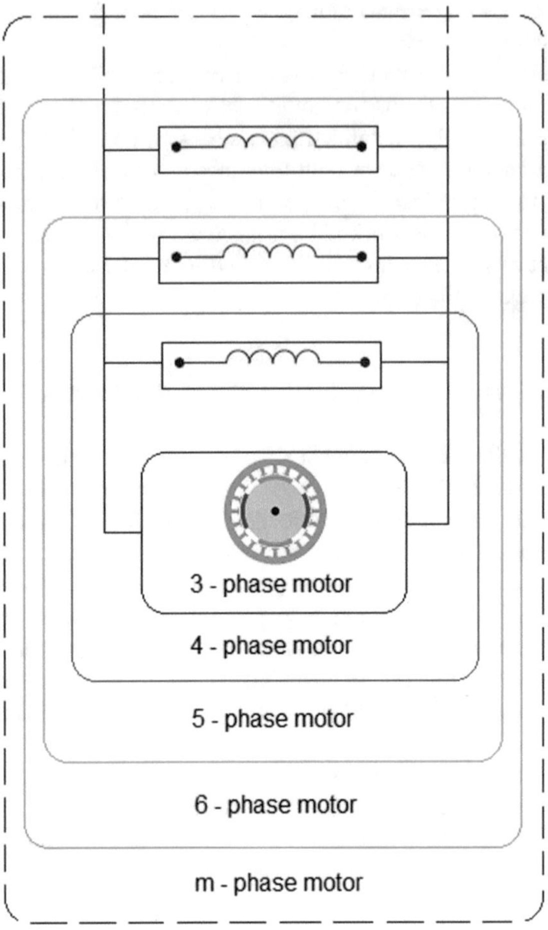

Fig. 12 Structural model of multiphase electric motor

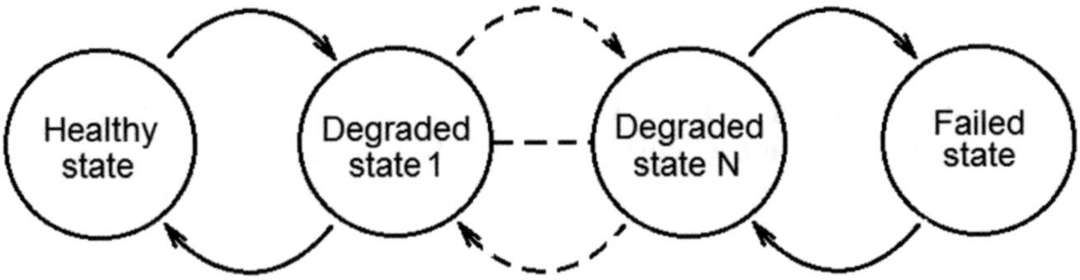

Fig. 13 State space diagram of MSSR MM for the fault tolerance estimation

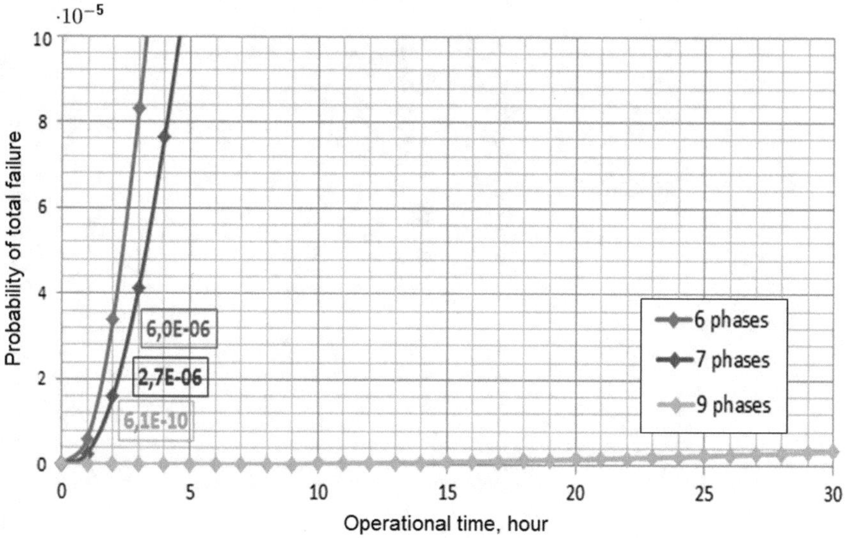

Fig. 14 Probability of total failure of the traction electric motor

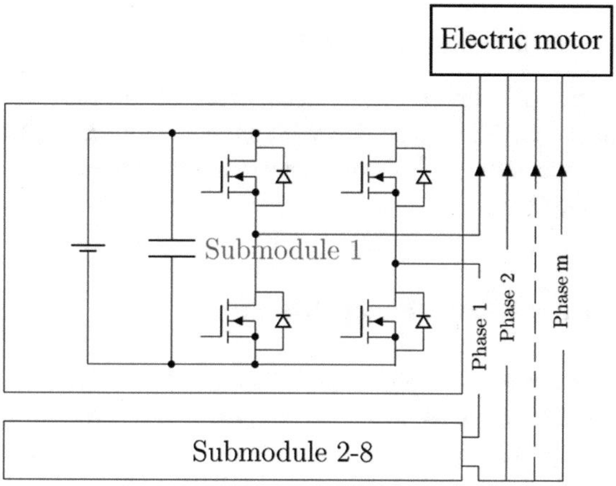

Fig. 15 17-level cascaded H-bridge inverter

3.4.3 The Choice of EM-PE Connection Topology

For the choice of the best topology of connecting 9-phase EM and PE, three well-known topologies were considered, presented in Fig. 17.

They are:

- the 3 × 3-phase system with three star connections (17a),
- the 1 × 9-phase system with one star connection (17b),

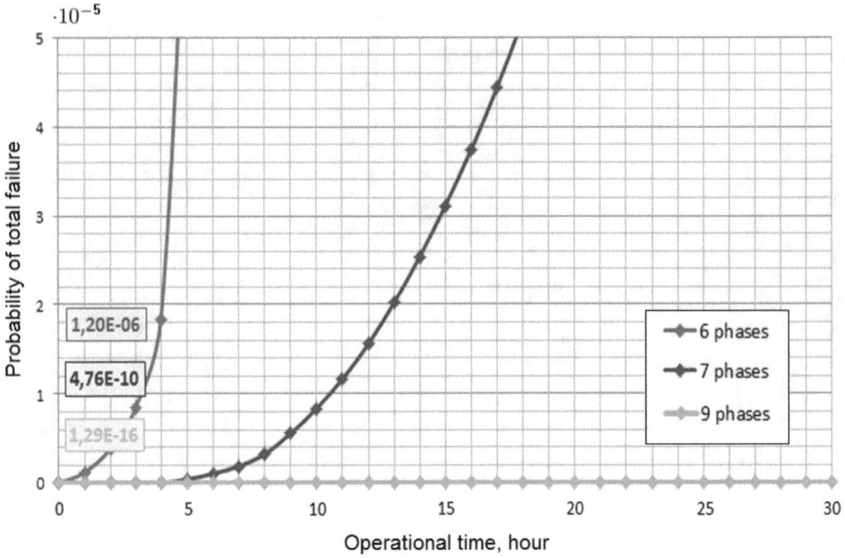

Fig. 16 Probability of total failure of one phase including the multilevel inverter

- the system with 9 galvanically separated phases (17c).

The results of the calculations, presented in Fig. 18, have shown that a modular topology of 9-phase electrical machine with galvanically uncoupled phases and a 17-level CHB inverter satisfies the requirements of the project on the fault tolerance for the propulsion system.

3.4.4 The Choice of Electric Energy Sources

The methods to analyze and improve the reliability of the electrical energy source and of the electric converter are discussed in [23, 27]. In order to meet the design requirements for reliability and fault tolerance as shown in [23], as electric energy sources it is advisable to apply the energy storage, with a matrix topology of battery or fuel cells with more than 22% battery cells' and more than 20% fuel cells' redundancy. The reliability characteristics of all units, taking into account the specific operational load conditions and aging processes, are advisable to be computed by means of the MSSR MM, as shown in [20, 21, 23, 24].

Considering the strict requirements on the level of fault tolerance of the helicopter's electric traction drive, it is advisable to use reconfigurable matrix topology of EES, shown in Fig. 19, which have higher reliability and fault tolerance indices than a conventional option.

Figure 20 presents the states-transitions diagram of the Markov model of EES consisting of NxM battery cells.

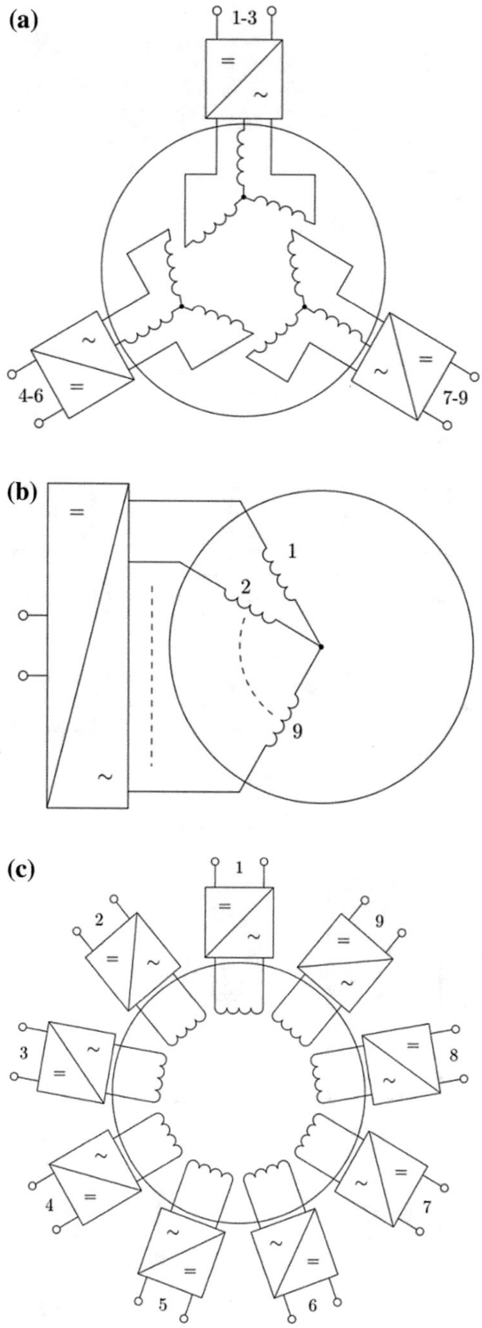

Fig. 17 Possible topologies of connecting EM and PE

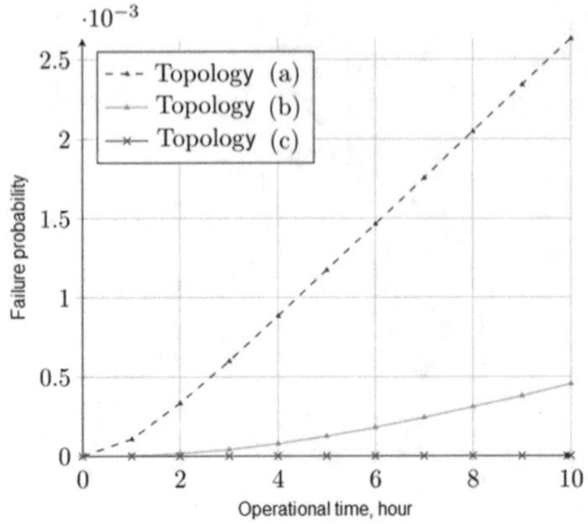

Fig. 18 Probabilities of total failure for different EM-PE topologies

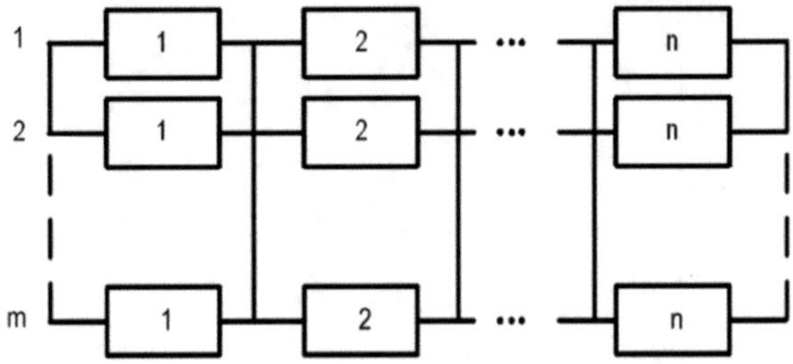

Fig. 19 Reconfigurable matrix topology of EES

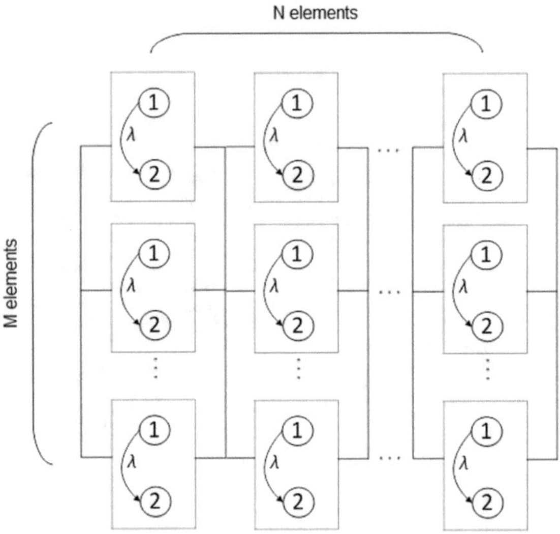

Fig. 20 Reliability block diagram

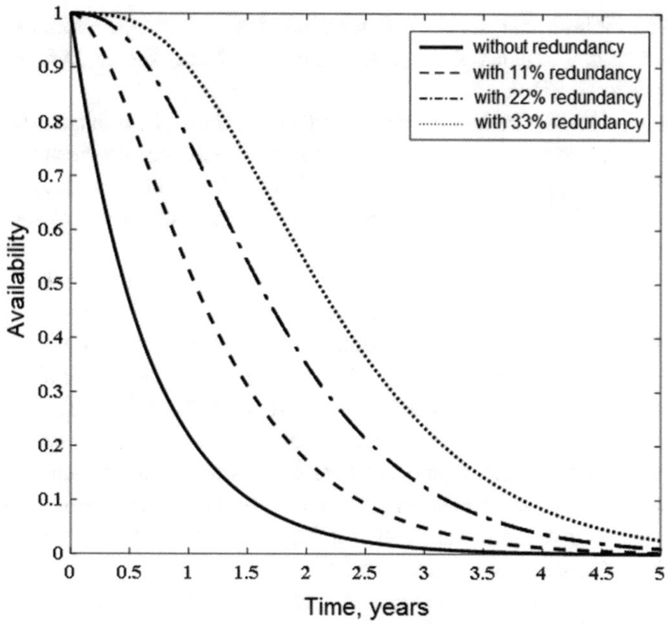

Fig. 21 Operational availability of EES based on Samsung cells

As it was mentioned, each battery cell is a device with two states of performance: a fully operational state with a nominal capacity and a total failure corresponding to a capacity of 0. According to the Markov method and the reliability block diagram (Fig. 20), the following system of differential equations has been constructed:

$$\begin{cases} \frac{dp_{i1}(t)}{dt} = -\lambda p_{i1}(t), \\ \frac{dp_{i2}(t)}{dt} = \lambda p_{i2}(t). \end{cases}$$

Initial conditions are: $p_{i1}(0) = 1,\ p_{i2}(0) = 0$.

Based on L_Z-transform method, the reliability function was calculated for the nominal load level, shown in Fig. 21.

3.5 *Subsystem Level*

At this level, the entire spectrum of technical tasks, which are related to the most important subsystem of an electric vehicle, is solved. The results of solving these problems will allow at higher levels to determine the financial equivalent of an important indicator of the level of excellence of an electric propulsion system—the sustainable functioning.

Such tasks include analysis and optimization of reliability, operational availability, fault tolerance, maintenance strategies, reliability associated cost, and performance of the propulsion system.

When analyzing the reliability characteristics at the SSL, it is necessary to take into account the operational load modes, the mutual influence between the units, the aging processes, the frequency, and the duration of maintenance and repairs, as well as the influence of structural and functional redundancy of the entire subsystem or its particular parts.

The required degree of redundancy of the electric propulsion system of the ice-breaker LNG tanker, depending on the requirements on safety and fault tolerance, can be achieved at the SSL by using multi power electric energy sources (MPEES) consisting of six diesel generator sets. The questions of features and the analysis of the reliability characteristics of MPEES are described in detail in [27, 28].

High survivability and fault tolerance of the electric propulsion system of LNG tanker are especially important in the extremely difficult ice conditions of the Arctic. In order to ensure the safe and sustainable navigation in ice conditions, on the SSL, it is necessary to provide the multi-motor electric drives with multi-phase electric motors, whose features are discussed in [27, 29].

The most comprehensive investigation of reliability indicators at the SSL is advisable to carry out by means of MSSR MM, MRM and MCS. Moreover, taking into account the high complexity of Markov models with a high number of states for the entire electric power system, it is proposed to perform the calculations using the new powerful L_z-transform method, described in detail in [20], which drastically simplified the solution of multiple differential equations.

3.5.1 Choice of the Number of Motors with Different Number of Phases

Performance of the whole propulsion system, i.e. shaft power, is 540 kW. In order to analyze the reliability features of electrical propulsion system of helicopter and to select the best one, four options of the traction drive topology were compared. The compared variants differ in the number of electrical machines and the number of phases of each motor. The structures of helicopter's electric traction drives are shown in Fig. 22 and are as follows:

- Six 3-phase motors, each generate 1/6 of total power;
- Three 6-phase motors, each generate 1/3 of the total power;
- Two 9-phase motors, each generate 1/2 of the total power;
- One 18-phase motors, which generate the total power.

Considering that the considered structures and their component parts (multiphase motors) of the traction drive are a multistate system, it is advisable to use L_z-transform method for their particular research, described in [20, 26, 28].

According to L_z-transform method, any j-component can have k_j different states, corresponding to different performances g_{ji}, represented by the set $\mathbf{g}_j = \{g_{j1}, \ldots, g_{jk_j}\}$, $j = \{1, \ldots, n\}; i = \{1, 2, \ldots, k_j\}$. The performance

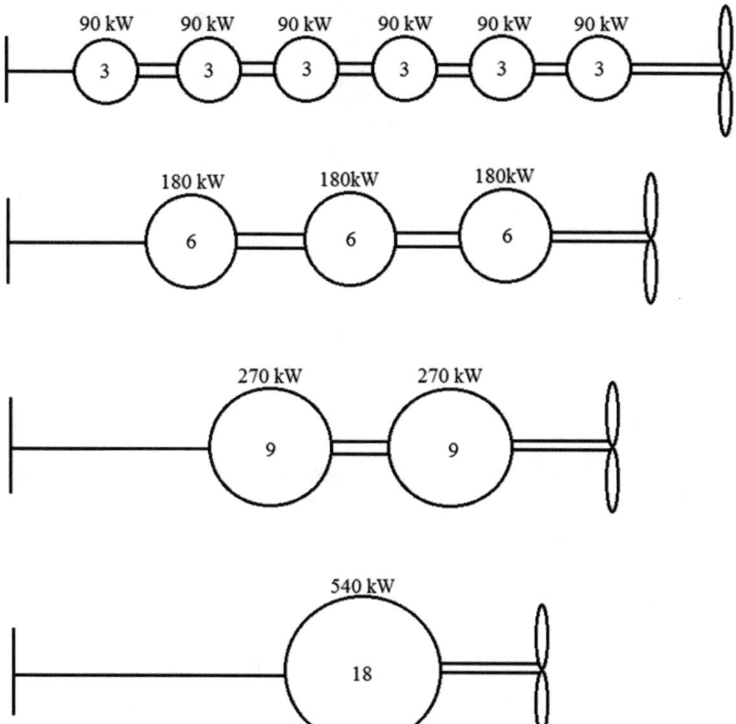

Fig. 22 Structures of various topologies of helicopter's electric drive

stochastic processes $G_j(t) \in \mathbf{g}_j$ and the system structure function $G(t) = f(G_1(t), \ldots, G_n(t))$, that produces the stochastic process corresponding to the output performance of the entire multi-state system, fully define the MSSR MM.

In this chapter, as an example of calculation, an electric drive scheme with three 6-phase traction electric motors is considered. The system's element, the 6-phase motor, has four states: fully working state with a performance of 180 kW, partial failure states with performances of 150 and 120 kW and full failure. The state-space diagram is presented in Fig. 23.

Using MATLAB® for numerical solution of the system of differential equations, it is possible to obtain the probabilities $p_1^{M_6}(t),\ p_2^{M_6}(t),\ p_3^{M_6}(t),\ p_4^{M_6}(t)$. Therefore, for such a system's element the output performance stochastic processes can be obtained as follows:

$$\begin{cases} \mathbf{g}^{M_6} = \left\{g_1^{M_6},\ g_2^{M_6},\ g_3^{M_6},\ g_4^{M_6}\right\} = \{180,\ 150,\ 120,\ 0\}, \\ \mathbf{p}^{M_6}(t) = \left\{p_1^{M_6}(t),\ p_2^{M_6}(t),\ p_3^{M_6}(t),\ p_4^{M_6}(t)\right\}. \end{cases}$$

Sets $\mathbf{g}^{M_6}$, $\mathbf{p}^{M_6}(t)$ define L_z-transforms for 6-phase motor as follows:

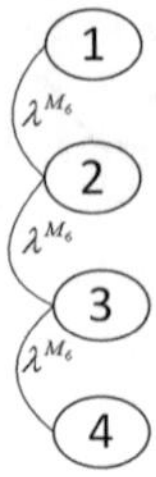

Fig. 23 State space diagram of 6-phase motor

$$\begin{aligned} L_z\{g^{M_6}(t)\} &= p_1^{M_6}(t)z^{g_1^{M6}} + p_2^{M_6}(t)z^{g_2^{M6}} + p_3^{M_6}(t)z^{g_3^{M6}} + p_4^{M_6}(t)z^{g_4^{M6}} \\ &= p_1^{M_6}(t)z^{180} + p_2^{M_6}(t)z^{150} + p_3^{M_6}(t)z^{120} + p_4^{M_6}(t)z^{0}. \end{aligned} \quad (3)$$

Multi-state model for the multi-motor electric traction drive may be presented as connected in parallel three 6-phase electric motors, shown in Fig. 24.

Therefore, the whole system L_Z-transform is as follows:

$$L_z\{G^{SysM_6}(t)\} = \Omega_{f_{par}}\left(L_z\{g^{M_6}(t)\}, L_z\{g^{M_6}(t)\}, L_z\{g^{M_6}(t)\}\right) \quad (4)$$

Using the composition operator $\Omega_{f_{par}}$, the L_Z-transform $L_z\{G^{SysM_6}(t)\}$ of the in-parallel-connected three identical 6-phase motors, can be obtained.

Table 2 presents the failure rates for the electric motors shown in Fig. 22.

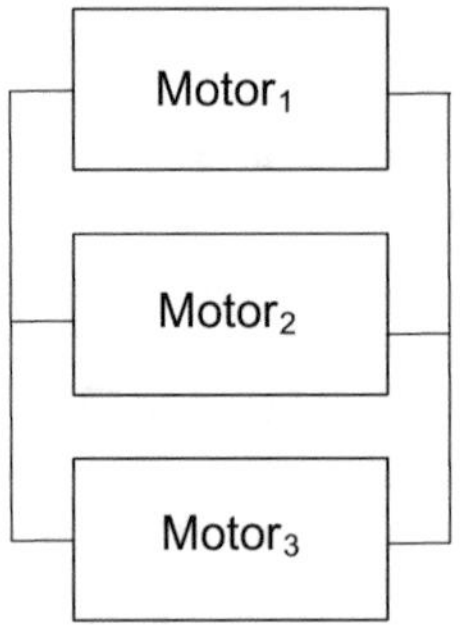

Fig. 24 Reliability block diagram of multi-motor drive with three 6-phase electric motors

Table 2 Failure rates of each element

	Failure rates ($year^{-1}$)
3-phase motor	0.09
6-phase motor	0.15
9-phase motor	0.21
18-phase motor	0.33

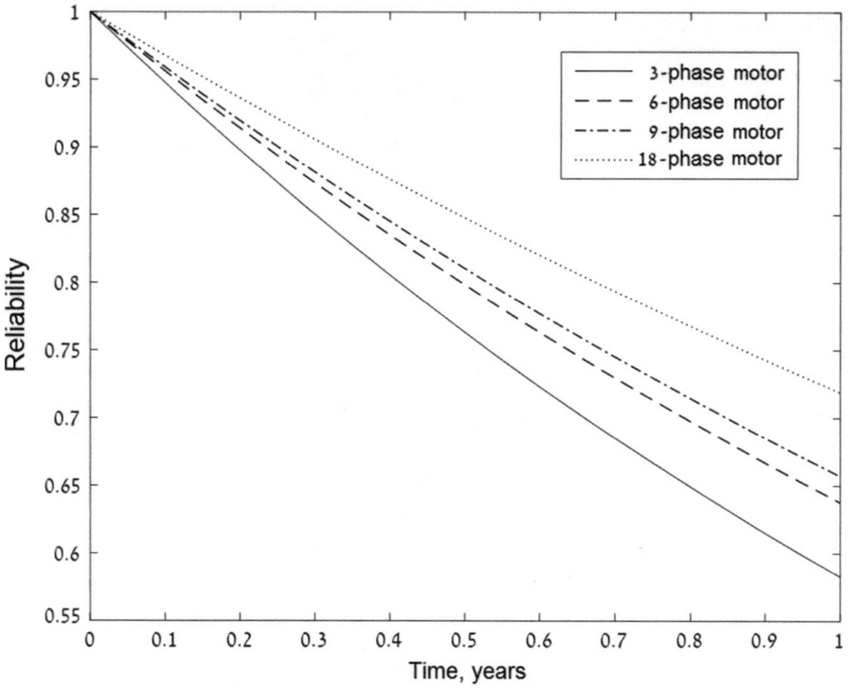

Fig. 25 Probability of the failure free operation for 100% load level

The obtained results of reliability calculation for the 100% load mode are shown in Fig. 25.

Such topologies allow to realize the required value of the performance and as a consequence, the high survivability of the helicopter (or other electric vehicles) with the possible occurrence of critical failures of the electric propulsion system.

3.5.2 The Model of the Whole Electric Propulsion System of the Helicopter

The resulting state space diagram of the MSSR MM of electric propulsion system of the helicopter, taking into account the impact of the human factor (HF), is shown in Fig. 26.

In Fig. 26 the blue state 0 of the graph corresponds to a full failure-free operation of all traction drive components. States 1–15 correspond to the partial failures of the elements of propulsion system with the partial loss of their functionality. The red state 16 represents the total failed electric traction drive and the inability of the helicopter to realize a safe flight.

Figure 27 presents the results of simulation on the Markov model, based on the state space diagram of Fig. 26.

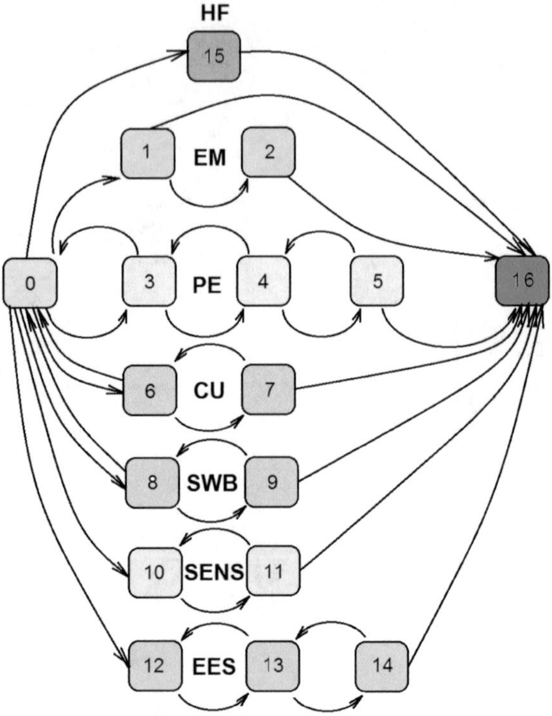

Fig. 26 State space diagram of helicopters' electric propulsion system

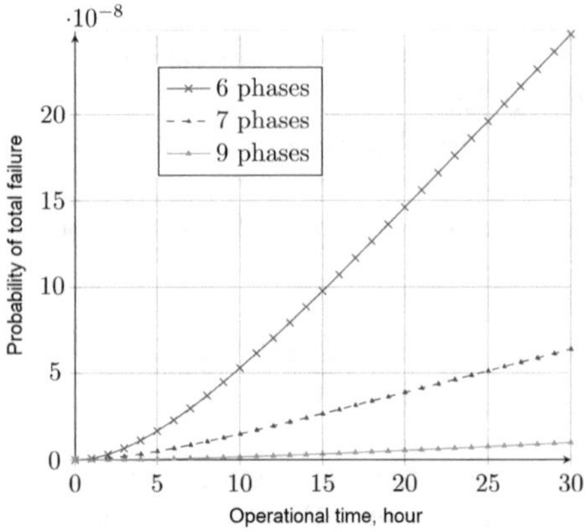

Fig. 27 Probability of total failure of electric propulsion system

For the considered topology of the helicopter's electric traction drive, which includes a multiphase traction motor, electric multilevel inverter, and 100% redundancy of SWB, SENS, and CU, the best option in terms of design requirements on the fault tolerance is the topology with a 9-phase electric motor, the 17-level CHB inverter and matrix topology of EES with partial redundancy.

Based on such MSSR MM the reliability characteristics of electric helicopter can be estimated and improved in accordance with the design requirements.

3.6 System Level

At the SL, the operation of the ship with electric propulsion subsystem as a whole system is considered, since this application is more informative at the SL compared to the helicopter. The operational conditions include several uncertainties and many random parameters. This fact has a significant influence on the comprehensive reliability characteristics of the Artic ship.

The objective function of the icebreaker LNG tanker is the safely, sustainable, and efficient shipping in the specified Arctic operating conditions. In accordance with this, the main objectives are to increase the carrying capacity of the tanker and to minimize the total operating costs and damages. The reliability characteristics of the icebreaker LNG tanker influence the values of both components of the objective function of the ship. In order to solve these problems, it is advisable to use MCS and MCDA, considering the random environment of the Arctic navigation conditions and the number of uncertainties, along with MSSR MM and MRM.

In this way, at the SL, it is recommendable to determine all reliability indicators of the whole tanker. Based on such reliability indices, the total cost can be calculated, which is needed to maintain sustainably the required level of performance during the operation of the tanker in real ice operating conditions. These are the operational availability, performance, deficiency of performance, maintainability, reliability associated cost, damages from unreliability, life cycle cost, risk probability, etc.

In order to improve the reliability and fault tolerance of the electric propulsion system and the LNG tanker as a whole, at this level, it is possible to use several autonomous electric drives with their own screws, to use the propulsion system of the gondola type with two screws, to optimize the maintenance and repair strategies of the power system of the tanker during navigation, and to use predictive reliability monitoring and a reliability control system of the ship electrical propulsion system.

In order to build the model of the LNG tanker life cycle at the SL, the process of the icebreaker LNG tanker operations is represented by a chain of the different operating modes. During the operation cycle depending on conditions of navigation, it is possible to distinguish four basic operating modes of an icebreaker LNG tanker. Each of them corresponds to a certain required number and power of the main engines. These operating modes are shown in Fig. 28 and defined as follows:

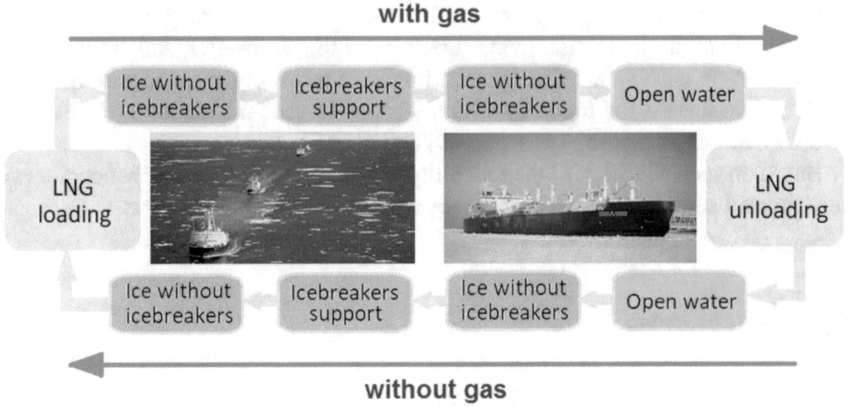

Fig. 28 Operational modes of icebreaker LNG tanker

- Loading and unloading of LNG at the terminal. Each of these two modes usually takes about 24 h. Sustainability of the loading and unloading process is determined by the reliability of onshore and ship gas liquefying and pumping systems.
- Navigation of a ship in ice-free water. The operation in this mode depends on the required velocity and needs the greater part of the operational time 50–80% of the nominal generated power.
- Autonomous movement in the ice without icebreaker support. The navigation in this mode depends on ice conditions and a wide power range from 50% up to 100% of the nominal power can be used.
- Navigation of a ship in heavy ice supported by icebreakers. In order to realize sustainable joint operation with icebreakers in this mode, electric propulsion system needs 80–100% of the nominal generated power.

Considering the abovementioned features of operational modes of the icebreaker LNG tanker propulsion system, three demand levels were chosen for calculation: 100, 80, and 50% of the main traction electric motors power.

For an accurate assessment of operational availability and performance of the electric propulsion system, it has been proposed to estimate the values separately for each of the above modes, followed by calculating the total impact on the value of the ship's operating speed and, accordingly, the amount of cargo transported per unit of time.

In order to analyze the reliability indicators at the system level of the MLHRM, the icebreaker LNG tanker power system—based on the decomposition principle—is presented in the form of four blocks: the electric energy source system (EES), the ship's electric propulsion system (EPS), the subsystem of the ship's consumers of electric energy (EEC) and LNG liquefaction and storage system (LSS). The simplified structure of the whole LNG tanker power system is shown in Fig. 29.

As a result of calculating the comprehensive reliability indices of each functional block, indicated in Fig. 29, based on the L_Z-transform method to solve the system

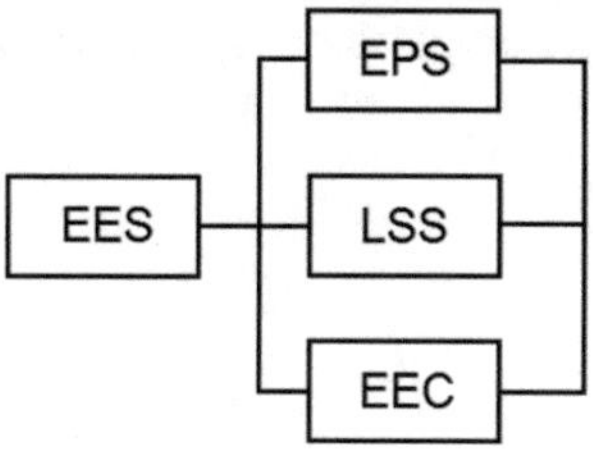

Fig. 29 Structure of the hybrid-electric power system of LNG tanker

of differential equations of MSSR MM, a schedule of operational availability of the power system of LNG tanker for different demands was simulated, which is presented in Fig. 30.

The graph of Fig. 28 demonstrates the ability of the tanker's power system to ensure sustainable functioning under the conditions of various operational demands. For this, the process of operating a fully loaded tanker during LNG delivery from the Sabetta terminal on the Russian Yamal Peninsula to the Chinese port of Shanghai was modeled. As can be seen from Fig. 28, the Arctic LNG tanker has high operational availability for the maximum levels of demand. Its value is equal to 85.82%. This

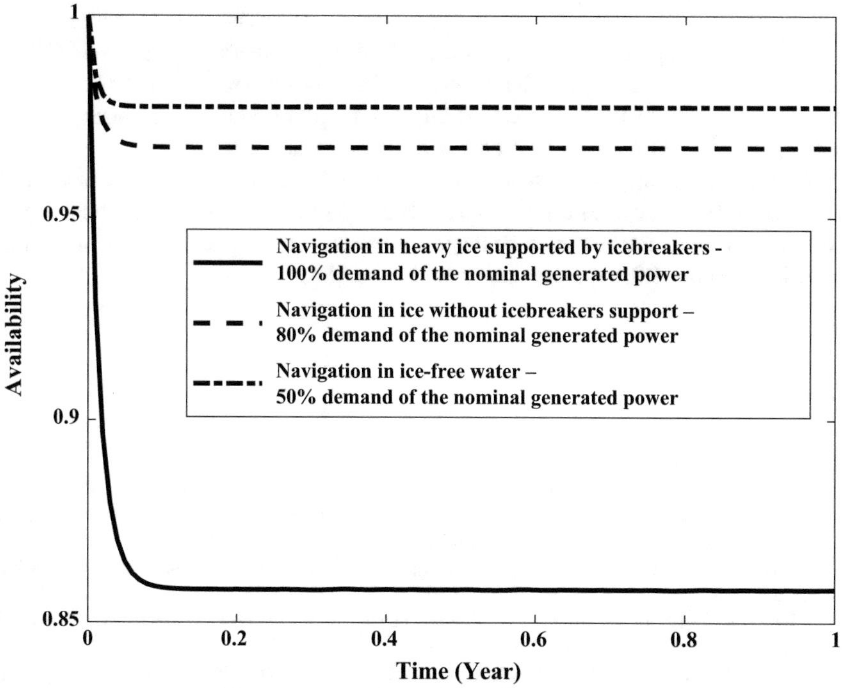

Fig. 30 Operational availability of the power system of LNG tanker for different demands

indicates that such multi-drive propulsion system closely related to the conditions of ice navigation.

4 Conclusions

It can be concluded that MLHRM and methodology of its application will allow to realize the overall analysis and estimation of comprehensive reliability characteristics of the vehicle electric propulsion systems at the design stage. It means to implement the so-called reliability oriented design of the traction electric drives. The suggested MLHRM of the vehicle's life cycle allows for each level to solve specific technical and technical-economical optimization tasks, such as optimization of the design of the electric machine, number of phases, number of electric motors, degree of fault tolerance, level of redundancy, maintenance strategy, topologies of electric converters, and electric energy sources.

The MLHRM approach allows to provide a quantitative comparative analysis of methods for improving the comprehensive reliability of the vehicle electric propulsion systems at each MLHRM level. In other words, in order to quantify the impact on the integrated reliability of the electric propulsion system and vehicle as whole, it is possible to use systems of diagnostics, fault detection, monitoring, fault prediction, varying degrees of redundancy of elements, and various maintenance strategies.

Two different application cases, namely, electric propulsion system for SAR helicopter and diesel-electric propulsion system of icebreaker Arctic LNG tanker, testify to the universality of the proposed MLHRM and appropriate methodology, as well as the possibility of its application for various technical systems.

In further studies, it is advisable to estimate the value of the reliability associated costs, as well as life cycle costs of Arctic LNG tanker for different operational routes by using different maintenance strategies, considering the gradual deterioration of the ship's icebreaking capacity during ice navigation.

References

1. Bani-Mustafa T, Pedroni N, Zio E, Vasseur D, Beaudouin F (2017) A hierarchical tree-based decision making approach for assessing the trustworthiness of risk assessment models. In: Proceedings of the international topical meeting on probabilistic safety assessment and analysis (PSA'17), 24th–28th Sept 2017, Pittsburgh, PA, pp 314–323
2. Saaty TL (2008) Decision making with the analytic hierarchy process. Int J Serv Sci (IJSSci) 1(1):83–98
3. Ziemba P (2019) Inter-criteria dependencies-based decision support in the sustainable wind energy management. Energies 12(749):1–29
4. Ganji SRS, Rassafi AA, Kordani AA (2018) Vehicle safety analysis based on a hybrid approach integrating DEMATEL, ANP and ER. KSCE J Civ Eng 22(11):4580–4592
5. Brown RE, Gupta S, Christie RD, Venkata SS, Fletcher R (1996) Distribution system reliability assessment using hierarchical markov modeling. IEEE Trans Power Delivery 11(4):1929–1934

6. Ataie E, Entezari-Maleki R, Rashidi L, Trivedi KS, Ardagna D, Movaghar A (2017) Hierarchical stochastic models for performance, availability, and power consumption analysis of IaaS clouds. IEEE Trans Cloud Comput 1–18
7. Nam T, Mavris DN (2018) Multistage reliability-based design optimization and application to aircraft conceptual design. J Aircr 1–15
8. Paulson EJ, Starkey RP (2013) Development of a multistage reliability-based design optimization method. J Mech Des 136(1):1–8
9. Wikström P, Terens LA, Kobi H (2000) Reliability, availability, and maintainability of high-power variable-speed drive systems. IEEE Trans Ind Appl 36(1):231–241
10. Bolvashenkov I, Kammermann J, Herzog H-G (2016) Research on reliability and fault tolerance of multi-phase traction electric motors based on markov models for multi-state systems. In: Proceedings of 23rd international IEEE symposium on power electronics, electrical drives, automation and motion (SPEEDAM), 22th–24th June 2016, Anacapri, Italy, pp 1–6
11. Linkov I, Fox-Lent C, Read L, et al (2018) Tiered approach to resilience assessment. J Risk Anal 1–9. https://doi.org/10.1111/risa.12991
12. Woo S, O'Neal DL (2019) Reliability design and case study of mechanical system like a hinge kit system in refrigerator subjected to repetitive stresses. Eng Fail Anal 99:319–329
13. Xiao N, Huang N-Z, Li Y, He L, Jin T (2011) Multiple failure modes analysis and weighted risk priority number evaluation in FMEA. Eng Fail Anal 18:1162–1170
14. Ding Y, Lin Y, Peng R, Zuo MJ (2019) Approximate reliability evaluation of large-scale multistate series-parallel systems. IEEE Trans Reliab 1–15
15. Xuy X, Liz Z, Chen N (2016) A hierarchical model for lithium-ion battery degradation prediction. IEEE Trans Reliab 65(1):310–325
16. Abbas M, Vachtsevanos GJ (2009) A system-level approach to fault progression analysis in complex engineering systems. In: Proceedings of annual conference of the prognostics and health management society, 27 Sept–1 Oct 2009, San Diego, CA, pp 1–7
17. Gomes JPP, Rodrigues LR, Galvão RKH, Yoneyama T (2013) System level RUL estimation for multiple-component systems. In: Proceedings of annual conference of the prognostics and health management society, 14th–17th Oct 2013, New Orleans, LA, USA, pp 1–9
18. Rodrigues LR (2017) Remaining useful life prediction for multiple-component systems based on a system-level performance indicator. IEEE/ASME Trans Mechatron 1–10
19. Bolvashenkov I, Kammermann J, Willerich S, Herzog H-G (2016) Comparative study of reliability and fault tolerance of multi-phase permanent magnet synchronous motors for safety-critical drive trains. In: Proceedings of the international conference on renewable energies and power quality (ICREPQ'16), 4th–6th May, Madrid, Spain, pp 1–6
20. Bolvashenkov I, Herzog H-G, Frenkel I, Khvatskin L, Lisnianski A (2018) Safety-critical electrical drives: topologies, reliability, performance. Springer, Berlin
21. Bolvashenkov I, Kammermann J, Willerich S, Herzog H-G (2015) Comparative study for the optimal choice of electric traction motors for a helicopter drive train. In: Proceedings of the 10th conference on sustainable development of energy, water and environment systems (SDEWES'15), 27th Sept–3rd Oct 2015, Dubrovnik, Croatia, pp 1–15
22. Kammermann J, Bolvashenkov I, Herzog H-G (2017) Reliability of induction machines: statistics, tendencies, and perspectives. In: Proceedings of 26th IEEE international symposium on industrial electronics (ISIE), 19th–21th June 2017, Edinburgh, UK, pp 1843–1847
23. Bolvashenkov I, Frenkel I, Kammermann J, Herzog HG (2017) Comparison of the battery energy storage and fuel cell energy source for the safety-critical drives considering reliability and fault tolerance. In: Proceedings of IEEE international conference on information and digital technologies (IDT), 5th–7th July 2017, Žilina, Slovakia, pp 63–70
24. Bolvashenkov I, Kammermann J, Herzog H-G (2016) Methodology for determining the transition probabilities for multi-state system markov models of fault tolerant electric vehicles. In: Proceedings of the Asian IEEE conference on energy, power and transportation electrification, 25th–27th Oct 2016, Singapore, pp 1–6
25. Bolvashenkov I, Kammermann J, Herzog HG, Frenkel I (2019) Operational availability and performance analysis of the multi-drive multi-motor electric propulsion system of an icebreaker

gas tanker for Arctic. In: Proceedings of IEEE 14th international conference on ecological vehicles and renewable energies (EVER'19), 8th–10th May 2019, Monaco, pp 1–6
26. Bolvashenkov I, Kammermann J, Herzog H-G, Frenkel I, Ikar E, Khvatskin L (2017) Investigation of reliability and fault tolerance of multiphase traction electric motor supplied with multi power source based on Lz-transform. In: Proceedings of IEEE international conference on system reliability and safety (ICSRS'17), 20th–22th Dec 2017, Milano, Italy, pp 303–309
27. Bolvashenkov I, Herzog H-G (2016) Use of stochastic models for operational efficiency analysis of multi power source traction drives. In: Proceedings of the second IEEE international symposium on stochastic models in reliability engineering, life science and operations management, (SMRLO), 15th–18th Feb 2016, Beer Sheva, Israel, pp 124–130
28. Frenkel I, Bolvashenkov I, Herzog H-G, Khvatskin L (2017) Operational sustainability assessment of multi power source traction drive. In: Ram M, Davim JP (eds) Mathematics applied to engineering. Elsevier, London, UK, pp 191–203
29. Bolvashenkov I, Kammermann J, Herzog H-G, Frenkel I (2017) Fault tolerance assessment of multi-motor electrical drives with multi-phase traction motors based on LZ-transform. In: Proceedings of IEEE 14th international conference on ecological vehicles and renewable energies (EVER'19), 8th–10th May 2019, Monaco, pp 1–6

Graph Theory Based Reliability Assessment Software Program for Complex Systems

Abdrabbi Bourezg and Hamid Bentarzi

Abstract Reliability is a conceptual term that means endurance, dependability, and good performance. However, in system engineering, it is more than a conceptual term; it can be measured and evaluated. Reliability means the ability of a system to perform the required task under the normal conditions during its age. A complexity in system reliability may be commonly arisen due to the interconnection of various elements in the form of a network that can be represented by graphs. The graph theory and computer programs are essential tools for analyzing large and complex systems. This chapter presents how a complexity of system reliability can be reduced through the use of computer programs based on a graph theory. The software program has been developed for reliability assessment of complex systems such as aircraft. It can handle any statistical distributions. It uses the inclusion-exclusion method for finding the minimal paths for directed acyclic graph using reliability block diagram (RBD). A system may be considered to operate if there exists a set of functioning components from source to target. So, at least one minimal path must function for the system operation. The probability of the union of all minimal paths can be used to find the reliability of the whole system.

Keywords Graph theory · Reliability · Failure distribution function · Reliability block diagram (RBD)

A. Bourezg
Higher Colleges of Technology, Dubai, UAE
e-mail: abourezg@hct.ac.ae

H. Bentarzi (✉)
Laboratory Signals and Systems, IGEE, University M'hamed Bougara Boumerdes, Boumerdes, Algeria
e-mail: lss@univ-boumerdes.dz

M. Ram and H. Pham (eds.), *Advances in Reliability Analysis and its Applications*, Springer Series in Reliability Engineering,
https://doi.org/10.1007/978-3-030-31375-3_6

1 Introduction

In the reliability engineering, reliability is more than an abstract term that means endurance, dependability, and good performance; it can be assessed and evaluated. It can be defined as the ability of a system (part or component) to perform the attended task under the specified conditions during specified period [1, 2].

Since a failure may occur in all industrial items and systems, adequate monitoring of it becomes crucial, and hence reliability analysis is an important step in all engineering fields especially in the most aware and dangerous one such as airspace and the nuclear industry. The concept of reliability is highly based on the probability theory.

A complexity of any systems reliabilities may be commonly arisen due to the interconnection of various elements in the form of a network or a graph. The use of graph theory and computer programs is important tool for analyzing these systems reliabilities as they become larger or complex [3–6]. A path enumeration method such as directed acyclic graph for representing the system reliability block diagram (RBD) associated with probability union law may be used to develop a Matlab graphical unit interface (GUI) application. The developed application named (Reliability_App.) can give to the user the most important parameters of reliability through the use of different statistical distribution functions for subsystems or units of any system.

2 Reliability Modeling

2.1 *What Is Reliability?*

Mathematical and statistical methods can be used for quantifying reliability (evaluation, prediction and measurement) and for analyzing reliability data. The reliability evaluation techniques first have been developed for the aerospace industry and military applications. Then, they have been followed by the nuclear industry, which requires now a high safety and hence reliability especially nuclear reactors. All these applications faced severe problems such as aerospace (Challenger space shuttle, in 1986; several commercial aircraft accidents such as recent ones such an Ethiopian Airlines Max 8 crash killing all 157 on board, less than five months before a Lion Air plane of the same type came down in Indonesia claiming 189 lives), nuclear (Three Mile Island, in 1979; Chernobyl, in 1986), electricity supply (USA blackout, in 2003), process plant (Bhopal, in 1984). In all these catastrophic events, severe social and environmental penalties have been resulted such as many deaths. These events have considerably increased pressure to objectively assess reliability, safety and risk [3].

It should be stated here, that the reliability theory is nothing more or less than a mathematic engineering judgment, that is, long-term engineering experience collected during the development of humankind, transformed into numerical models. This transformation would never be possible without:

- Mathematical statistics and probability theory,
- Thinking in terms of systems,
- Introduction of time as an additional dimension.

These three fundamental features of the reliability theory and its applications should continually be kept in mind by all who want to master the reliabilistic approach to the problems of constructed facilities [4].

The probability ($T > t$) that is the time to failure T is greater than a specified time t is given by the reliability function [5]:

$$\boldsymbol{R(t) = P(T > t)} \tag{1}$$

Also referred to as survival function, The reliability function is a monotonic non-increasing function, always unity at the beginning of age [$(0) = 1$, $R(\infty) = 0$].

It is linked with unreliability function or the cumulative distribution function (CDF) (t) (or $Q(t)$) of the time to failure by [5]:

$$\text{Reliability} = 1 - \text{probability of failure}$$

$$\mathrm{R}(\boldsymbol{t}) = \boldsymbol{1} - \boldsymbol{F(t)}, \tag{2}$$

If T is the time to failure, $F(t)$ gives the probability $P(T \le t)$ that the system or component will fail before time t.

The probability density function (PDF) of the time to failure is denoted by (t). It describes how the failure probability is distributed as function of time. In the infinitesimal interval [t, $t + dt$], the probability of failure is f(t) dt [5].

The probability of failure in any specified time interval $t_1 \le T \le \boldsymbol{t_2}$ can be written as follows:

$$P(t_1 \le T \le t_2) = \int_{t_1}^{t_2} f(t)dt \tag{3}$$

Basic properties of the PDF are: (i) it is always non-negative and (ii) the total area beneath f(t) is always equal to one (the item will certainly fail). The cumulative distribution function of the time to failure and the probability density function are related to each other by (see Fig. 1) [5]:

$$\mathrm{f}(\boldsymbol{t}) = \boldsymbol{dF(t)/dt} \tag{4}$$

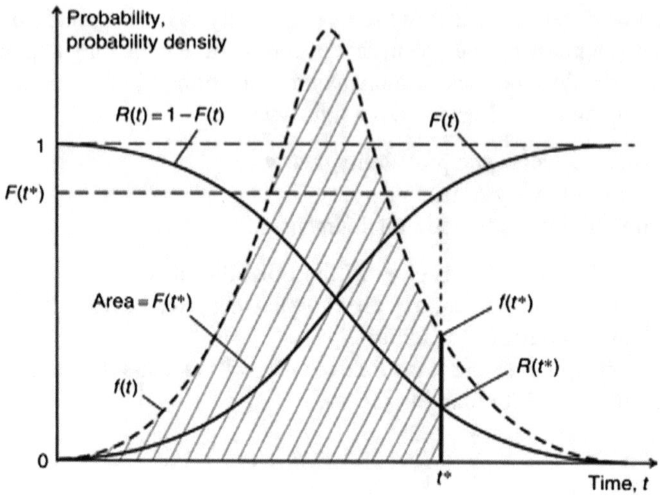

Fig. 1 Reliability function, cumulative distribution of the time to failure and failure density function

and,

$$F(t) = \int_{-\infty}^{t} f(\xi)d\xi \tag{5}$$

Usually, in reliability, Eq. (5) can simply be written as:

$$F(t) = \int_{0}^{t} f(\xi)d\xi \tag{6}$$

Since the system should not fail before time = 0.

2.2 *Statistical Distributions*

In mathematical reliability analyses, various types of probability or statistical distributions function may be used as illustrated in Fig. 1.

In all practical cases, the appropriate probability distribution can be determined from sample testing or from a data collection scheme associated with the operation of the components, devices or systems [3].

These distributions are needed to formulate [2]:

- The probability that a component will start (fail) on demand,

- The probability that a component will run for a period of time given a successful start,
- The impact of repair on these probabilities,
- The frequency of initiating events.

3 Graph Theory

Graph theory has drawn increased interest of scientists and engineers in the last decades due to its demonstrated ability to solve problems of a variety of fields. Graphs have been found highly useful for modeling systems such as physical science, engineering, social sciences and even economic problems as well as reliability engineering due to their intuitive diagrammatic representation [7].

The application of the graph theory for network reliability evaluation was proposed by Misra and Rao in 1970, then many studies have been performed, and a number of algorithms, techniques and approaches have been proposed in the literature. In fact, today, the use of graph theory becomes inevitable in network reliability evaluation [7].

A graph $G = (V, E)$ consists of a set of vertices $|V|$ (or nodes$n$) and a set of edges $|E|$ (or links$e$) as illustrated in Fig. 2. If an edge connects two vertices i and j; is said to be adjacent to i [7]. In the obtained graph G, each edge has an orientation. Obviously, a source node would not have any edge incidents on it whereas a destination node (target) would not have any edge emerging out of it. An arc from node i to node j is represented as an ordered pair (i, j), where "i" is called the tail and "j" is called the head of the arc [7].

In order to describe the graph G, a Boolean matrix can be used as shown in Fig. 3, the adjacency matrix is a matrix of size $n \times n$, [7]:

$$a_{ij} = 1 \text{ if there is an edge from i to vertex } j \tag{7a}$$

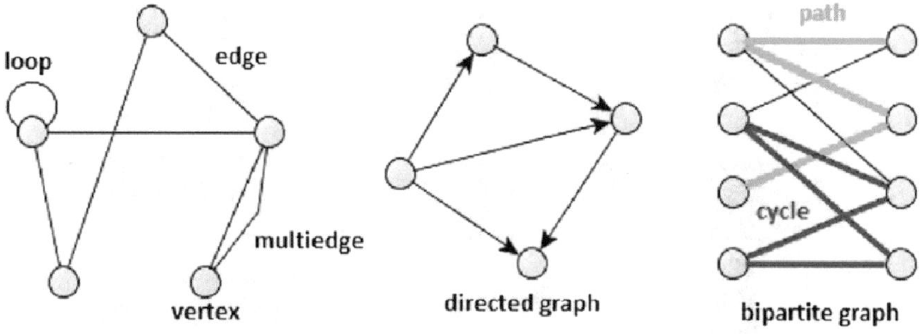

Fig. 2 Graph representation of system reliability

Fig. 3 Graph G with adjacency matrix of complex system

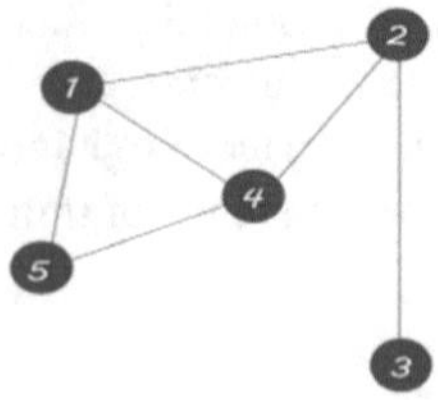

$$A = \begin{pmatrix} 0 & 1 & 0 & 1 & 1 \\ 1 & 0 & 1 & 1 & 0 \\ 0 & 1 & 0 & 0 & 0 \\ 1 & 1 & 0 & 0 & 1 \\ 1 & 0 & 0 & 1 & 0 \end{pmatrix}$$

$$a_{ij} = 0 \text{ if there is no edge from i to vertex } j \quad (7b)$$

Different approaches require that all possible path sets are first enumerated and then the reliability expression is driven. Thus, a path is a subset of components, so that the system can work if all elements of the subset are on.

A path *P* is a minimal path (min path) if it does not exist a subset of P or a path (i.e. it does not have proper sub paths). Figure 4 shows Reliability Block Diagram of a system represented by its minimal path subsets.

A single component is considered in a Reliability Block Diagram (RBD), a "flow" is moving from a source, through the component, to a target. The component is supposed to be functioning if the flow can pass through it without checking. However, if the component has failed, the flow is barred to reach the target [8].

The "flow" concept can be extended to a whole system. A system is considered to function if there exists a set of functioning components that permits the flow from source to target. In an RBD, a path is defined as a set of functioning components that guarantees a functioning system. Since a directed acyclic graph has been selected to

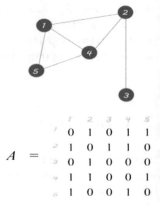

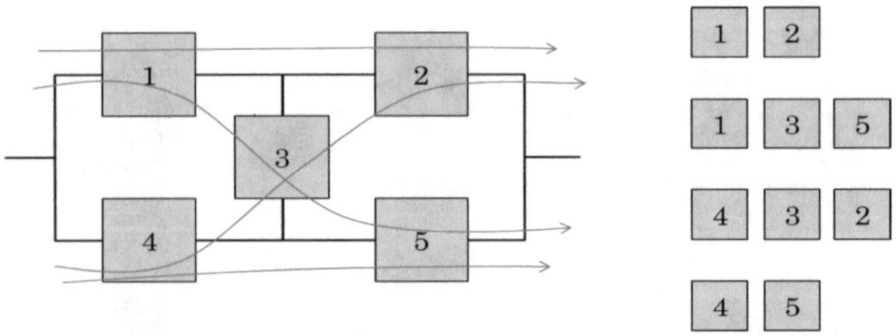

Minimal path set indicated by the arrows.

Minimal path subsets

Fig. 4 Reliability block diagram of a system represented by its minimal path subsets

be employed to represent a system's RBD, all paths are minimal. It can be noted, all components are distinct in a path [8].

After obtaining the minimal paths of a system's RBD, the principle of inclusion-exclusion for probability evaluation can be used to generate an accurate analytical expression for reliability. Let $\{A_1, A_2, \ldots, A_M\}$ be the set of all minimal paths of a system. At least one minimal path must function in order the system to operate. The reliability of the system R_s as the probability of the union of all minimal paths can be written as follows [8]:

$$R_s = P\left\{\bigcup_{i=1}^{M} A_i\right\} \quad (8)$$

4 Reliability Assessment Software Program

A system's RBD can be represented by a directed acyclic graph (DAG) with a source vertex s and a target vertex t. Matlab GUI application (Reliability_App.m) can be developed to give an exact analytical expression for the reliability and provide the user of some useful reliability information about the system.

Figure 5 shows an algorithm that may be used to calculate the reliability of a complex system using DAG starting by the determination of the reliability of every path. The steps of this algorithm are:

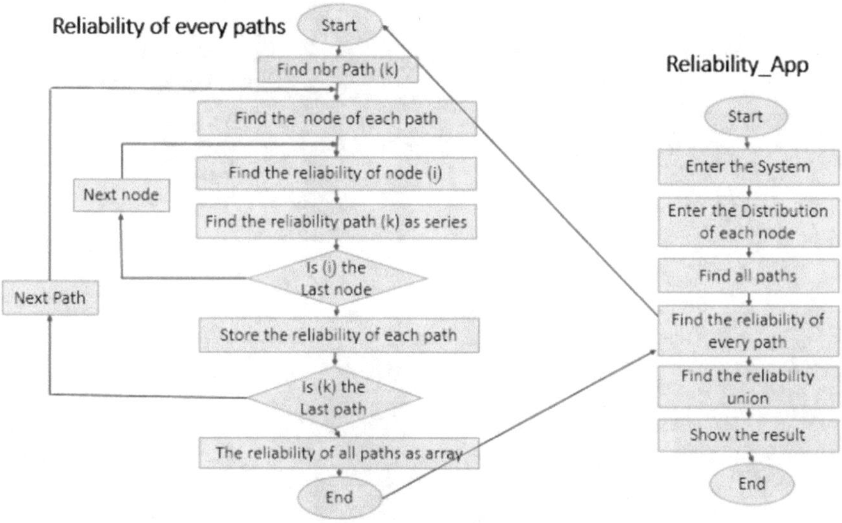

Fig. 5 Flowchart of the algorithm used to calculate reliability of complex system

(a) Compute the reliability of each node by taking into consideration its distribution,
(b) Compute the path reliability as series configuration,
(c) Finally, find the reliability of the whole system as the probability of the union of all minimal paths.

Reliability_Network_Eval_Funct.m (found in toolbox) can be used to proceed all steps that mentioned previously. In step one, finding all paths in a directed acyclic graph G between two arbitrary vertices, $t \in V$. Where s is referred as the source and as the target. Since a directed acyclic graph is selected, all paths are simple (all vertices in a path are distinct). It visits a vertex and then recursively visit all of its neighbors. Then, the current vertex to the target is compared. If it does not match, it continues to traverse the graph. Otherwise, the target has been reached and the path for later output is recorded. For a given directed acyclic graph, a source vertex s, and a target vertex *t*, Find_Paths_Funct.m returns a matrix its rows are all paths connecting s to t [8].

The application includes also a weakness elements algorithm based on intersection method to define which element is more sensitive, also its handles the computational of the MTTF by using Simpson's rules. and elapsed time function to evaluate the running time for Path Finding and system reliability. The profiler proved execution time for all executed functions, as proved in this application.

Figure 6 shows the GUI which contains three panels labeled (1) Your System, (2) Distribution and (3) Reliability.

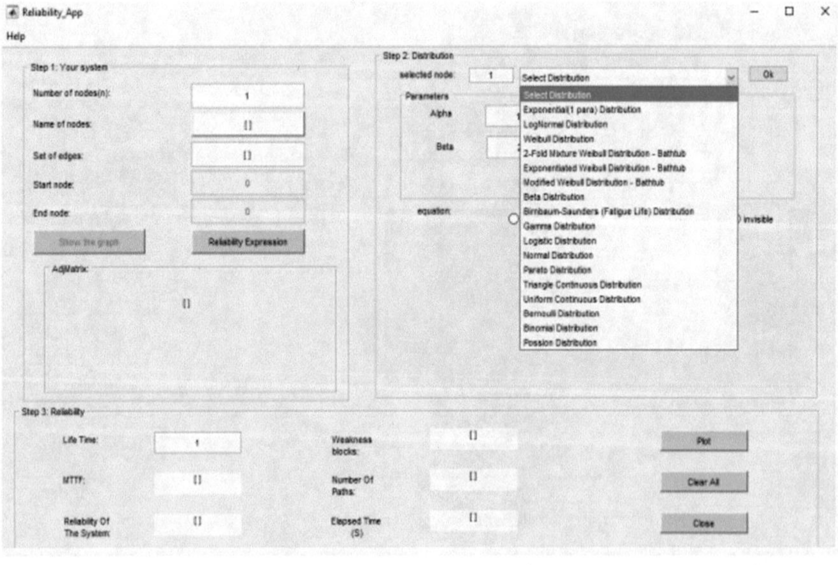

Fig. 6 GUI application front view

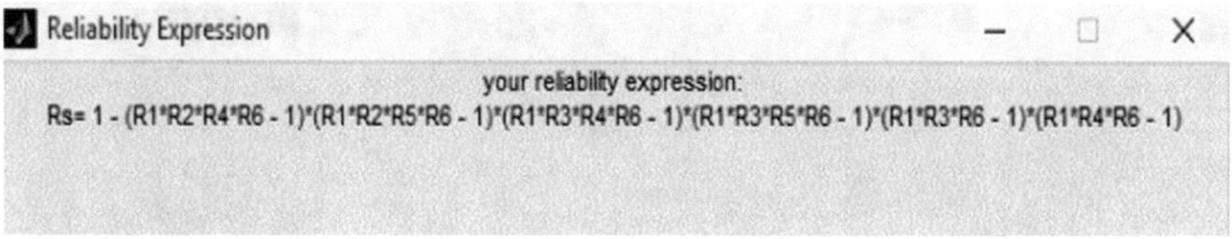

Fig. 7 Reliability expression

Panel (1) Your System

In this panel, the user can enter the number of nodes n and name every node, after that he can enter the connecting edge set that is a matrix of size n $\times$ 2, and start node and end node for the system/subsystem that concerned by reliability analysis. It shows the reliability block diagram using Biographic function of Matlab Bioinformatics Toolbox, also gives reliability expression as illustrated in Fig. 7.

Panel (2) Distribution

In this panel, the user can select a distribution function from the list that is adequate for his network's subsystem/unit/component. Then, he can introduce the corresponding distribution parameters values based on the described distribution equations (CDF, PDF or Hazard).

Panel (3) Reliability

In this panel, the user can specify the life time that is chosen for the reliability computation. After executing the application, it will give the reliability of the system at that time; also give MTTF value of the system, weakness blocks, the number of paths and the elapsed time for Reliability_Network_Eval_Funct.m function.

By depressing the plot pushbutton, the four important plots of reliability analysis such as PDF, CDF, the survival function and the hazard will be displayed.

A generic system composed of six subsystems has been used in the developed software program and the reliability relationships among them are known. In addition, the underlying statistical distributions and parameters used to model the subsystems' reliabilities are given in Table 1 [8]. The application can be used to create the system's RBD as illustrated in Fig. 8.

Table 1 Subsystems distribution parameters [8]

	Distribution	Parameters
Subsystem1	Weibull	$\alpha = 3.44, \beta = 2.21$
Subsystem2	Exponential	$\lambda = 1.06$
Subsystem3	Exponential	$\lambda = 1.38$
Subsystem4	Weibull	$\alpha = 2.25, \beta = 2.68$
Subsystem5	Exponential	$\lambda = 0.94$
Subsystem6	Weibull	$\alpha = 4.36, \beta = 3.97$

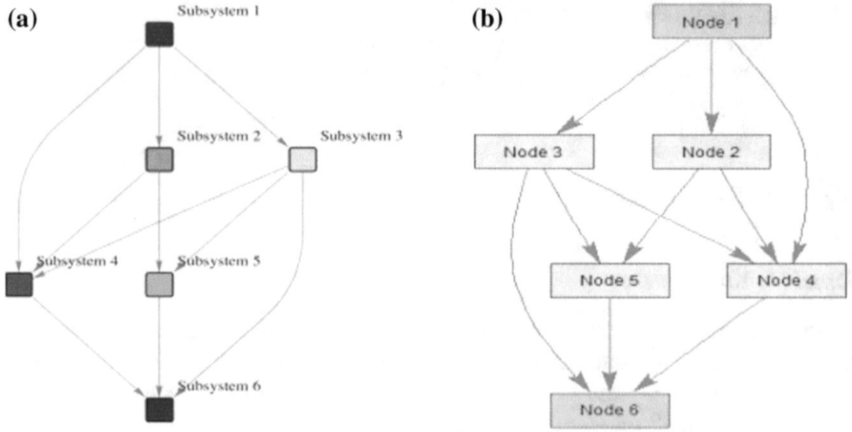

Fig. 8 **a** RBD diagram of the example, **b** the graph generated by GUI application

5 Reliability Assessment Application for Aircraft

5.1 Aircraft System

The goal of this study is to assess the current reliability of complex General Aviation (GA) Aircraft Systems. For providing pertinent information concerning GA aircraft reliability that is helpful to the engineering goal of ensuring development of a reasonable, advanced single pilot transportation aircraft, it is necessary to include airplanes that have the characteristics of future aircraft design [9].

The used approach for studying the reliability is to define the complex GA subsystems of aircraft, collect data of failures from a random sample of complex aircraft, and then, analyze these data in order to estimate the reliability. To accomplish this, complex GA Aircraft may be divided into the five subsystems indicating their primary functions [9]:

A. ***Airframe***: It consists of any element or structure that is necessary to the structural integrity of the aircraft. The interior upholstery, the aircraft paint and the static wicks are also considered part of the Airframe subsystem even they are not part of its structural integrity.
B. ***Cockpit Instrumentation***: it is the required minimum instrumentation for general aviation aircraft flying for satisfying conditions of Instrument Flight Rules (IFR) as defined in Federal Aviation.
C. ***Part-91.Control***: it deals with any component that controls the aircraft's attitude, heading, and altitude or changes the aerodynamic characteristics of the aircraft in the air or on the ground. This system is composed of two primary systems: Flight Control and Ground Control.
D. ***Electrical circuit***: it contains the lighting system and any components involved in the source and distribution of electrical power.

E. ***Power plant***: any component or system that is essential to developing thrust for the aircraft (engine, fuel…). (Except the heating and ventilation system).

5.2 *Reliability Aircraft Analysis*

The Weibull distribution function (with two-parameters, β and α) is used to estimate Airplane System reliability. The fitting tests and the bias tests show that the Weibull distribution function fits well the aircraft data. Besides, the used sample is not significantly biased. The obtained results indicate that the random sample of aircraft, used along with the Weibull distribution function, is appropriate for estimating the system reliabilities for the Complex GA Aircraft [9].

NASA report [9] that contains the system definition and information of the airplane cockpit, failure rates of system parts may be used to model an airplane.

The airplane needs both wings, one engine out of two on each wing, and the control system to function. The control system is shown in Fig. 9, with the mapping to component names in Table 1. Based on these subsystems, the whole system structure can be modeled, which is shown in Fig. 9. All components have lifetime's distribution that fit a Weibull distribution function. The parameters are presented in Table 2 [10].

By introducing these data in Reliability_App for reliability analysis and comparing the obtained results with reference calculation [10], Figs. 10, 11, 12 and 13 show the obtained results displayed by the GUI. The probability of the airplane surviving a six-hour flight is given in Table 3.

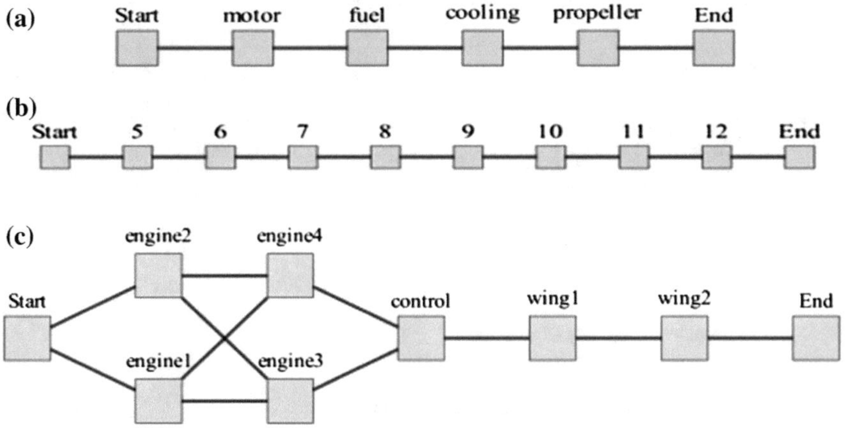

Fig. 9 **a** System structure for an airplane engine, **b** system structure for airplane controls, **c** system structure for airplane [10]

Table 2 Weibull function parameters for airplane [10]

No	Component	α (hours)	β
1	Motor	4830	1.5
2	Fuel	5130	1.44
3	Cooling	4190	1.60
4	Propeller	3740	1.63
5	Directional	4729	1.85
6	Longitudinal	4718.22	1.57
7	Lateral	5843.58	2.25
8	Flaps	3956	0.95
9	Trims	2672.1	0.73
10	Hydraulics	3977.39	1.14
11	Landing gear	2895.62	0.92
12	Steering	3994.78	1.65
13	Wing	4250	1.79

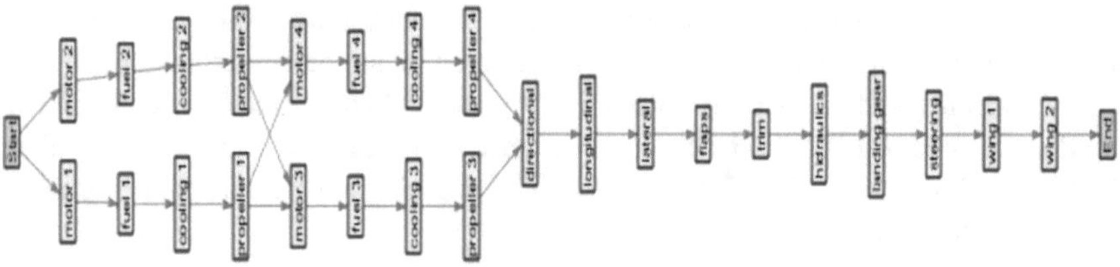

Fig. 10 RBD of airplane system generated by GUI application

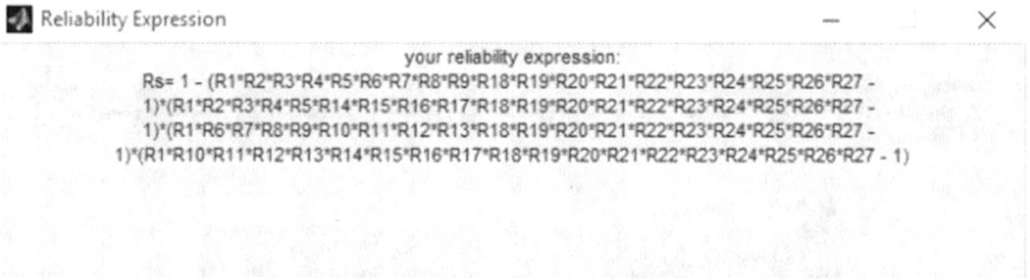

Fig. 11 The obtained reliability expression

5.3 Simulation Results and Discussion

The approach used in this study to estimate the current reliability of Complex GA Aircraft Systems (i.e., Airframe, Electrical, Power plant, Flight Control, and Ground Control) has utilized a random sample that reflects the actual aircraft-operating environment.

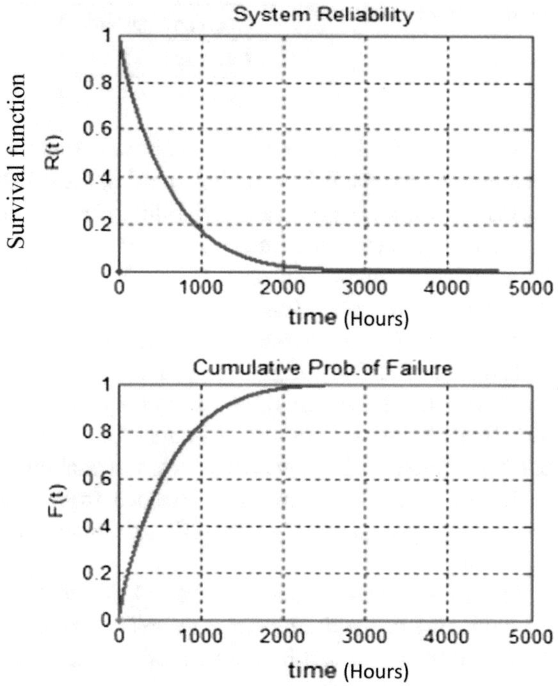

Fig. 12 Plots of GUI application: survival function (top), CDF (bottom)

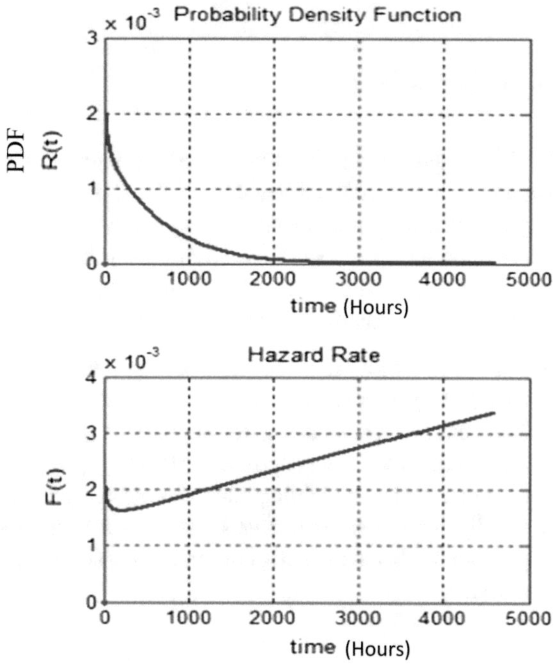

Fig. 13 Plots of GUI application: PDF (top), and hazard function (bottom)

Table 3 Result comparison

Proposed Reliabilty_App	Result 0.98234 ≈ 98.2%
The reference [10]	Result 0.982331 ≈ 98.2%

The exponential distribution is not the only function that may be employed to find system failure rates. It is commonly used in the reliability field and provide a good means for estimating a system reliability. However, if wear-out or infant mortality is considered, then, other function may be used in the determination of distribution function which fits well the data. The Weibull distribution function is widely used in engineering and it can be used to model both the increase and decrease of failure rates. This analysis shows that the current general aviation cockpit has little redundancy in its design. Currently, flight safety and success relies heavily on pilot training and situational awareness. This analysis predicts that a current general aviation aircraft, on a trip taking approximately six-hours, would have a 98.2% probability to complete this trip without losing any cockpit instrumentation.

The of Reliability_App plots, obtained from results as shown in Figs. 12 and 13, describe properties of system lifetime assuming that the components are non-repairable.

From the hazard function of Fig. 13, it can be noted that a burn-in period at the beginning of the lifetime of the airplane and the system will be unreliable after 2000 h. It may even fail before that, this lifetime can be extended by maintenance and monitoring the friction and corrosion.

6 Conclusions

In this chapter, the reliability theoretical background and graph theory have been presented. After that, the developed Matlab GUI application based on graph theory for the reliability assessment of complex systems has been discussed. Besides, this application has been used for studying the reliability of the complex system such as aircraft system and the obtained results are very encouraging.

The conclusions that may be drawn from this work are:

- For the reliability assessment of any system (complex or simple), the developed Matlab GUI application based on graph theory may be used.
- The obtained results using this application present insight into current general aviation reliability. The proposed approach may be used to assist in the development of future system reliability such GA aircraft and safety requirements.
- Finally, the reliability analysis can minimize risk and safe lives, it can improve the product quality which an economic necessity to survive in the global competitive markets.

References

1. Chowdhury A, Koval D (2009) Power distribution system reliability: practical methods and applications (IEEE Press series on power engineering). Wiley, Hoboken, New Jersey
2. Gunawa I (2014) Fundamentals of reliability engineering: applications in multistage interconnection networks. Scrivener Publishing LLC, Hoboken, New Jersey, Wiley, Salem, Massachusetts
3. Billinton R, Ronald NA (1992) Reliability evaluation of engineering systems: concepts and techniques, 2nd edn. Springer Science + Business Media New York
4. Tichý M (1993) Applied methods of structural reliability. Springer Science + Business Media, B.V.
5. Todinov M (2016) Reliability and risk models: setting reliability requirements, 2nd edn. Wiley, Atrium, Southern Gate, UK
6. Bourezg A, Meglouli H (2015) Reliability assessment of power distribution systems using disjoint path-set algorithm. J Ind Eng Int 45–57
7. Henley EJ, Williams RA (1973) Graph theory in modern engineering: computer aided design, control, optimization, reliability analysis. Academic Press, New York
8. Silvestri T (2014) Complex system reliability: a graph theory approach. Math J
9. Pettit D, Turnbull A (2001) General aviation aircraft reliability study. NASA, Hampton, Virginia
10. Rhodin MLJ (2011) Reliability calculations for complex systems. Sweden Linköping

Reliability and Vacation: The Critical Issue

Chandra Shekhar, Shreekant Varshney and Amit Kumar

Abstract Predicting and understanding the machining system performance is a constant challenge in the process industries like manufacturing and production systems, computer and communication systems, just-in-time (JIT) service systems. The reliability modeling supports the decision-making process from early to the optimal state-of-the-art design of the machining system. Reliability measures account the performance of different preventive, predictive, corrective, zero-hours, and periodic maintenance strategies. For all strategies, the most anterior arrangement is the availability of the service facility as and when required to maintain the high grade or efficient quality of service (QoS). The permanent service facility may increase cost, idleness, deterioration in quality. To reduce the wastage of valuable resources like time, money, quality, etc., vacation is a prominent idea for the service facility. The vacation time is a period of time of not doing the usual service or activities. In this time, the server may take rest to rejuvenate, to reduce idle time at the station, to diminish the expected cost incurred in service. The long vacation time also wastes the valuable resources substantially due to the long waiting queue of failed machines. The vacation time period is a critical issue and needs to analyze judgementally. In this chapter, a comparative study of different vacation policies on the reliability characteristics of the machining system is presented. For that purpose, the queueing-theoretic approach is employed, and the Markovian models are developed for various types of vacation policy namely N-policy, single vacation, multiple vacations, Bernoulli vacation, working vacation, vacation interruption, etc. For all vacation policies, the reliability and mean-time-to-failure ($MTTF$) of the system are compared, and results are depicted in the graphs for quick insights. From this study, readers get a glance to understand about vacation, researchers get a concrete platform to choose appropriate assumptions for their research in machining/service

C. Shekhar (✉) · S. Varshney · A. Kumar
Department of Mathematics, Birla Institute of Technology and Science Pilani, Pilani Campus, Pilani 333 031, Rajasthan, India
e-mail: chandrashekhar@pilani.bits-pilani.ac.in

S. Varshney
e-mail: skvarshney91@gmail.com

A. Kumar
e-mail: amitk251@gmail.com

M. Ram and H. Pham (eds.), *Advances in Reliability Analysis and its Applications*, Springer Series in Reliability Engineering,
https://doi.org/10.1007/978-3-030-31375-3_7

system or system analyst may opt suitable vacation policy as per limitation of the system.

Keywords Machine repair problem · Quasi birth-death process · Markovian queue · Vacation policies · Reliability · $MTTF$

1 Introduction

In today's highly technological environment, there is a requirement of a specific machine for almost each and every task. Our dependence on machines is not only clearly observed, but it is also increasing with the requirements of our day-to-day life. But over the time, all machines with their components will experience breakdowns, i.e. machines are unreliable in nature. Unreliability leads to unexpected failures, expensive repairs, costly replacement of their components and even caused the significant interruption in the business of any industry. Also, the increasing complexity of machining systems, as well as the cumulative cost incurred due to the failure of the operation have brought the issue of reliability to the forefront. Hence, it creates a challenge for system analysts and engineers for the better reliability of machining systems and as a result, researchers need to pay special attention to modeling and analysis. Besides this, time variability and random failure events also lead to interrupt the working of the machining system. Over the last few decades, the reliability of machining systems has become a more challenging problem because the highly technological industries include more complex engineering systems with increasing levels of sophistication. Mathematically, reliability is the probability that the system will work properly without interruption over the time interval $[0, t)$ under certain operating conditions. Reliability is a much narrower concept than dependability and has a mathematical formulation in contrast to availability.

Under the preventive-maintenance policy, to avoid any loss of production, hindrance in functioning, data losses, monetary losses, the system designer always keeps some standby machines as safety measure so that a standby machine can immediately act as a substitute when an operating machine fails. In many critical applications of computer, communication, and manufacturing systems, the provision of standby redundancy besides active redundancy has been an essential architectural attribute for achieving high reliability and grade of service (GoS). The standby machines can be classified into three types, hot-standby, warm-standby and cold-standby. If the failure rate of the standby machine is similar to the failure rate of the operating machine then it is known as hot-standby. A warm-standby is the one whose failure rate is less than that of the operating machine and greater than zero. However, the failure rate of a cold-standby is zero. The standbys to be considered in the present investigation are warm-standby and can be extended for another type just by setting the failure rate as per need. Queueing modeling is being used tremendously and effectively in congestion problems encountered due to the failure of machines in day-to-day life as well as industrial scenario including manufacturing systems, computer systems, web services, and communication networks. In this chapter, we consider a basic machine repair problem where a group of a finite number of identical and unreliable operating machines functions in the parallel, is under the supervision of a repairman in the repair facility as corrective maintenance measure and provision of a finite number of standby machines as a preventive maintenance measure.

Under the strategic-maintenance policy, in many practical machining/service systems, the repairman may become unavailable deliberately for some time due to a variety of reasons like minimizing the idle period, utilizing the efficiency of repairman for secondary services, etc. A vacation is defined as the random period when a repairman is unavailable for a random period of time in a repair facility. The duration of the vacation of the repairman is a critical issue in the machining system. Longer vacation period increases the loss of production/revenue and cost incurred due to the failed machines in the queue, while the shorter vacation period increases idleness and the cost of the system. Optimal vacation policies for the repairman balance among the loss of production, idleness of repairman, loss of revenue and extra cost. Vacation policies like N-policy, single vacation, multiple vacations, Bernoulli vacation, and working vacation et cetera are some important strategies to leave the repair facility for the vacation of random length. Each vacation policy differs from other policies either on starting or terminating strategy. Besides the expected total cost, all vacation policies directly affect the reliability characteristics also.

In the forthcoming sections, we analyze the effect of different vacation policies on reliability characteristics of the machining system via queueing-theoretic approach. Based on the rule of the resumption of service after the vacation, the machine repair models have been analyzed in different categories which are given below. For that purpose, we first introduce a basic machine repair problem, and then we implement the machine repair model with several types of vacation policies.

2 Machine Repair Problem (MRP)

Notations

M: Number of operating machines in the system
S: Number of standby machines in the system
m: Minimum number of machines required in short mode
λ: Mean arrival rate
μ: Mean service rate

Machine repair problem (MRP) is a typical example of a real-time finite population queueing model, where the machines represent the population of prospective customers, an arrival corresponds to a failure of the machine in the system, and the caretaker who provides the repair to the failed machines is known as server/repairman. Now, for the analysis purpose, we have considered a finite population machine repair problem consisting of M identical operating machines and S warm standby machines under the care of a single repairman. When an operating machine breakdowns, it is immediately replaced perfectly by an available standby machine with the negligible switch-over time and has the same failure characteristics as of an operating machine. In the normal mode, the system requires M operating machines, but it continues to function properly until there are at least m $(1 \leq m < M)$ machines remain in the system, i.e., the system also works in short mode with an increased likelihood of the failure. Therefore, for the well functioning of the system, only $K = M + S - m + 1$ machines are allowed to fail. For the modeling purpose, we assume that the time-to-failure of an operating machine as well as standby machine follow the exponential distribution with failure rates λ & ν $(0 < \nu < \lambda)$ and the time-to-repair of a failed machine is identically and independently exponentially distributed random variate with rate μ. Hence, the overall state-dependent failure rate of machines is represented as

$$\lambda_n = \begin{cases} M\lambda + (S-n)\nu; & 0 \leq n < S \\ (M+S-n)\lambda; & S \leq n < K \end{cases}$$

Some good reference books for machine repair problem via queueing-theoretic approach have been authored by following learners [1–7]. During the last few decades, the MRP with several essential and/or optional queueing terminologies has been investigated by many researchers (cf. [8–23], and [24]).

For the reliability analysis at time instant t, we define the state probabilities of the system at time t as $P_n(t) = \Pr[N(t) = n;\ n = 0, 1, \ldots, K]$, which represent that there are n failed machines in the system at time t. For the better understanding of the governing model, the transition-state diagram for continuous-time Markov chain (CTMC) involved in machine repair model is depicted in Fig. 1.

Using the concept of quasi-birth and death (QBD) process and balancing the input and output flow in Fig. 1, the governing time-dependent Chapman-Kolmogorov differential-difference equations are constructed as follows.

$$\frac{dP_0(t)}{dt} = -\lambda_0 P_0(t) + \mu P_1(t)$$

$$\frac{dP_n(t)}{dt} = -(\lambda_n + \mu) P_n(t) + \lambda_{n-1} P_{n-1}(t) + \mu P_{n+1}(t); \quad 1 \leq n \leq K-2$$

$$\frac{dP_{K-1}(t)}{dt} = -(\lambda_{K-1} + \mu) P_{K-1}(t) + \lambda_{K-2} P_{K-2}(t)$$

$$\frac{dP_K(t)}{dt} = \lambda P_{K-1}(t)$$

Now, one can easily transform the above-mentioned system of differential-difference equation into the matrix form as follows

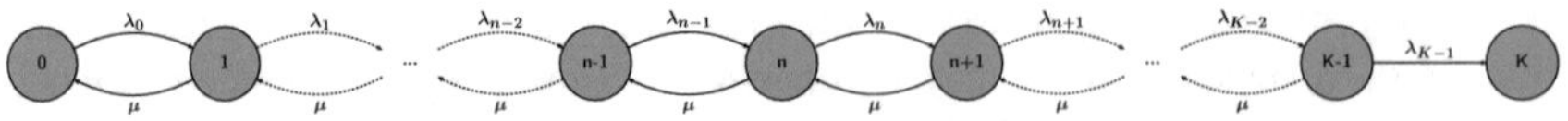

Fig. 1 State transition diagram of a basic machine repair problem

$$\mathbf{DX} = \mathbf{Q}_1\mathbf{X}$$

where **X** is the column vector of all the time-dependent state probabilities and **DX** is the derivative of the probability vector **X** while $\mathbf{Q}_1$ be the block-square matrix which is obtained by using the tri-diagonal characteristics of the above system of differential-difference equations. The structure of the block-square matrix $\mathbf{Q}_1$ is represented as follows.

$$\mathbf{Q}_1 = \begin{bmatrix} \mathbf{A}_1 & \mathbf{B}_1 \\ \mathbf{C}_1 & \mathbf{D}_1 \end{bmatrix}$$

where $\mathbf{A}_1$ is the scalar matrix, $\mathbf{B}_1$ & $\mathbf{C}_1$ are the vectors of order $(1 \times K)$ & $(K \times 1)$ respectively and $\mathbf{D}_1$ is the tri-diagonal square matrix of order $(K \times K)$. The structures of these block sub-matrices are given as follow.

$$\mathbf{A}_1 = [-\lambda_0]$$

$$\mathbf{B}_1 = [\lambda_0, 0, 0, \ldots, 0]$$

$$\mathbf{C}_1 = [\mu, 0, 0, \ldots, 0]^T$$

and

$$\mathbf{D}_1 = \begin{bmatrix} u_1^1 & v_1^1 & 0 & \cdots & 0 & 0 \\ w_2^1 & u_2^1 & v_2^1 & \cdots & 0 & 0 \\ 0 & w_3^1 & u_3^1 & \cdots & 0 & 0 \\ \vdots & \vdots & \vdots & \vdots & \vdots & \vdots \\ 0 & 0 & 0 & \cdots & u_{K-1}^1 & v_{K-1}^1 \\ 0 & 0 & 0 & \cdots & w_K^1 & u_K^1 \end{bmatrix}$$

where

$$u_n^1 = \begin{cases} -(\lambda_n + \mu); & 1 \le n \le K-1 \\ 0; & \text{otherwise} \end{cases}$$

$$v_n^1 = \begin{cases} \lambda_n; & 1 \le n \le K-1 \\ 0; & \text{otherwise} \end{cases}$$

$$w_n^1 = \begin{cases} \mu; & 2 \le n \le K-1 \\ 0; & \text{otherwise} \end{cases}$$

The system performance measures are scale to calculate the efficiency of the system. These performance measures may either be quantitative or qualitative and help the system designers to rank the complex machining/service systems. Following are some basic queueing/reliability measures which are required to analyze the machine repair problem.

- Expected number of failed machines in the system at time t

$$E_N(t) = \sum_{n=0}^{K} n P_n(t)$$

- Throughput of the system at time t

$$\tau_P(t) = \sum_{n=1}^{K-1} \mu P_n(t)$$

- Probability that the repairman is on vacation/idle

$$P_V(t) = P_0(t)$$

- Reliability of the system

$$R_Y(t) = 1 - P_K(t)$$

- Mean time-to-failure of the system

$$MTTF = \int_0^\infty R_Y(t)dt$$

- Failure frequency of the system at time t

$$FF(t) = \lambda_{K-1} P_{K-1}(t)$$

For the comparative analysis purpose, first we fix the default values of several system parameters as $M = 10$, $S = 5$, $m = 2$, $\nu = 0.2$, and vary the values of λ from 1.1 to 1.5 with $\mu = 10$ and μ from 5 to 7 with $\lambda = 0.8$ for all the test instances. We consider same default values for other MRP with different vacation policies in forthcoming sections also. In practice, the incremental changes in failure rate and service rate show the decreasing and increasing effects on the reliability characteristics of the machining systems. From Fig. 2, the obvious results are observed that on increasing the value of λ, the reliability of the system and $MTTF$ decreases. Sim-

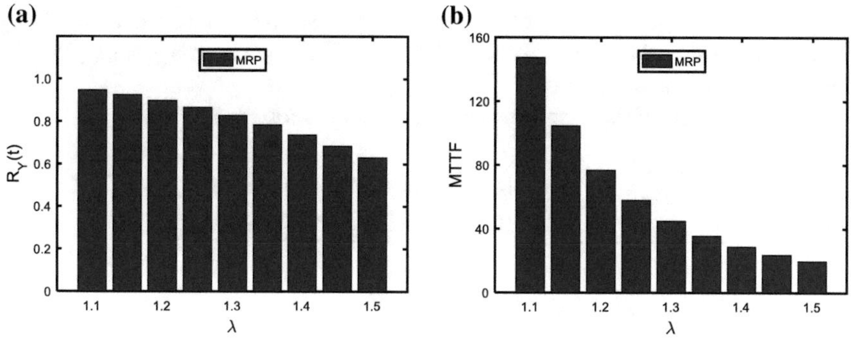

Fig. 2 Effect of failure rate λ on reliability and $MTTF$ of the system respectively for MRP

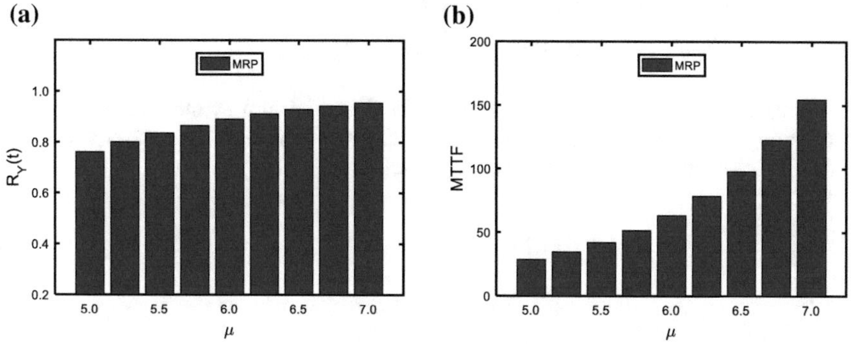

Fig. 3 Effect of repair rate μ on reliability and $MTTF$ of the system respectively for MRP

ilarly, in accordance with Fig. 3, the reliability and $MTTF$ of the system increase correspondingly with incremental change in service rate μ.

In the following sections, we develop the queueing models for MRP with different vacation policies and comparative analysis of reliability characteristics are done. For that purpose, we consider the pre-define assumptions, notations, performance measures, and default parameters value for forthcoming MRP models also.

3 MRP with N-Policy

Notations

N: Threshold value for N-policy vacation

In the machining systems, if the repairman becomes unavailable at the end of a busy period and resumes repairing of the failed machines instantly when the queue length of failed machines reaches a critical number N, then it is known as N-policy vacation of the repairman. In the past, many of the researches have been done using the concept of N-policy in the queueing literature. For a quick glance, refer [25–35], and [36] and references therein.

Define the state of the repairman and state probabilities as follows

$$I(t) = \begin{cases} 0; & \text{Repairman is in vacation state} \\ 1; & \text{Repairman is in working state} \end{cases}$$

and

$$\begin{aligned} P_{0,n}(t) &= \Pr[I(t) = 0 \text{ and } N(t) = n;\ n = 0, 1, \cdots, N-1] \\ P_{1,n}(t) &= \Pr[I(t) = 1 \text{ and } N(t) = n;\ n = 0, 1, \cdots, K-1] \\ P_K(t) &= \Pr[\text{The system is in failed state}] \end{aligned}$$

Using the assumptions and notations of MRP model in previous Sect. 2 and state transition diagram in Fig. 4, following governing differential-difference equations in terms of transient-state probabilities are developed.

$$\begin{aligned} \frac{dP_{0,0}(t)}{dt} &= -\lambda_0 P_{0,0}(t) + \mu P_{1,1}(t) \\ \frac{dP_{0,n}(t)}{dt} &= -\lambda_n P_{0,n}(t) + \lambda_{n-1} P_{0,n-1}(t); \quad 1 \le n \le N-1 \\ \frac{dP_{1,1}(t)}{dt} &= -(\lambda_1 + \mu) P_{1,1}(t) + \mu P_{1,2}(t) \\ \frac{dP_{1,n}(t)}{dt} &= -(\lambda_n + \mu) P_{1,n}(t) + \lambda_{n-1} P_{n-1}(t) + \mu P_{1,n+1}(t); \quad 2 \le n \le N-1 \\ \frac{dP_{1,N}(t)}{dt} &= -(\lambda_N + \mu) P_{1,N}(t) + \lambda_{N-1} P_{1,N-1}(t) + \lambda_{N-1} P_{0,N-1}(t) + \mu P_{1,N+1}(t) \\ \frac{dP_{1,n}(t)}{dt} &= -(\lambda_n + \mu) P_{1,n}(t) + \lambda_{n-1} P_{1,n-1}(t) + \mu P_{1,n+1}(t); \quad N+1 \le n \le K-2 \\ \frac{dP_{1,K-1}(t)}{dt} &= -(\lambda_{K-1} + \mu) P_{K-1}(t) + \lambda_{K-2} P_{1,K-2}(t) \\ \frac{dP_K(t)}{dt} &= \lambda_{K-1} P_{K-1}(t) \end{aligned}$$

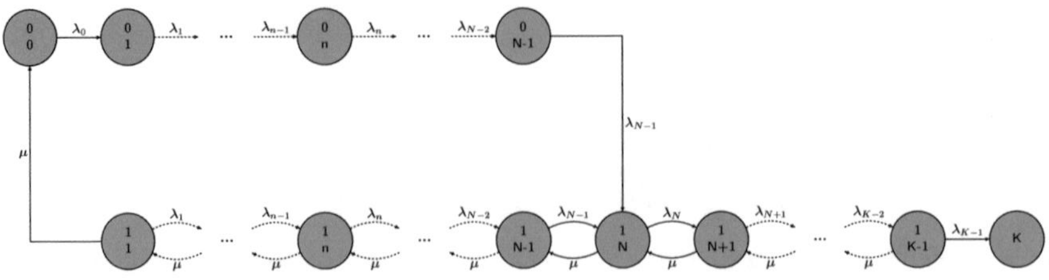

Fig. 4 State transition diagram of the machine repair model with N-policy

The generator matrix $\mathbf{Q}_2$ of the quasi-birth-death process for MRP with N-policy is as follows

$$\mathbf{Q}_2 = \begin{bmatrix} \mathbf{A}_2 & \mathbf{B}_2 \\ \mathbf{C}_2 & \mathbf{D}_2 \end{bmatrix}$$

where $\mathbf{A}_2$, $\mathbf{B}_2$, $\mathbf{C}_2$, and $\mathbf{D}_2$ are the block matrices of order $(N \times N)$, $(N \times K)$, $(K \times N)$, and $(K \times K)$ respectively and have the following matrix form.

$$\mathbf{A}_2 = \begin{bmatrix} x_1^2 & y_1^2 & 0 & \cdots & 0 & 0 \\ 0 & x_2^2 & y_2^2 & \cdots & 0 & 0 \\ 0 & 0 & x_3^2 & \cdots & 0 & 0 \\ \vdots & \vdots & \vdots & \vdots & \vdots & \vdots \\ 0 & 0 & 0 & \cdots & x_{N-1}^2 & y_{N-1}^2 \\ 0 & 0 & 0 & \cdots & 0 & x_N^2 \end{bmatrix}$$

$$x_n^2 = \begin{cases} -\lambda_{n-1}; & 1 \le n \le N \\ 0; & \text{otherwise} \end{cases}$$

$$y_n^2 = \begin{cases} \lambda_{n-1}; & 1 \le n \le N-1 \\ 0; & \text{otherwise} \end{cases}$$

and

$$\mathbf{B}_2[N, N] = \lambda_{N-1}$$

$$\mathbf{C}_2[1, 1] = \mu$$

$$\mathbf{D}_2 = \begin{bmatrix} u_1^2 & v_1^2 & 0 & \cdots & 0 & 0 \\ w_2^2 & u_2^2 & v_2^2 & \cdots & 0 & 0 \\ 0 & w_3^2 & u_3^2 & \cdots & 0 & 0 \\ \vdots & \vdots & \vdots & \vdots & \vdots & \vdots \\ 0 & 0 & 0 & \cdots & u_{K-1}^2 & v_{K-1}^2 \\ 0 & 0 & 0 & \cdots & w_K^2 & u_K^2 \end{bmatrix}$$

$$u_n^2 = \begin{cases} -(\lambda_n + \mu); & 1 \le n \le K-1 \\ 0; & \text{otherwise} \end{cases}$$

$$v_n^2 = \begin{cases} \lambda_n; & 1 \le n \le K-1 \\ 0; & \text{otherwise} \end{cases}$$

$$w_n^2 = \begin{cases} \mu; & 2 \le n \le K-1 \\ 0; & \text{otherwise} \end{cases}$$

Using the transient-state probabilities obtained by above differential-differential equations on employing the Runge-Kutta method of the fourth order, performance measures for MRP with N-policy corresponds to MRP model are as follows.

$$E_N(t) = \sum_{n=0}^{N-1} n P_{0,n}(t) + \sum_{n=1}^{K-1} n P_{1,n}(t)$$

$$\tau_p(t) = \sum_{n=1}^{K-1} \mu P_{1,n}(t)$$

$$P_V(t) = \sum_{n=0}^{N-1} P_{0,n}(t)$$

$$R_Y(t) = 1 - P_K(t)$$

$$MTTF = \int_0^\infty R_Y(t)dt$$

$$FF(t) = \lambda_{K-1} P_{1,K-1}(t)$$

For the graphical analysis, we fix $N = 4$ besides default value of governing parameters in previous section. From Fig. 5, it is observed that on increasing the value of λ, the value of $R_Y(t)$ and $MTTF$ decreases but the observed values of these indices in the case of N-policy vacation queueing system is lesser than the basic machine repair model. Similarly, if we increase the value of μ, the corresponding increasing values of $R_Y(t)$ and $MTTF$ are observed which validates the mathematical formulation (Fig. 6).

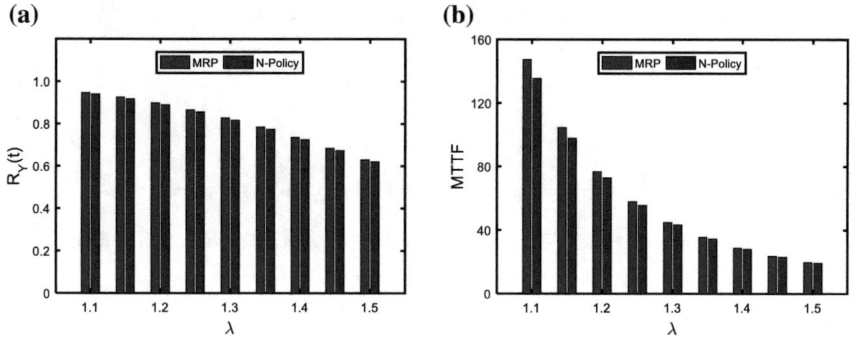

Fig. 5 Effect of failure rate λ on reliability and $MTTF$ respectively wrt (i) MRP and (ii) N-policy

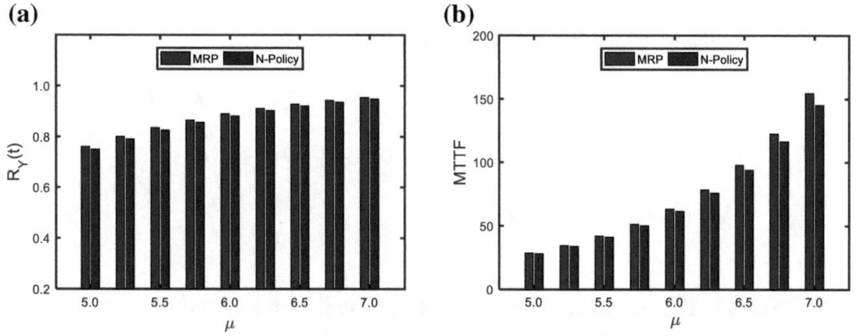

Fig. 6 Effect of repair rate μ on reliability and $MTTF$ respectively wrt (i) MRP and (ii) N-policy

4 MRP with Bernoulli Vacation Policy (BV)

Notations

p: Probability that the repairman chooses the vacation
θ: Vacation rate of the server

The repairing system in which the repairman decides whether to leave for a vacation of random duration with some certain probability p or continue to repair the next failed machine, if any, with the complementary probability $1 - p$ after inspecting the system state is known as Bernoulli scheduled vacation of the repairman. The duration-of-vacation follows an exponential distribution with mean time $1/\theta$. To deal MRP and queueing problem with Bernoulli vacation, many studies have been done in the past also (cf. [37–45]).

For the Markovian analysis, we consider identical assumptions and notations for arrival process and service process as in previous sections. By using the appropriate transitions and balancing the input-output flow in Fig. 7, the governing system of differential-difference equations in the transient state are constructed as follows.

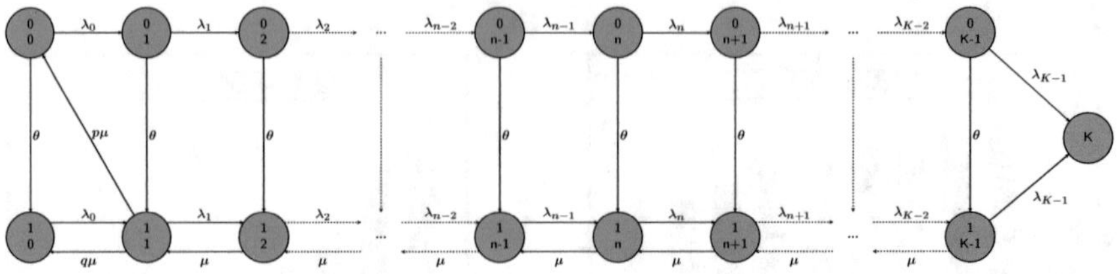

Fig. 7 State transition diagram of the machine repair model with Bernoulli vacation policy

$$\frac{dP_{0,0}(t)}{dt} = -(\lambda_0 + \theta)P_{0,0}(t) + p\mu P_{1,1}(t)$$
$$\frac{dP_{0,n}(t)}{dt} = -(\lambda_n + \theta)P_{0,n}(t) + \lambda_{n-1}P_{0,n-1}(t); \quad 1 \le n \le K-1$$
$$\frac{dP_{1,0}(t)}{dt} = -\lambda_0 P_{1,0}(t) + q\mu P_{1,1}(t) + \theta P_{0,0}(t)$$
$$\frac{dP_{1,n}(t)}{dt} = -(\lambda_n + \mu)P_{1,n}(t) + \lambda_{n-1}P_{1,n-1}(t) + \mu P_{1,n+1}(t) + \theta P_{0,n}(t); \quad 1 \le n \le K-2$$
$$\frac{dP_{1,K-1}(t)}{dt} = -(\lambda_{K-1} + \mu)P_{1,K-1}(t) + \lambda_{K-2}P_{1,K-2}(t) + \theta P_{0,K-1}(t)$$
$$\frac{dP_K(t)}{dt} = \lambda_{K-1}P_{0,K-1}(t) + \lambda_{K-1}P_{1,K-1}(t)$$

To calculate the transient probabilities from the flow balance equations, we construct the generator matrix $\mathbf{Q}_3$ in the following form

$$\mathbf{Q}_3 = \begin{bmatrix} \mathbf{A}_3 & \mathbf{B}_3 & \mathbf{E}_3 \\ \mathbf{C}_3 & \mathbf{D}_3 & \mathbf{F}_3 \\ \mathbf{G}_3 & \mathbf{H}_3 & \mathbf{I}_3 \end{bmatrix}$$

where $\mathbf{A}_3$, $\mathbf{B}_3$, $\mathbf{C}_3$, and $\mathbf{D}_3$ are the square matrices of order $(K \times K)$, both $\mathbf{E}_3$ and $\mathbf{F}_3$ are the column vectors of order $(K \times 1)$ whereas the vectors $\mathbf{G}_3$, $\mathbf{H}_3$, $\mathbf{I}_3$ are the null vectors respectively.

$$\mathbf{A}_3 = \begin{bmatrix} x_1^3 & y_1^3 & 0 & \cdots & 0 & 0 \\ 0 & x_2^3 & y_2^3 & \cdots & 0 & 0 \\ 0 & 0 & x_3^3 & \cdots & 0 & 0 \\ \vdots & \vdots & \vdots & \vdots & \vdots & \vdots \\ 0 & 0 & 0 & \cdots & x_{K-1}^3 & y_{K-1}^3 \\ 0 & 0 & 0 & \cdots & 0 & x_K^3 \end{bmatrix}$$

$$x_n^3 = \begin{cases} -(\lambda_{n-1} + \theta); & 1 \le n \le K \\ 0; & \text{otherwise} \end{cases}$$

$$y_n^3 = \begin{cases} \lambda_{n-1}; & 1 \le n \le K-1 \\ 0; & \text{otherwise} \end{cases}$$

and

$$\mathbf{B}_3 = diag[\theta, \theta, \cdots, \theta]$$

$$\mathbf{C}_3[2, 1] = p\mu$$

$$\mathbf{D}_3 = \begin{bmatrix} u_1^3 & v_1^3 & 0 & \cdots & 0 & 0 \\ w_2^3 & u_2^3 & v_2^3 & \cdots & 0 & 0 \\ 0 & w_3^3 & u_3^3 & \cdots & 0 & 0 \\ \vdots & \vdots & \vdots & \vdots & \vdots & \vdots \\ 0 & 0 & 0 & \cdots & u_{K-1}^3 & v_{K-1}^3 \\ 0 & 0 & 0 & \cdots & w_K^3 & u_K^3 \end{bmatrix}$$

$$u_n^3 = \begin{cases} -\lambda_0; & n = 1 \\ -(\lambda_{n-1} + \mu); & 2 \le n \le K \\ 0; & \text{otherwise} \end{cases}$$

$$v_n^3 = \begin{cases} \lambda_{n-1}; & 1 \le n \le K-1 \\ 0; & \text{otherwise} \end{cases}$$

$$w_n^3 = \begin{cases} q\mu; & n = 2 \\ \mu; & 3 \le n \le K \\ 0; & \text{otherwise} \end{cases}$$

$$\mathbf{E}_3 = \mathbf{F}_3 = [0, 0, \cdots, 0, \lambda_{K-1}]^T$$

$$\mathbf{G}_3 = \mathbf{H}_3 = [0, 0, \cdots, 0]^T$$

$$\mathbf{I}_3 = [0]$$

On the basis of computed transient-state probabilities, the expressions for system performance indices, $E_N(t)$, $P_V(t)$, $\tau_p(t)$, $FF(t)$, $R_Y(t)$ and $MTTF$, are delineated as follows.

$$E_N(t) = \sum_{j=0}^{1}\sum_{n=0}^{K-1} nP_{j,n}(t) + KP_{K}(t)$$

$$\tau_p(t) = \sum_{n=1}^{K-1} \mu P_{1,n}(t)$$

$$P_V(t) = \sum_{n=0}^{K-1} P_{0,n}(t)$$

$$R_Y(t) = 1 - P_{K}(t)$$

$$MTTF = \int_0^{\infty} R_Y(t)dt$$

$$FF(t) = \lambda_{K-1}P_{0,K-1}(t) + \lambda_{K-1}P_{1,K-1}(t)$$

For observing the comparative reliability characteristics for different vacation policies, we fix $p = 0.5$ & $\theta = 0.5$ and other default parameters as same as in the previous section and depict the results in Figs. 8 and 9 by varying the values of failure rate λ and service rate μ.

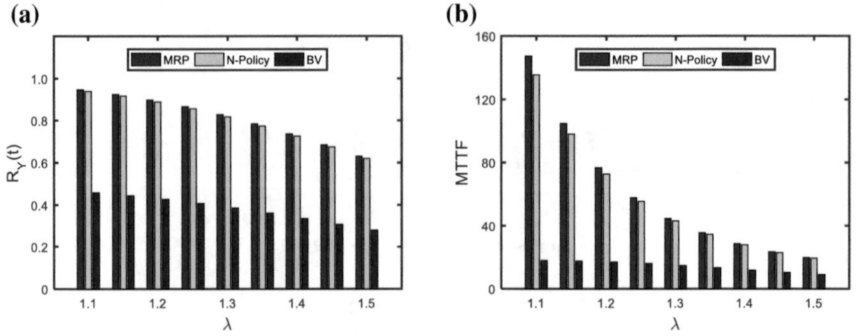

Fig. 8 Effect of failure rate λ on reliability and $MTTF$ respectively wrt (i) MRP, (ii) N-policy, and (iii) BV

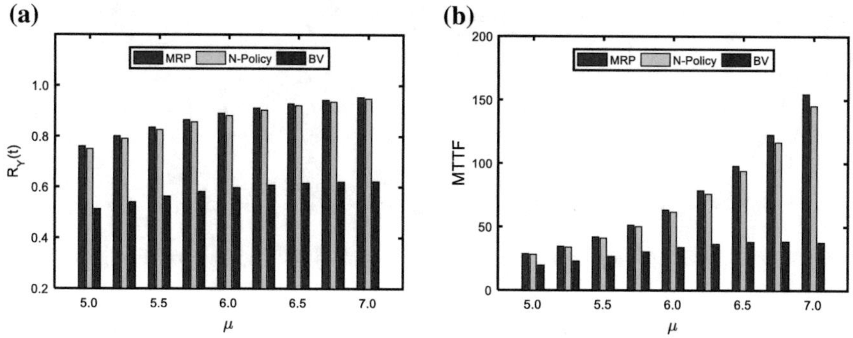

Fig. 9 Effect of repair rate μ on reliability and $MTTF$ respectively wrt (i) MRP, (ii) N-policy, and (iii) BV

For the varied values of λ & μ range from 1.1 to 1.5 & 5.0 to 7.0, respectively, we depict that the observed values of $R_Y(t)$ and $MTTF$ follow the obvious trend. For Bernoulli vacation policy, these reliability characteristics are quite lesser than the N-policy vacation.

5 MRP with Multiple Vacation Policy (MV)

Notations

θ: Vacation rate of the server

In this section, we study the MRP with multiple vacation policy. According to this policy, a repairman leaves for the vacation of random duration when there is no failed machine for repair in the repair station. On return from the vacation, if the repairman finds no waiting failed machine in the queue, he takes another vacation and continues this process until he finds at least one failed machine awaiting for repair. Numerous problems on machine repair model with multiple vacation policy have been proposed by many researchers in the queueing literature (cf. [31, 36, 46–52]).

The basic assumptions and notations are identical as for previous sections. Referring to the state-transition diagram for the MRP with the multiple vacation policy shown in Fig. 10, we have following time-dependent Chapman Kolmogrove equations in terms of state probabilities.

$$\frac{dP_{0,0}(t)}{dt} = -\lambda_0 P_{0,0}(t) + \mu P_{1,1}(t)$$
$$\frac{dP_{0,n}(t)}{dt} = -(\lambda_n + \theta) P_{0,n}(t) + \lambda_{n-1} P_{0,n-1}(t); \quad 1 \le n \le K-1$$
$$\frac{dP_{1,1}(t)}{dt} = -(\lambda_1 + \mu) P_{1,1}(t) + \mu P_{1,2}(t) + \theta P_{0,1}(t)$$

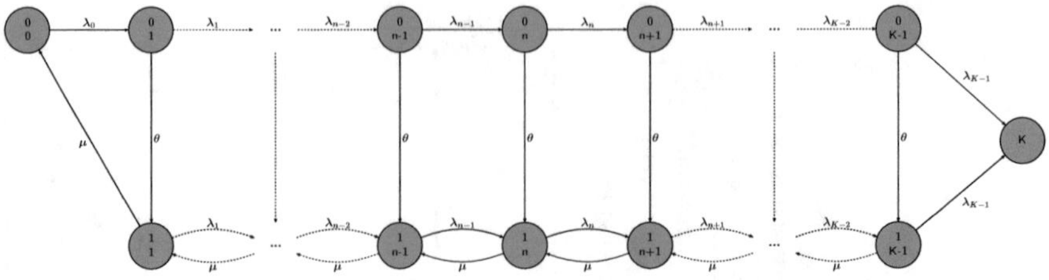

Fig. 10 State transition diagram of the machine repair model with multiple vacation policy

$$\frac{dP_{1,n}(t)}{dt} = -(\lambda_n + \mu)P_{1,n}(t) + \lambda_{n-1}P_{1,n-1}(t) + \mu P_{1,n+1}(t) + \theta P_{0,n}(t); \quad 2 \le n \le K-2$$

$$\frac{dP_{1,K-1}(t)}{dt} = -(\lambda_{K-1} + \mu)P_{1,K-1}(t) + \lambda_{K-2}P_{1,K-2}(t) + \theta P_{0,K-1}(t)$$

$$\frac{dP_K(t)}{dt} = \lambda_{K-1}P_{0,K-1}(t) + \lambda_{K-1}P_{1,K-1}(t)$$

For employing the Runge-Kutta numerical method to compute the state probabilities, the corresponding matrix representation is provided for the set of differential equations for the MRP with multiple vacation policy

$$\mathbf{Q}_4 = \begin{bmatrix} \mathbf{A}_4 & \mathbf{B}_4 & \mathbf{E}_4 \\ \mathbf{C}_4 & \mathbf{D}_4 & \mathbf{F}_4 \\ \mathbf{G}_4 & \mathbf{H}_4 & \mathbf{I}_4 \end{bmatrix}$$

where $\mathbf{A}_4$, $\mathbf{B}_4$, $\mathbf{C}_4$, and $\mathbf{D}_4$ are the matrices of order $(K \times K)$, $(K \times K-1)$, $(K-1 \times K)$, and $(K-1 \times K-1)$ respectively, $\mathbf{E}_4$ and $\mathbf{F}_4$ are the column vectors of order $(K \times 1)$ and $(K-1 \times 1)$ respectively while $\mathbf{G}_4$, $\mathbf{H}_4$, and $\mathbf{I}_4$ are the null vectors. The structures of the block matrices are represented as

$$\mathbf{A}_4 = \begin{bmatrix} x_1^4 & y_1^4 & 0 & \cdots & 0 & 0 \\ 0 & x_2^4 & y_2^4 & \cdots & 0 & 0 \\ 0 & 0 & x_3^4 & \cdots & 0 & 0 \\ \vdots & \vdots & \vdots & \vdots & \vdots & \vdots \\ 0 & 0 & 0 & \cdots & x_{K-1}^4 & y_{K-1}^4 \\ 0 & 0 & 0 & \cdots & 0 & x_K^4 \end{bmatrix}$$

$$x_n^4 = \begin{cases} -\lambda_0; & n = 1 \\ -(\lambda_{n-1} + \theta); & 2 \le n \le K \\ 0; & \text{otherwise} \end{cases}$$

$$y_n^4 = \begin{cases} \lambda_{n-1}; & 1 \le n \le K-1 \\ 0; & \text{otherwise} \end{cases}$$

and

$$\mathbf{B}_4[l, l-1] = \theta; \quad l = 2, 3, \cdots, K$$

$$\mathbf{C}_4[1, 1] = \mu$$

$$\mathbf{D}_4 = \begin{bmatrix} u_1^4 & v_1^4 & 0 & \cdots & 0 & 0 \\ w_2^4 & u_2^4 & v_2^4 & \cdots & 0 & 0 \\ 0 & w_3^4 & u_3^4 & \cdots & 0 & 0 \\ \vdots & \vdots & \vdots & \vdots & \vdots & \vdots \\ 0 & 0 & 0 & \cdots & u_{K-2}^4 & v_{K-2}^4 \\ 0 & 0 & 0 & \cdots & w_{K-1}^4 & u_{K-1}^4 \end{bmatrix}$$

$$u_n^4 = \begin{cases} -(\lambda_n + \mu); & 1 \le n \le K-1 \\ 0; & \text{otherwise} \end{cases}$$

$$v_n^4 = \begin{cases} \lambda_n; & 1 \le n \le K-2 \\ 0; & \text{otherwise} \end{cases}$$

$$w_n^4 = \begin{cases} \mu; & 2 \le n \le K-1 \\ 0; & \text{otherwise} \end{cases}$$

$$\mathbf{E}_4 = [0, 0, \cdots, 0, \lambda_{K-1}]^T$$
$$\mathbf{F}_4 = [0, 0, \cdots, 0, \lambda_{K-1}]^T$$

$$\mathbf{G}_4 = [0, 0, \cdots, 0]^T$$
$$\mathbf{H}_4 = [0, 0, \cdots, 0]^T$$

$$\mathbf{I}_4 = [0]$$

Using transient-state probabilities derived from the above generator matrix Q_4 on employing Runge-Kutta method of the fourth-order, the explicit expression of different system performance measures in closed form are as follow.

$$E_N(t) = \sum_{j=0}^{1} \sum_{n=j}^{K-1} n P_{j,n}(t) + K P_{K}(t)$$

$$\tau_p(t) = \sum_{n=1}^{K-1} \mu P_{1,n}(t)$$

$$P_V(t) = \sum_{n=0}^{K-1} P_{0,n}(t)$$

$$R_Y(t) = 1 - P_{K}(t)$$

$$MTTF = \int_0^{\infty} R_Y(t)dt$$

$$FF(t) = \lambda_{K-1} P_{0,K-1}(t) + \lambda_{K-1} P_{1,K-1}(t)$$

The effects of λ and μ on $R_Y(t)$ and $MTTF$ are shown in Figs. 11 and 12. For that purpose, we fix $\theta = 0.5$ and other parameters as in previous section as default

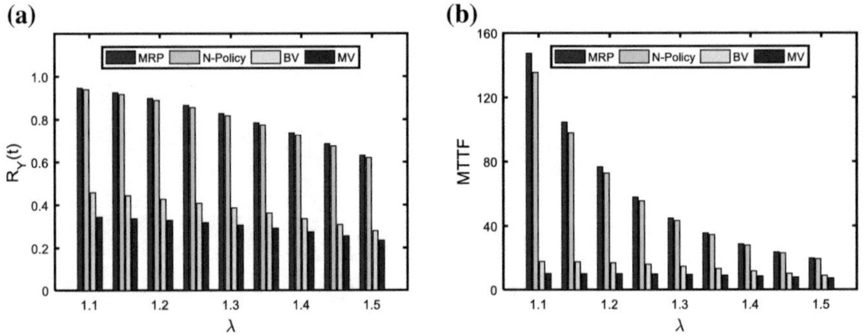

Fig. 11 Effect of failure rate λ on reliability and $MTTF$ respectively wrt (i) MRP, (ii) N-policy, (iii) BV, and (iv) MV

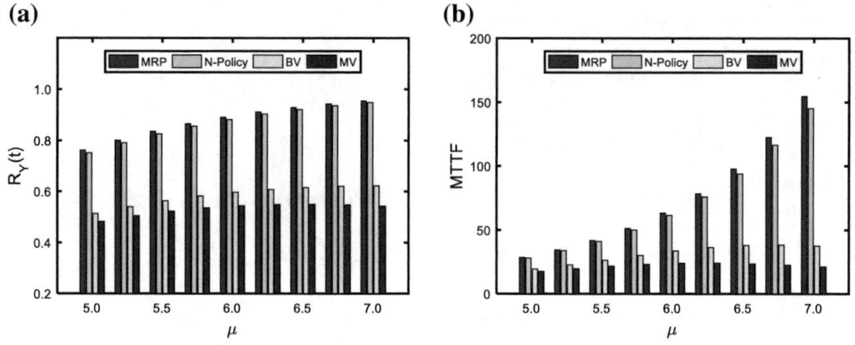

Fig. 12 Effect of repair rate μ on reliability and $MTTF$ respectively wrt (i) MRP, (ii) N-policy, (iii) BV, and (iv) MV

value. These figures reveal that the values of reliability and $MTTF$ of the system are almost constant for increasing values of λ and μ for the multiple vacation case. Also, for a fixed value of λ and μ, we see that the values of $R_Y(t)$ and $MTTF$ are lesser as compared to other vacation policies as expected.

6 MRP with Single Vacation Policy (SV)

Notations

θ: Vacation rate of the server

In the single vacation policy, the repairman takes exactly one vacation i.e. at the end of a vacation of random duration, if the repairman finds no failed machine in the repair station for repair, he then waits idly in the system unless a prospect failed machine arrives in the repair station for the seek of repair. Over the last few decades, vacation queues have been emerged as a significant area of research in queueing literature. A number of researchers including [53–59] and many more studied the vacation queues with single vacation policy.

Now, connecting the states of the system according to the transitions in Fig. 13, we obtain the following set of a transient system of differential equations.

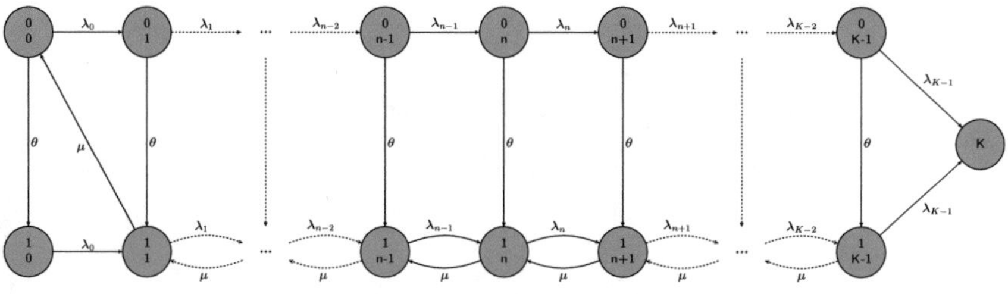

Fig. 13 State transition diagram of the machine repair model with single vacation policy

$$\frac{dP_{0,0}(t)}{dt} = -(\lambda_0 + \theta)P_{0,0}(t) + \mu P_{1,1}(t)$$
$$\frac{dP_{0,n}(t)}{dt} = -(\lambda_n + \theta)P_{0,n}(t) + \lambda_{n-1}P_{0,n-1}(t); \quad 1 \le n \le K-1$$
$$\frac{dP_{1,0}(t)}{dt} = -\lambda_0 P_{1,0}(t) + \theta P_{0,0}(t)$$
$$\frac{dP_{1,n}(t)}{dt} = -(\lambda_n + \mu)P_{1,n}(t) + \lambda_{n-1}P_{1,n-1}(t) + \mu P_{1,n+1}(t) + \theta P_{0,n}(t); \quad 1 \le n \le K-2$$
$$\frac{dP_{1,K-1}(t)}{dt} = -(\lambda_{K-1} + \mu)P_{1,K-1}(t) + \lambda_{K-2}P_{1,K-2}(t) + \theta P_{0,K-1}(t)$$
$$\frac{dP_K(t)}{dt} = \lambda_{K-1}P_{0,K-1}(t) + \lambda_{K-1}P_{1,K-1}(t)$$

A matrix-form is presented to compute the state probabilities through Runge-Kutta method from the resulting system of differential equations mentioned earlier. Using the results given by Neuts [60], the generator matrix $\mathbf{Q}_5$ of the Markov chain in MRP with single vacation is partitioned in the following form

$$\mathbf{Q}_5 = \begin{bmatrix} \mathbf{A}_5 & \mathbf{B}_5 & \mathbf{E}_5 \\ \mathbf{C}_5 & \mathbf{D}_5 & \mathbf{F}_5 \\ \mathbf{G}_5 & \mathbf{H}_5 & \mathbf{I}_5 \end{bmatrix}$$

where $\mathbf{A}_5$, $\mathbf{B}_5$, $\mathbf{C}_5$, and $\mathbf{D}_5$ are the square matrices of order $(K \times K)$, $\mathbf{E}_5$ and $\mathbf{F}_5$ are the column vectors of order $(K \times 1)$, and $\mathbf{G}_5$, $\mathbf{H}_5$ & $\mathbf{I}_5$ are the zero vectors.

$$\mathbf{A}_5 = \begin{bmatrix} x_1^5 & y_1^5 & 0 & \cdots & 0 & 0 \\ 0 & x_2^5 & y_2^5 & \cdots & 0 & 0 \\ 0 & 0 & x_3^5 & \cdots & 0 & 0 \\ \vdots & \vdots & \vdots & \vdots & \vdots & \vdots \\ 0 & 0 & 0 & \cdots & x_{K-1}^5 & y_{K-1}^5 \\ 0 & 0 & 0 & \cdots & 0 & x_K^5 \end{bmatrix}$$

$$x_n^5 = \begin{cases} -(\lambda_{n-1} + \theta); & 1 \le n \le K \\ 0; & \text{otherwise} \end{cases}$$

$$y_n^5 = \begin{cases} \lambda_{n-1}; & 1 \le n \le K-1 \\ 0; & \text{otherwise} \end{cases}$$

and

$$\mathbf{B}_5 = diag[\theta, \theta, \cdots, \theta]$$

$$\mathbf{C}_5[2,1] = \mu$$

$$\mathbf{D}_5 = \begin{bmatrix} u_1^5 & v_1^5 & 0 & \cdots & 0 & 0 \\ w_2^5 & u_2^5 & v_2^5 & \cdots & 0 & 0 \\ 0 & w_3^5 & u_3^5 & \cdots & 0 & 0 \\ \vdots & \vdots & \vdots & \vdots & \vdots & \vdots \\ 0 & 0 & 0 & \cdots & u_{K-1}^5 & v_{K-1}^5 \\ 0 & 0 & 0 & \cdots & w_K^5 & u_K^5 \end{bmatrix}$$

$$u_n^5 = \begin{cases} -\lambda_0; & n = 1 \\ -(\lambda_{n-1} + \mu); & 2 \le n \le K \\ 0; & \text{otherwise} \end{cases}$$

$$v_n^5 = \begin{cases} \lambda_{n-1}; & 1 \le n \le K - 1 \\ 0; & \text{otherwise} \end{cases}$$

$$w_n^5 = \begin{cases} \mu; & 3 \le n \le K \\ 0; & \text{otherwise} \end{cases}$$

$$\mathbf{E}_5 = \mathbf{F}_5 = [0, 0, \ldots, 0, \lambda_{K-1}]^T$$

$$\mathbf{G}_5 = \mathbf{H}_5 = [0, 0, \ldots, 0]^T$$

$$\mathbf{I}_5 = [0]$$

In this chapter, we mainly deal with the reliability and $MTTF$ of the machining system. For in-depth analysis, we require other queueing and reliability characteristics also. Hence, the explicit expressions in the closed form of some other system performance measures are derived in terms of state probabilities as follow.

$$E_N(t) = \sum_{j=0}^{1} \sum_{n=0}^{K-1} nP_{j,n}(t) + KP_{_K}(t)$$

$$\tau_p(t) = \sum_{n=1}^{K-1} \mu P_{1,n}(t)$$

$$P_V(t) = \sum_{n=0}^{K-1} P_{0,n}(t)$$

$$R_Y(t) = 1 - P_{_K}(t)$$

$$MTTF = \int_0^{\infty} R_Y(t)dt$$

$$FF(t) = \lambda_{_{K-1}} P_{_{0,K-1}}(t) + \lambda_{_{K-1}} P_{_{1,K-1}}(t)$$

For the study purpose, we set the parameters as same as in previous sections. Figures 14 and 15 display the variation of reliability characteristics with respect to failure rate and repair rate respectively. It is observed that the single vacation is better than multiple vacation for enhancing the reliability measures.

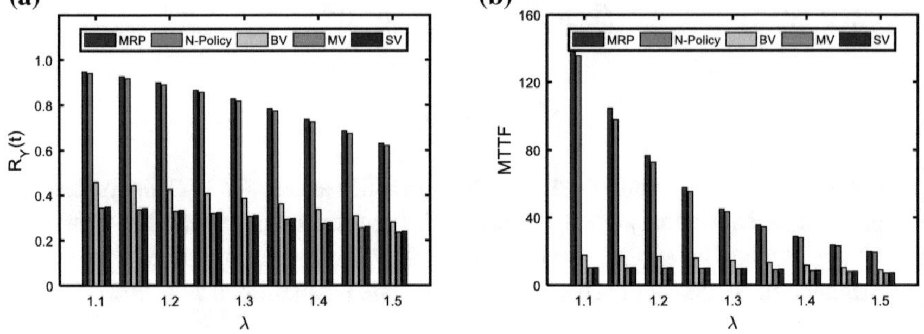

Fig. 14 Effect of failure rate λ on reliability and $MTTF$ respectively wrt (i) MRP, (ii) N-policy, (iii) BV, (iv) MV, and (v) SV

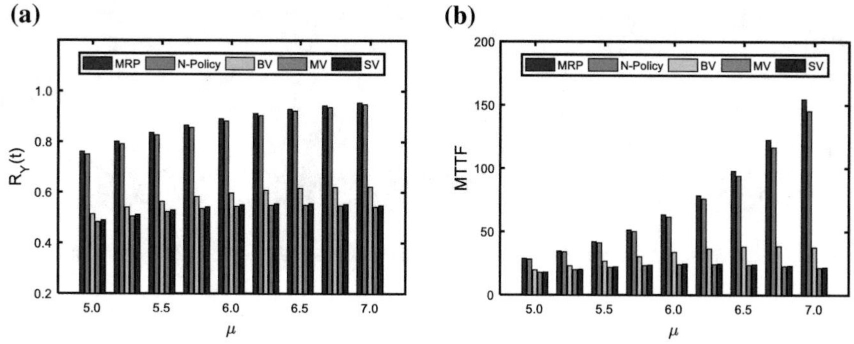

Fig. 15 Effect of repair rate μ on reliability and $MTTF$ respectively wrt (i) MRP, (ii) N-policy, (iii) BV, (iv) MV, and (v) SV

7 MRP with Multiple Working Vacation Policy (MWV)

Notations

μ_b: Mean service rate during a busy period
μ_v: Mean service rate during the working vacation period
θ: Vacation rate of the server

The working vacation policy was firstly introduced by Servi and Finn [61], in which the repairman works with a slower rate rather than completely terminating the repair in the vacation period. The repairman contiunes the repair with the slower rate till his random vacation period ends. There are mainly two special cases of the working vacation policy depending on vacation terminating criterion, first one is multiple working vacation policy and the second is a single working vacation policy. In the multiple working vacation policy, at the end of the vacation period, if the repairman does not find any failed machine in the system for repair then he repeatedly takes the working vacation of random duration otherwise provides the repair to the failed machine if any present in the system with normal mean rate. Several researchers investigated the queueing problems with the multiple working vacation policy in the past (cf. [62–66]).

The basic assumptions and notations are simillar as in previous section. The time-to-repair in busy mode and vacation mode follow an exponential distribution with mean rate μ_b and μ_v respectively. The following transient-state differential equations are obtained by balancing the transitions at time instant t in Fig. 16.

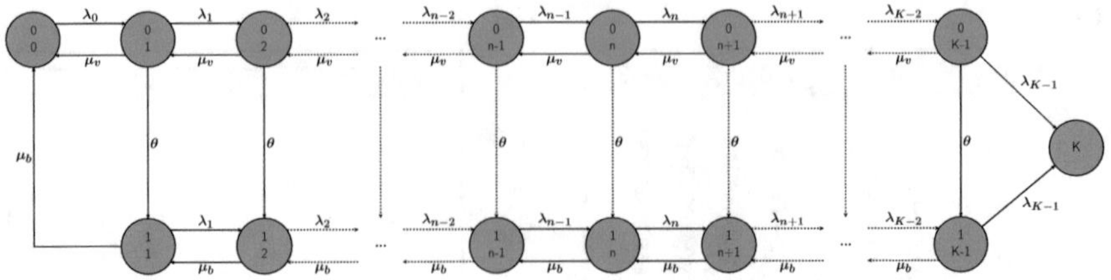

Fig. 16 State transition diagram of the machine repair model with multiple working vacation policy

$$\frac{dP_{0,0}(t)}{dt} = -\lambda_0 P_{0,0}(t) + \mu_v P_{0,1}(t) + \mu_b P_{1,1}(t)$$
$$\frac{dP_{0,n}(t)}{dt} = -(\lambda_n + \mu_v + \theta)P_{0,n}(t) + \lambda_{n-1} P_{0,n-1}(t) + \mu_v P_{0,n+1}(t); \quad 1 \le n \le K-2$$
$$\frac{dP_{0,K-1}(t)}{dt} = -(\lambda_{K-1} + \mu_v + \theta)P_{0,K-1}(t) + \lambda_{K-2} P_{0,K-2}(t)$$
$$\frac{dP_{1,1}(t)}{dt} = -(\lambda_1 + \mu_b)P_{1,1}(t) + \mu_b P_{1,2}(t) + \theta P_{0,1}(t)$$
$$\frac{dP_{1,n}(t)}{dt} = -(\lambda_n + \mu_b)P_{1,n}(t) + \lambda_{n-1} P_{1,n-1}(t) + \mu_b P_{1,n+1}(t) + \theta P_{0,n}(t); \quad 2 \le n \le K-2$$
$$\frac{dP_{1,K-1}(t)}{dt} = -(\lambda_{K-1} + \mu_b)P_{1,K-1}(t) + \lambda_{K-2} P_{1,K-2}(t) + \theta P_{0,K-1}(t)$$
$$\frac{dP_K(t)}{dt} = \lambda_{K-1} P_{0,K-1}(t) + \lambda_{K-1} P_{1,K-1}(t)$$

Suppose that $\mathbf{Q}_6$ denotes the corresponding transition rate matrix, then $\mathbf{Q}_6$ can be expressed in block structures as

$$\mathbf{Q}_6 = \begin{bmatrix} \mathbf{A}_6 & \mathbf{B}_6 & \mathbf{E}_6 \\ \mathbf{C}_6 & \mathbf{D}_6 & \mathbf{F}_6 \\ \mathbf{G}_6 & \mathbf{H}_6 & \mathbf{I}_6 \end{bmatrix}$$

where the block-matrices $\mathbf{A}_6$, $\mathbf{B}_6$, $\mathbf{C}_6$, and $\mathbf{D}_6$ are of order $(K \times K)$, $(K \times K-1)$, $(K-1 \times K)$ and $(K-1 \times K-1)$ respectively, $\mathbf{E}_6$ and $\mathbf{F}_6$ are the column vectors of order $(K \times 1)$ and $(K-1 \times 1)$ respectively, and $\mathbf{G}_6$, $\mathbf{H}_6$, $\mathbf{I}_6$ are the zero vectors of appropriate order. These sub-matrices are given as follow.

$$\mathbf{A}_6 = \begin{bmatrix} x_1^6 & y_1^6 & 0 & \cdots & 0 & 0 \\ z_2^6 & x_2^6 & y_2^6 & \cdots & 0 & 0 \\ 0 & z_3^6 & x_3^6 & \cdots & 0 & 0 \\ \vdots & \vdots & \vdots & \vdots & \vdots & \vdots \\ 0 & 0 & 0 & \cdots & x_{K-1}^6 & y_{K-1}^6 \\ 0 & 0 & 0 & \cdots & z_K^6 & x_K^6 \end{bmatrix}$$

$$x_n^6 = \begin{cases} -\lambda_0; & n = 1 \\ -(\lambda_{n-1} + \mu_v + \theta); & 2 \le n \le K \\ 0; & \text{otherwise} \end{cases}$$

$$y_n^6 = \begin{cases} \lambda_{n-1}; & 1 \le n \le K - 1 \\ 0; & \text{otherwise} \end{cases}$$

$$z_n^6 = \begin{cases} \mu_v; & 2 \le n \le K \\ 0; & \text{otherwise} \end{cases}$$

and

$$\mathbf{B}_6[l, l-1] = \theta; \quad l = 2, 3, \ldots, K$$

$$\mathbf{C}_6[1, 1] = \mu_b$$

$$\mathbf{D}_6 = \begin{bmatrix} u_1^6 & v_1^6 & 0 & \cdots & 0 & 0 \\ w_2^6 & u_2^6 & v_2^6 & \cdots & 0 & 0 \\ 0 & w_3^6 & u_3^6 & \cdots & 0 & 0 \\ \vdots & \vdots & \vdots & \vdots & \vdots & \vdots \\ 0 & 0 & 0 & \cdots & u_{K-2}^6 & v_{K-2}^6 \\ 0 & 0 & 0 & \cdots & w_{K-1}^6 & u_{K-1}^6 \end{bmatrix}$$

$$u_n^6 = \begin{cases} -(\lambda_n + \mu_b); & 1 \le n \le K - 1 \\ 0; & \text{otherwise} \end{cases}$$

$$v_n^6 = \begin{cases} \lambda_n; & 1 \le n \le K - 2 \\ 0; & \text{otherwise} \end{cases}$$

$$w_n^6 = \begin{cases} \mu_b; & 2 \le n \le K - 1 \\ 0; & \text{otherwise} \end{cases}$$

$$\mathbf{E}_6 = [0, 0, \ldots, 0, \lambda_{K-1}]^T$$
$$\mathbf{F}_6 = [0, 0, \ldots, 0, \lambda_{K-1}]^T$$

$$\mathbf{G}_6 = [0, 0, \ldots, 0]^T$$
$$\mathbf{H}_6 = [0, 0, \ldots, 0]^T$$

$$\mathbf{I}_6 = [0]$$

The system performance measures of the machine repair model with multiple working vacation policy are given by

$$E_N(t) = \sum_{j=0}^{1} \sum_{n=j}^{K-1} n P_{j,n}(t) + K P_K(t)$$

$$\tau_p(t) = \sum_{n=1}^{K-1} \mu_v P_{0,n}(t) + \sum_{n=1}^{K-1} \mu_b P_{1,n}(t)$$

$$P_{WV}(t) = \sum_{n=0}^{K-1} P_{0,n}(t)$$

$$R_Y(t) = 1 - P_K(t)$$

$$MTTF = \int_0^{\infty} R_Y(t) dt$$

$$FF(t) = \lambda_{K-1} P_{0,K-1}(t) + \lambda_{K-1} P_{1,K-1}(t)$$

For the graphical comparative purpose, we fix the value of system parameters as in the last section along with $\mu_b = \mu = 10$ & $\mu_v = 8$ for the Fig. 17 wrt failure rate (λ) and $\mu_b = \mu = 6$ & $\mu_v = 4$ for Fig. 18 wrt service rate (μ or μ_b) respectively. Figures 17 and 18 portray the changing trend of reliability measures with respect to breakdown rate and repair rate. The inferences from these figures are obvious that reliability and mean-time-to-failure are decreasing on the increased value of breakdown rate. To achieve the high reliability of the machining system, high repair

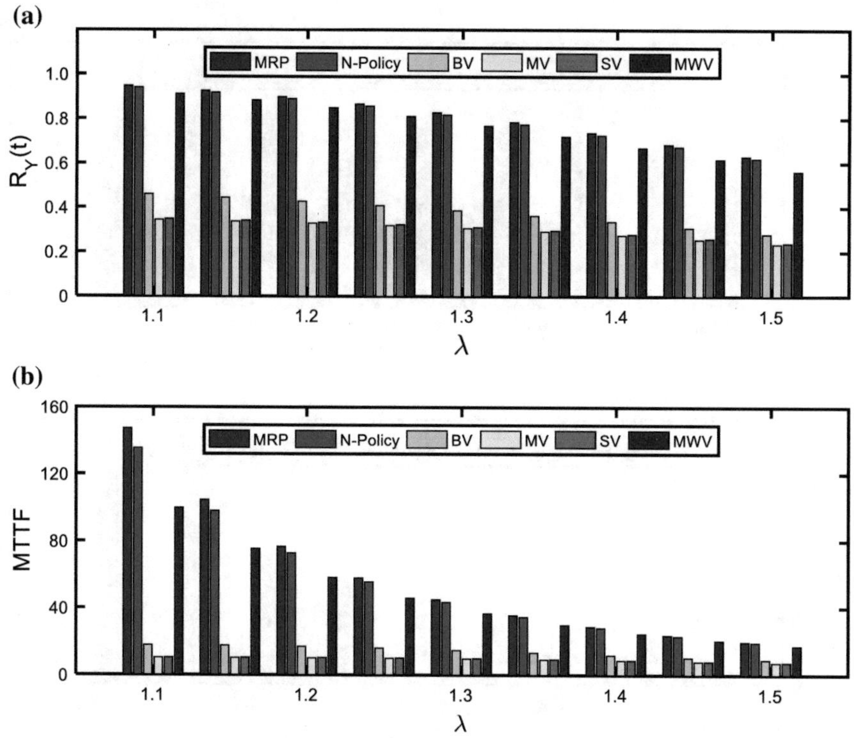

Fig. 17 Effect of failure rate λ on reliability and $MTTF$ respectively wrt (i) MRP, (ii) N-policy, (iii) BV, (iv) MV, (v) SV, and (vi) MWV

rate should be maintained. It also observed that multiple working vacation is better than the multiple vacation for the seek of the high level of reliability measures.

8 MRP with Single Working Vacation Policy (SWV)

Notations

μ_b: Mean service rate during a busy period
μ_v: Mean service rate during the working vacation period
θ: Vacation rate of the server

In the single working vacation policy, a working vacation starts when there is no failed machine in the system at service completion instant. When a working vacation ends, if there is no failed machine for repair in the service station, the repairman stays in system idle and is ready for serving the newly arriving failed machines. Otherwise, after returning from the vacation if he finds the non-empty system, he immediately

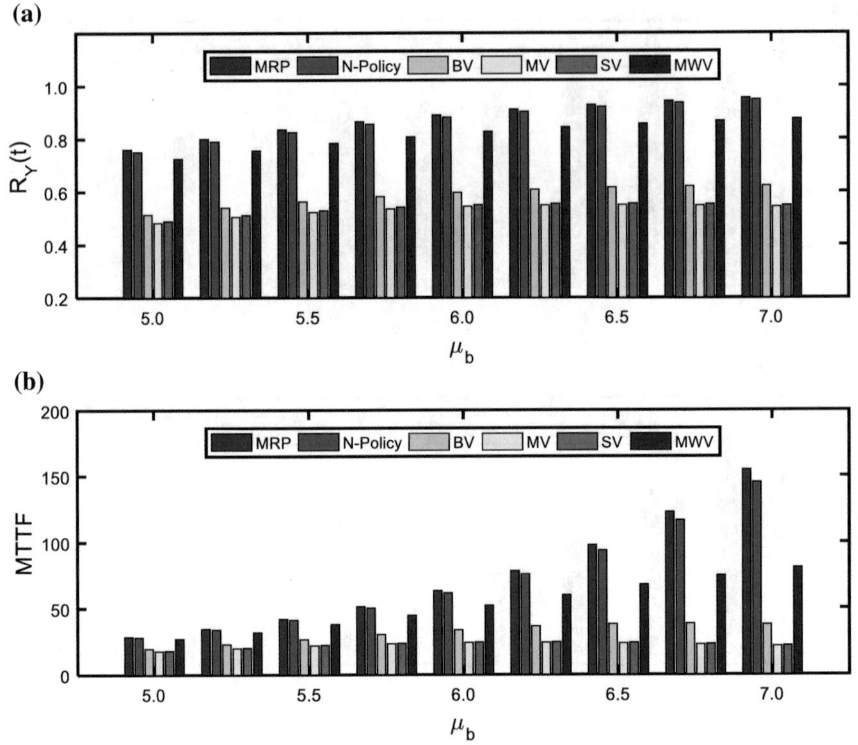

Fig. 18 Effect of repair rate μ_b on reliability and $MTTF$ respectively wrt (i) MRP, (ii) N-policy, (iii) BV, (iv) MV, (v) SV, and (vi) MWV

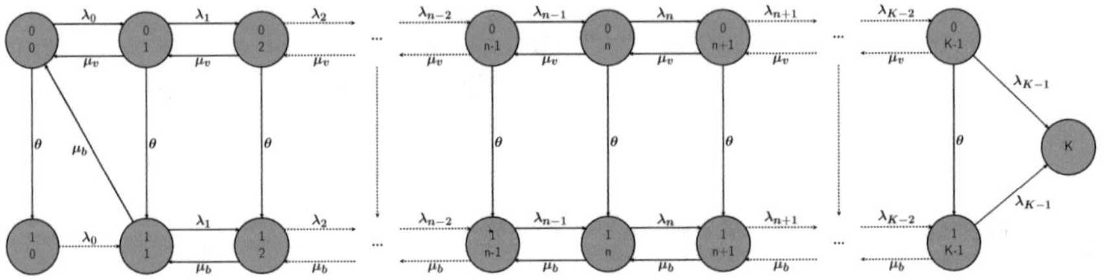

Fig. 19 State transition diagram of the machine repair model with single working vacation policy

switches to the normal service rate. A number of research papers have appeared in the queueing literature in which the MRP and several Makov/Non-Markov queueing models with single working vacation are considered (cf. [65, 67–71]).

Figure 19 represents the state-transition diagram of a MRP with single working vacation policy. For the modeling purpose, we take assumptions and notations similar as in the previous section. The governing set of differential-difference equations is given as follows.

$$\frac{dP_{0,0}(t)}{dt} = -(\lambda_0 + \theta)P_{0,0}(t) + \mu_v P_{0,1}(t) + \mu_b P_{1,1}(t)$$

$$\frac{dP_{0,n}(t)}{dt} = -(\lambda_n + \mu_v + \theta)P_{0,n}(t) + \lambda_{n-1}P_{0,n-1}(t) + \mu_v P_{0,n+1}(t); \quad 1 \le n \le K-2$$

$$\frac{dP_{0,K-1}(t)}{dt} = -(\lambda_{K-1} + \mu_v + \theta)P_{0,K-1}(t) + \lambda_{K-2}P_{0,K-2}(t)$$

$$\frac{dP_{1,0}(t)}{dt} = -\lambda_0 P_{1,0}(t) + \theta P_{0,0}(t)$$

$$\frac{dP_{1,n}(t)}{dt} = -(\lambda_n + \mu_b)P_{1,n}(t) + \lambda_{n-1}P_{1,n-1}(t) + \mu_b P_{1,n+1}(t) + \theta P_{0,n}(t); \quad 1 \le n \le K-2$$

$$\frac{dP_{1,K-1}(t)}{dt} = -(\lambda_{K-1} + \mu_b)P_{1,K-1}(t) + \lambda_{K-2}P_{1,K-2}(t) + \theta P_{0,K-1}(t)$$

$$\frac{dP_K(t)}{dt} = \lambda_{K-1}P_{0,K-1}(t) + \lambda_{K-1}P_{1,K-1}(t)$$

Using the lexicographical sequence of the states of the system, the generator matrix $\mathbf{Q}_7$ in the terms of block-matrices is represented as

$$\mathbf{Q}_7 = \begin{bmatrix} \mathbf{A}_7 & \mathbf{B}_7 & \mathbf{E}_7 \\ \mathbf{C}_7 & \mathbf{D}_7 & \mathbf{F}_7 \\ \mathbf{G}_7 & \mathbf{H}_7 & \mathbf{I}_7 \end{bmatrix}$$

where the matrices $\mathbf{A}_7$, $\mathbf{B}_7$, $\mathbf{C}_7$, and $\mathbf{D}_7$ are the block matrices of order $(K \times K)$, $\mathbf{E}_7$ and $\mathbf{F}_7$ are the column vectors of order $(K \times 1)$, and $\mathbf{G}_7$, $\mathbf{H}_7$, $\mathbf{I}_7$ are the null vectors.

$$\mathbf{A}_7 = \begin{bmatrix} x_1^7 & y_1^7 & 0 & \cdots & 0 & 0 \\ z_2^7 & x_2^7 & y_2^7 & \cdots & 0 & 0 \\ 0 & z_3^7 & x_3^7 & \cdots & 0 & 0 \\ \vdots & \vdots & \vdots & \vdots & \vdots & \vdots \\ 0 & 0 & 0 & \cdots & x_{K-1}^7 & y_{K-1}^7 \\ 0 & 0 & 0 & \cdots & z_K^7 & x_K^7 \end{bmatrix}$$

$$x_n^7 = \begin{cases} -(\lambda_0 + \theta); & n = 1 \\ -(\lambda_{n-1} + \mu_v + \theta); & 2 \le n \le K \\ 0; & \text{otherwise} \end{cases}$$

$$y_n^7 = \begin{cases} \lambda_{n-1}; & 1 \le n \le K-1 \\ 0; & \text{otherwise} \end{cases}$$

$$z_n^7 = \begin{cases} \mu_v; & 2 \le n \le K \\ 0; & \text{otherwise} \end{cases}$$

and

$$\mathbf{B}_7 = diag[\theta, \theta, \ldots, \theta]$$

$$\mathbf{C}_7[2, 1] = \mu_b$$

$$\mathbf{D}_7 = \begin{bmatrix} u_1^7 & v_1^7 & 0 & \cdots & 0 & 0 \\ w_2^7 & u_2^7 & v_2^7 & \cdots & 0 & 0 \\ 0 & w_3^7 & u_3^7 & \cdots & 0 & 0 \\ \vdots & \vdots & \vdots & \vdots & \vdots & \vdots \\ 0 & 0 & 0 & \cdots & u_{K-1}^7 & v_{K-1}^7 \\ 0 & 0 & 0 & \cdots & w_K^7 & u_K^7 \end{bmatrix}$$

$$u_n^7 = \begin{cases} -\lambda_0; & n = 1 \\ -(\lambda_{n-1} + \mu_b); & 2 \le n \le K \\ 0; & \text{otherwise} \end{cases}$$

$$v_n^7 = \begin{cases} \lambda_{n-1}; & 1 \le n \le K - 1 \\ 0; & \text{otherwise} \end{cases}$$

$$w_n^7 = \begin{cases} \mu_b; & 3 \le n \le K \\ 0; & \text{otherwise} \end{cases}$$

$$\mathbf{E}_7 = \mathbf{F}_7 = [0, 0, \ldots, 0, \lambda_{K-1}]^T$$

$$\mathbf{G}_7 = \mathbf{H}_7 = [0, 0, \ldots, 0]^T$$

$$\mathbf{I}_7 = [0]$$

Following are some system performance measures on which the overall mathematical analysis of the machine repair model with single working vacation is depended.

$$E_N(t) = \sum_{j=0}^{1}\sum_{n=0}^{K-1} nP_{j,n} + KP_K(t)$$

$$\tau_p(t) = \sum_{n=1}^{K-1} \mu_v P_{0,n}(t) + \sum_{n=1}^{K-1} \mu_b P_{1,n}(t)$$

$$P_{WV}(t) = \sum_{n=0}^{K-1} P_{0,n}(t)$$

$$R_Y(t) = 1 - P_K(t)$$

$$MTTF = \int_0^{\infty} R_Y(t)dt$$

$$FF(t) = \lambda_{K-1} P_{0,K-1}(t) + \lambda_{K-1} P_{1,K-1}(t)$$

On fixing the default parameters as same as for the previous section, we observe from Figs. 20 and 21 that single working vacation is superior to multiple working vacation and single vacation as well for the high demand of better reliability.

9 MRP with Vacation Interruption Policy (VI)

Notations

T: Threshold for vacation interruption
μ_b: Mean service rate during a busy period
μ_v: Mean service rate during the working vacation period
θ: Vacation rate of the server

In the working vacation policy, it is assumed that the server provides the service with the slower service rate in the vacation period, but in a real-time scenario, this assumption seems more restrictive. Therefore, to overcome this limitation, Li and Tian [72] proposed a vacation interruption policy for a $M/M/1$ queue. According to

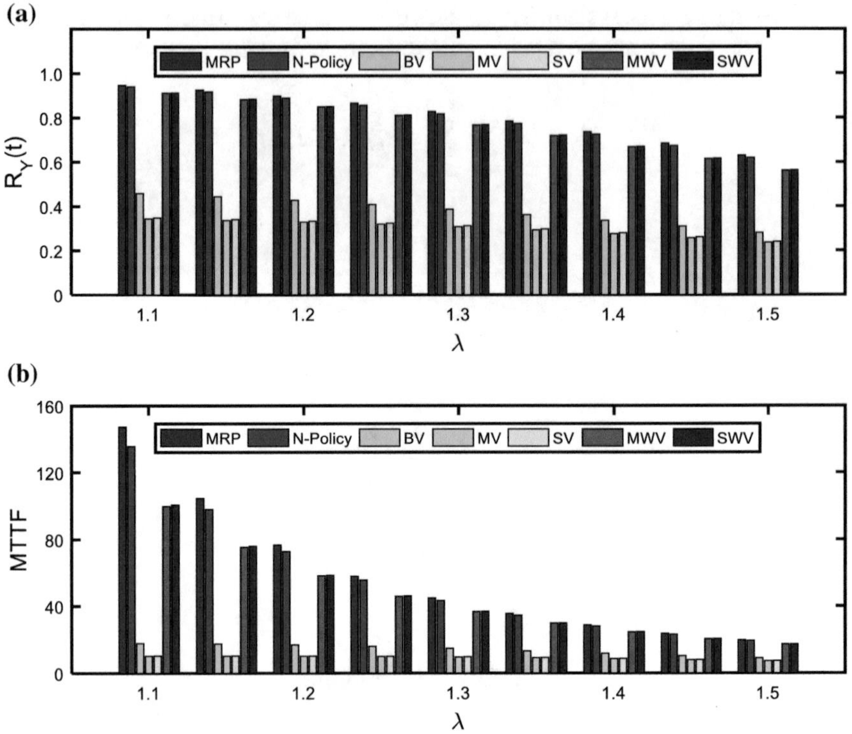

Fig. 20 Effect of failure rate λ on reliability and $MTTF$ respectively wrt (i) MRP, (ii) N-policy, (iii) BV, (iv) MV, (v) SV, (vi) MWV, and (vii) SWV

this policy, if there are more than a certain pre-specified number of failed machines in the system (T) waiting for repair after the service completion instant during the vacation period, the repairman ends his vacation and resumes the normal working attribute. Machining systems with vacation interruption and other features in different contexts have been investigated by many researchers (cf. [73–77]).

The assumptions and notions are identical to respective assumptions/notations in the previous section. Using the transition-state diagram depicted in Fig. 22 of a machine repair model with vacation interruption policy, the following differential-difference equations governing the system are developed.

$$\frac{dP_{0,0}(t)}{dt} = -(\lambda_0 + theta)P_{0,0}(t) + \mu_v P_{0,1}(t) + \mu_b P_{1,1}(t)$$

$$\frac{dP_{0,n}(t)}{dt} = -(\lambda_n + \mu_v + \theta)P_{0,n}(t) + \lambda_{n-1}P_{0,n-1}(t) + \mu_v P_{0,n+1}(t); \quad 1 \le n \le T-1$$

$$\frac{dP_{0,n}(t)}{dt} = -(\lambda_n + \mu_v + \theta)P_{0,n}(t) + \lambda_{n-1}P_{0,n-1}(t); \quad T \le n \le K-1$$

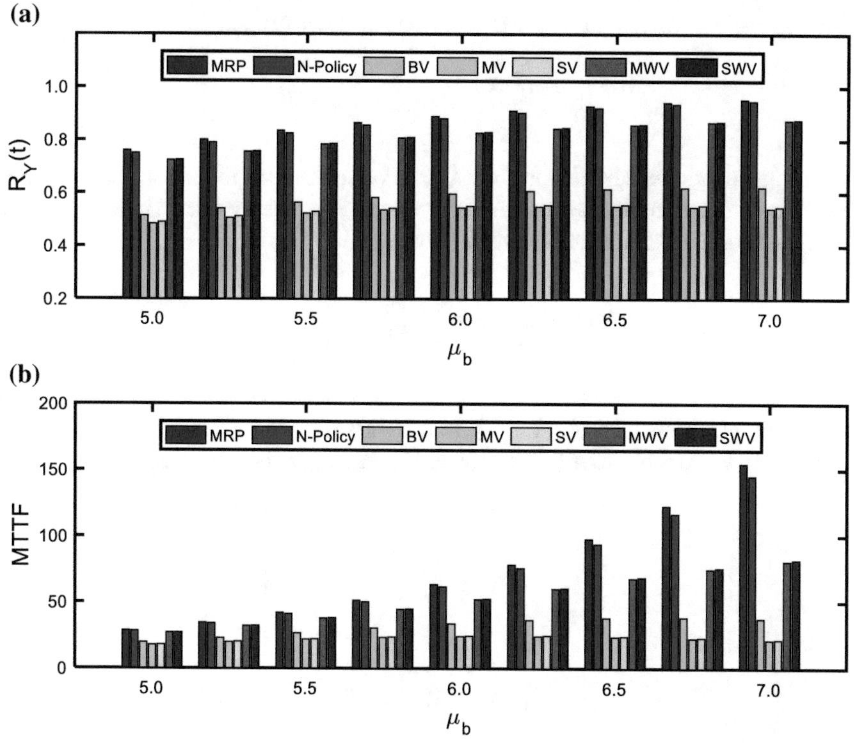

Fig. 21 Effect of repair rate μ_b on reliability and $MTTF$ respectively wrt (i) MRP, (ii) N-policy, (iii) BV, (iv) MV, (v) SV, (vi) MWV, and (vii) SWV

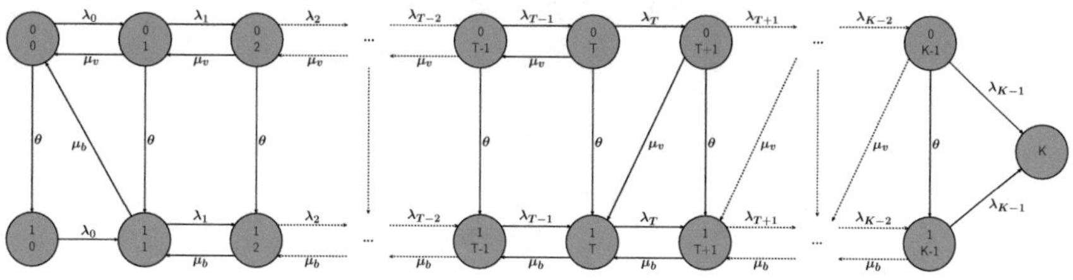

Fig. 22 State transition diagram of the machine repair model with vacation interruption policy

$$\frac{dP_{1,0}(t)}{dt} = -\lambda_0 P_{1,0}(t) + \theta P_{0,0}(t)$$

$$\frac{dP_{1,1}(t)}{dt} = -(\lambda_1 + \mu_b)P_{1,1}(t) + \lambda_0 P_{1,0}(t) + \mu_b P_{1,2}(t) + \theta P_{0,1}(t)$$

$$\frac{dP_{1,n}(t)}{dt} = -(\lambda_n + \mu_b)P_{1,n}(t) + \lambda_{n-1} P_{1,n-1}(t) + \mu_b P_{1,n+1}(t) + \theta P_{0,n}(t); \quad 2 \le n \le T-1$$

$$\frac{dP_{1,n}(t)}{dt} = -(\lambda_n + \mu_b)P_{1,n}(t) + \lambda_{n-1} P_{1,n-1}(t) + \mu_b P_{1,n+1}(t) + \theta P_{0,n}(t) + \mu_v P_{0,n+1}(t);$$

$$T \le n \le K-2$$

$$\frac{dP_{1,K-1}(t)}{dt} = -(\lambda_{K-1} + \mu_b)P_{1,K-1}(t) + \lambda_{K-2}P_{1,K-2}(t) + \theta P_{0,K-1}(t)$$
$$\frac{dP_K(t)}{dt} = \lambda_{K-1}P_{0,K-1}(t) + \lambda_{K-1}P_{1,K-1}(t)$$

The generator matrix, denoted by $\mathbf{Q}_8$, is composed by the block matrices obtained by corresponding transitions between states of the system. The structure of the generator matrix is expressed as follows

$$\mathbf{Q}_8 = \begin{bmatrix} \mathbf{A}_8 & \mathbf{B}_8 & \mathbf{E}_8 \\ \mathbf{C}_8 & \mathbf{D}_8 & \mathbf{F}_8 \\ \mathbf{G}_8 & \mathbf{H}_8 & \mathbf{I}_8 \end{bmatrix}$$

where, the block matrices $\mathbf{A}_8$, $\mathbf{B}_8$, $\mathbf{C}_8$, and $\mathbf{D}_8$ are of order $(K \times K)$, $\mathbf{E}_8$ and $\mathbf{F}_8$ are the column vectors of order $(K \times 1)$, and $\mathbf{G}_8$, $\mathbf{H}_8$, $\mathbf{I}_8$ are the zero vectors.

$$\mathbf{A}_8 = \begin{bmatrix} x_1^8 & y_1^8 & 0 & \cdots & 0 & 0 \\ z_2^8 & x_2^8 & y_2^8 & \cdots & 0 & 0 \\ 0 & z_3^8 & x_3^8 & \cdots & 0 & 0 \\ \vdots & \vdots & \vdots & \vdots & \vdots & \vdots \\ 0 & 0 & 0 & \cdots & x_{K-1}^8 & y_{K-1}^8 \\ 0 & 0 & 0 & \cdots & z_K^8 & x_K^8 \end{bmatrix}$$

$$x_n^8 = \begin{cases} -(\lambda_0 + \theta); & n = 1 \\ -(\lambda_{n-1} + \mu_v + \theta); & 2 \le n \le K \\ 0; & \text{otherwise} \end{cases}$$

$$y_n^8 = \begin{cases} \lambda_{n-1}; & 1 \le n \le K-1 \\ 0; & \text{otherwise} \end{cases}$$

$$z_n^8 = \begin{cases} \mu_v; & 2 \le n \le T+1 \\ 0; & \text{otherwise} \end{cases}$$

and

$$\mathbf{B}_8 = \begin{bmatrix} k_1^8 & 0 & 0 & \cdots & 0 & 0 \\ m_2^8 & k_2^8 & 0 & \cdots & 0 & 0 \\ 0 & m_3^8 & k_3^8 & \cdots & 0 & 0 \\ \vdots & \vdots & \vdots & \vdots & \vdots & \vdots \\ 0 & 0 & 0 & \cdots & k_{K-1}^8 & 0 \\ 0 & 0 & 0 & \cdots & m_K^8 & k_K^8 \end{bmatrix}$$

$$k_n^8 = \begin{cases} \theta; & 1 \le n \le K \\ 0; & \text{otherwise} \end{cases}$$

$$m_n^8 = \begin{cases} \mu_v; & T+2 \le n \le K \\ 0; & \text{otherwise} \end{cases}$$

$$\mathbf{C}_8[2,1] = \mu_b$$

$$\mathbf{D}_8 = \begin{bmatrix} u_1^8 & v_1^8 & 0 & \cdots & 0 & 0 \\ w_2^8 & u_2^8 & v_2^8 & \cdots & 0 & 0 \\ 0 & w_3^8 & u_3^8 & \cdots & 0 & 0 \\ \vdots & \vdots & \vdots & \vdots & \vdots & \vdots \\ 0 & 0 & 0 & \cdots & u_{K-1}^8 & v_{K-1}^8 \\ 0 & 0 & 0 & \cdots & w_K^8 & u_K^8 \end{bmatrix}$$

$$u_n^8 = \begin{cases} -\lambda_0; & n = 1 \\ -(\lambda_{n-1} + \mu_b); & 2 \le n \le K \\ 0; & \text{otherwise} \end{cases}$$

$$v_n^8 = \begin{cases} \lambda_{n-1}; & 1 \le n \le K-1 \\ 0; & \text{otherwise} \end{cases}$$

$$w_n^8 = \begin{cases} \mu_b; & 3 \le n \le K \\ 0; & \text{otherwise} \end{cases}$$

$$\mathbf{E}_8 = \mathbf{F}_8 = [0, 0, \ldots, 0, \lambda_{K-1}]^T$$

$$\mathbf{G}_8 = \mathbf{H}_8 = [0, 0, \ldots, 0]^T$$

$$\mathbf{I}_8 = [0]$$

The closed-form expressions for the expected number of failed machines in the system $E_N(t)$, throughput of the system $\tau_p(t)$, probability that the repairman is on working vacation $P_{WV}(t)$, probability that the vacation of the repairman is interrupted $P_{VI}(t)$, reliability of the system $R_Y(t)$, mean time-to-failure of the system $MTTF$, and failure frequency $FF(t)$ of the system are obtained respectively as follows.

$$E_N(t) = \sum_{j=0}^{1} \sum_{n=0}^{K-1} n P_{j,n}(t) + K P_{_K}(t)$$

$$\tau_p(t) = \sum_{n=1}^{K-1} \mu_v P_{0,n}(t) + \sum_{n=1}^{K-1} \mu_b P_{1,n}(t)$$

$$P_{WV}(t) = \sum_{n=0}^{K-1} P_{0,n}(t)$$

$$P_{VI}(t) = \sum_{T+1}^{K-1} P_{0,n}(t)$$

$$R_Y(t) = 1 - P_{_K}(t)$$

$$MTTF = \int_0^{\infty} R_Y(t)dt$$

$$FF(t) = \lambda_{_{K-1}} P_{_{0,K-1}}(t) + \lambda_{_{K-1}} P_{_{1,K-1}}(t)$$

For the numerical illustration, we consider the values of governing threshold as $T = 8$ and other system parameters as same as in the previous section. The variation

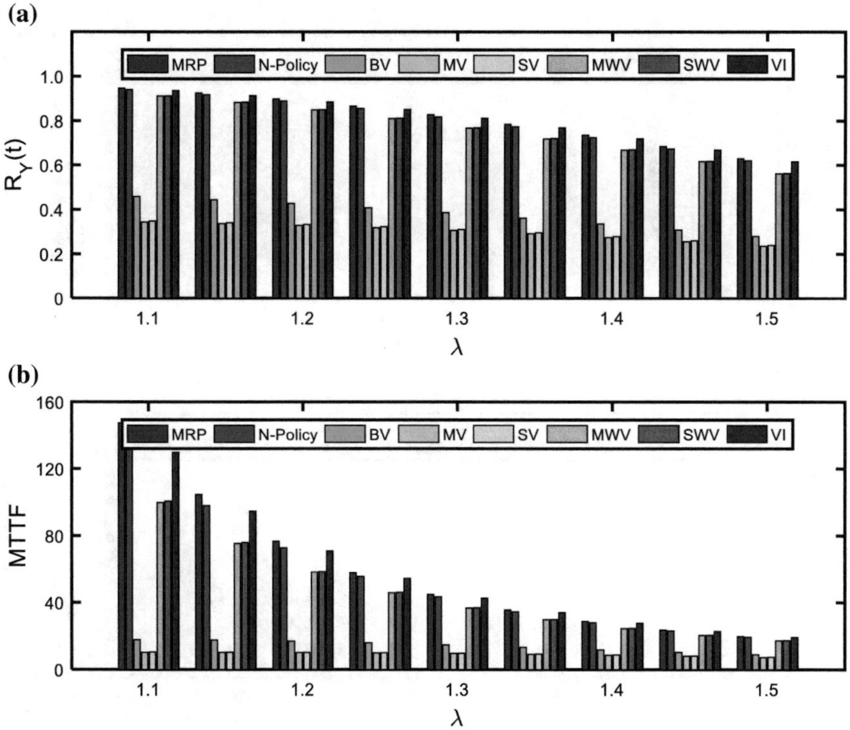

Fig. 23 Effect of failure rate λ on reliability and $MTTF$ respectively wrt (i) MRP, (ii) N-policy, (iii) BV, (iv) MV, (v) SV, (vi) MWV, (vii) SWV, and (viii) VI

of reliability measures is depicted in Figs. 23 and 24 for the varied value of failure rate λ and service rate μ respectively. It is clearly observed that vacation interruption always be a better solution for getting the better reliability of the machining system.

10 Discussion

In this chapter, a comparative analysis of different vacation policies on the reliability characteristics of the machining system is presented. For that purpose, the queueing-theoretic approach has been used and Markovian models are developed. The governing system of Chapman Kolmogrove differential-difference equations has been derived with some assumptions. We employ the numerical scheme, namely Runge-Kutta method of the fourth order, and program the code in MATLAB (2018b) software for computing the transient-state probabilities and henceforth system performance measures. For illustrative purpose, we present the numerical experiment

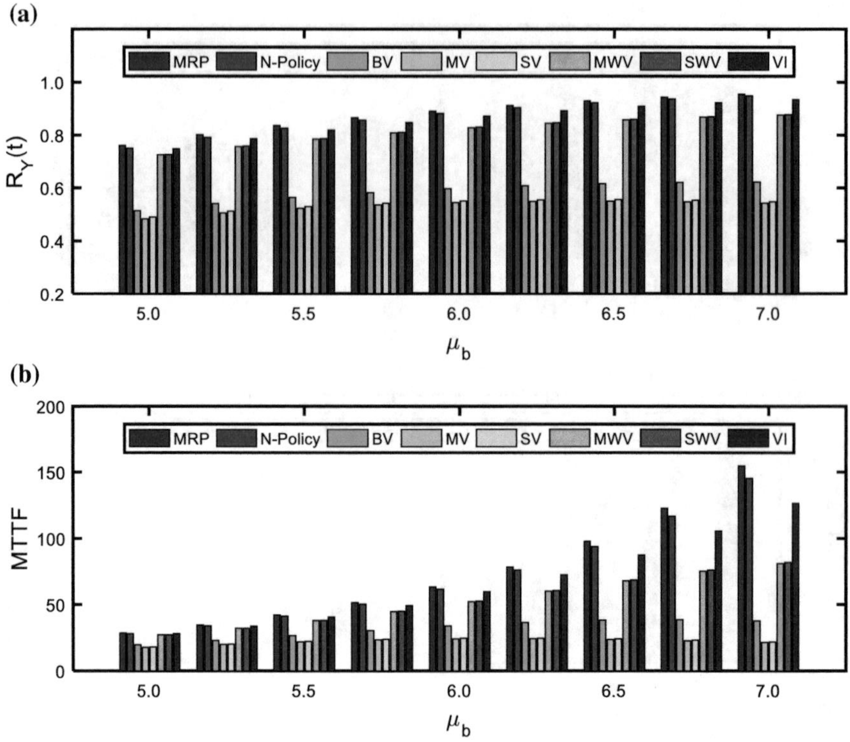

Fig. 24 Effect of repair rate μ_b on reliability and $MTTF$ respectively wrt (i) MRP, (ii) N-policy, (iii) BV, (iv) MV, (v) SV, (vi) MWV, (vii) SWV, and (viii) VI

for some hypothetical default parameters and results are depicted with bar graphs for comparision at glance.

The vacation of the repairman teminates when either number of failed machines exceeds the pre-specified threshold (e.g. N-policy), random duration ends (e.g. Bernoulli vacation, multiple/single vacation, multiple/single working vacation), or due to interruption (e.g. vacation interruption). In general, the vacation of repairman starts when there is no failed machine in the system to be repaired. But, in Bernoulli vacation, the repairman may opt the vacation in the presence of failed machines to be repaired probabilistically. In single vacation, the repairman may remain idle even when there is no failed machine to be repaired in the system after the completion of the vacation period. All vacation policies directly affect the reliability of the system. Under all vacations, reliability of the system and $MTTF$ reduce.

Next, for the comparative analysis purpose, we range the values of failure rate (λ) and service rate ($\mu = \mu_b$) from 1.1 to 1.5 and 5.0 to 7.0 respectively. Figures 23 and 24 depict the combined illustration of all the studied vacation policies on the reliability ($R_Y(t)$) and $MTTF$ of the system. Figure 23 reveals that as the value of λ increases, correspondingly the reliability and $MTTF$ of the system decreases which is quite obvious. Figure 24 exhibits the reverse trend of variation for the increasing rate of service. For any test instance, if we fix the value of λ and μ, it is easily observed that the value of reliability and $MTTF$ is higher for working vacation and vacation interruption policies in comparison of fully vacation policies. Therefore, the selection of the working vacation policies is a much better choice for the system analyst and engineers instead of opting the full vacation policies. So, as the conclusive remark, the system analyst and engineers may opt the working vacation and vacation interruption policies in order to achieve the maximal reliability of the system.

References

1. Cox DR, Miller HD (1965) The theory of stochastic process. Chapman & Hall, London
2. Bhat UN (1972) Elements of applied stochastic process. Wiley, New York
3. Cooper RB (1972) Introduction to queueing theory. MacMillan, New York
4. Medhi J (1982) Stochastic processes. Wiley
5. Gross D, Harris C (1998) Fundamentals of queueing theory. Wiley Series in Probability and Statistics
6. Medhi J (2002) Stochastic models in queueing theory. Academic Press, Amsterdam
7. Gross D, Shortle JF, Thompson JM, Harris CM (2008) Fundamentals of queueing theory. Wiley, New York
8. Balagurusamy E, Misra KB (1976) Reliability calculation of redundant systems with non-identical units. Microelectron Reliab 15(5):376–377
9. Trivedi KS, Dugan JB, Geist R, Smotherman M (1984) Hybrid reliability modeling of fault-tolerant computer systems. Comput Electr Eng 11(2–3):87–108
10. Sztrik J, Bunday BD (1993) Machine interference problem with a random environment. Eur J Oper Res 65(2):259–269
11. Levantesi R, Matta A, Tolio T (2003) Performance evaluation of continuous production lines with machines having different processing times and multiple failure modes. Perform Eval 51(2–4):247–268
12. Haque L, Armstrong MJ (2007) A survey of the machine interference problem. Eur J Oper Res 179(2):469–482
13. Dimitriou I, Langaris C (2010) A repairable queueing model with two-phase service, start-up times and retrial customers. Comput Oper Res 37(7):1181–1190
14. Ke JC, Lin CH (2010) Maximum entropy approach to machine repair problem. Int J Serv Oper Inform 5(3):197–208
15. Lv S, Yue D, Li J (2010) Transient reliability of machine repairable system. J Inf Comput Sci 7(13):2879–2885
16. Jain M, Sharma GC, Pundhir RS (2010) Some perspectives of machine repair problem. IJE Trans B: Appl 23(3–4):253–268
17. Wu Q, Wu S (2011) Reliability analysis of two-unit cold standby repairable systems under Poisson shocks. Appl Math Comput 218(1):171–182
18. Ruiz CJE, Li QL (2011) Algorithm for a general discrete K-out-of-N: G system subject to several types of failure with an indefinite number of repair-persons. Eur J Oper Res 211(1):97–111
19. Nourelfath M, Chatelet E, Nahas N (2012) Joint redundancy and imperfect preventive maintenance optimization for series-parallel multi-state degraded systems. Reliab Eng Syst Saf 103:51–60
20. El-Damcese MA, Shama MS (2013) Reliability and availability analysis of a standby repairable system with degradation facility. Int J Res Rev Appl Sci 16(3):501–507

21. Wells CE (2014) Reliability analysis of a single warm-standby system subject to repairable and non-repairable failures. Eur J Oper Res 235(1):180–186
22. Kuo C, Sheu S, Ke JC, Zhang ZG (2014) Reliability-based measures for a retrial system with mixed standby components. Appl Math Model 38(19–20):4640–4651
23. Ke JC, Liu TH, Wu CH (2015) An optimum approach of profit analysis on the machine repair system with heterogeneous repairmen. Appl Math Comput 253(15):40–51
24. Gonzaleza PP, Viagasa VF, Garciab MZ, Framinan JM (2019) Constructive heuristics for the unrelated parallel machines scheduling problem with machine eligibility and setup times. Comput Ind Eng 131:131–145
25. Jain M (1997) Optimal N-policy for single server Markovian queue with breakdown, repair and state dependant arrival rate. Int J Manag Syst 13(3):245–260
26. Ushakumari PV, Krishnamoorthy A (1998) k-out-of-n system with general repair: the N-policy. In: Proceedings of the II international symposium on semi-markov process, UTC, Compiegne, France
27. Gupta SM (1999) N-policy queueing system with finite source and warm spares. Opsearch 36(3):189–217
28. Krishnamoorthy A, Ushakumari PV, Lakshmi B (2002) k-out-of-n system with repair: the N-policy. Asia Pac J Oper Res 19(1):47–61
29. Ushakumari PV, Krishnamoorthy A (2004) k-out-of-n system with repair: the max(N, T)-policy. Perform Eval 57(2):221–234
30. Yue D, Yue W, Qi H (2008) Analysis of a machine repair system with warm spares and N-policy Vacations. In: The 7th international symposium on operations research and its applications (ISORA'08), pp 190–198
31. Jain M, Upadhyaya S (2009) Threshold N-policy for degraded machining system with multiple types of spares and multiple vacations. Qual Technol Quant Manag 6(2):185–203
32. Wang TY, Yang DY, Li MJ (2010) Fuzzy analysis for the N-policy queues with infinite capacity. Int J Inf Manag Sci 21:41–56
33. Singh CJ, Jain M, Kumar B (2013) Analysis of queue with two phases of service and m phases of repair for server breakdown under N-policy. Int J Serv Oper Manag 16(3):373–406
34. Jain M, Shekhar C, Shukla S (2016) A time-shared machine repair problem with mixed spares under N-policy. J Ind Eng Int 12(2):145–157
35. Chen WL, Wang KH (2018) Reliability analysis of a retrial machine repair problem with warm standbys and a single server with N-policy. Reliab Eng Syst Saf 180:476–486
36. Ayyappan G, Karpagam S (2019) Analysis of a bulk queue with unreliable server, immediate feedback, N-policy, Bernoulli schedule multiple vacation and stand-by server. Ain Shams Eng J. https://doi.org/10.1016/j.asej.2019.03.008
37. Blanc JPC, Mei RDV (1995) Optimization of polling systems with Bernoulli schedules. Perform Eval 22(2):139–158
38. Choudhury G, Madan KC (2004) A two-phase batch arrival queueing system with a vacation time under Bernoulli schedule. Appl Math Comput 149(2):337–349
39. Choudhury G, Madan KC (2005) A two-stage batch arrival queueing system with a modified Bernoulli schedule vacation. Math Comput Model 42(1–2):71–85
40. Choudhury G, Deka M (2012) A single server queueing system with two phases of service subject to server breakdown and Bernoulli vacation. Appl Math Model 36(12):6050–6060
41. Liu TH, Ke JC (2014) On the multi-server machine interference with modified Bernoulli vacations. J Ind Manag Optim 10(4):1191–1208
42. Shrivastava RK, Mishra AK (2014) Analysis of queuing model for machine repairing system with Bernoulli vacation schedule. Int J Math Trends Technol 10(2):85–92
43. Rajadurai P, Chandrasekaran VM, Saravanarajan MC (2015) Analysis of an $M^{[X]}/G/1$ unreliable retrial G-queue with orbital search and feedback under Bernoulli vacation schedule. OPSEARCH 53(1):197–223
44. Singh CJ, Jain M, Kumar B (2016) $M^X/G/1$ unreliable retrial queue with option of additional service and Bernoulli vacation. Ain Shams Eng J 7(1):415–429

45. Choudhury G, Deka M (2018) A batch arrival unreliable server delaying repair queue with two phases of service and Bernoulli vacation under multiple vacation policy. Qual Technol Quant Manag 15(2):157–186
46. Thomo L (1997) A multiple vacation model $M^X/G/1$ with balking. Nonlinear Amlysis, Theory, Methods Appl 30(4):2025–2030
47. Zhang ZG, Vickson R, Love E (2001) The optimal service policies in an $M/G/1$ queueing system with multiple vacation types. Inf Syst Oper Res 39(4):357–366
48. Ke JC, Wang KH (2007) Vacation policies for machine repair problem with two type spares. Appl Math Model 31(5):880–894
49. Ke JC, Wu CH (2012) Multi-server machine repair model with standbys and synchronous multiple vacation. Comput Ind Eng 62(1):296–305
50. Yuan L (2012) Reliability analysis for a system with redundant dependency and repairmen having multiple vacations. Appl Math Comput 218(24):11959–11969
51. Liou CD (2015) Optimization analysis of the machine repair problem with multiple vacations and working breakdowns. J Ind Manag Optim 11(1):83–104
52. Sun W, Li S, Guo CE (2016) Equilibrium and optimal balking strategies of customers in Markovian queues with multiple vacations and N-policy. Appl Math Model 40(1):284–301
53. Lee HW, Lee SS, Chae KC, Nadarajan R (1992) On a batch service queue with single vacation. Appl Math Model 16(1):36–42
54. Madan KC, Al-Rub AZA (2004) On a single server queue with optional phase type server vacations based on exhaustive deterministic service and a single vacation policy. Appl Math Comput 149(3):723–734
55. Xu X, Zhang ZG (2006) Analysis of multi-server queue with a single vacation (e, d)-policy. Perform Eval 63(8):825–838
56. Wang HX, Xu GQ (2012) A cold system with two different components and a single vacation of the repairman. Appl Math Comput 219(5):2634–2657
57. Wu W, Tang Y, Yu M, Jiang Y (2014) Reliability analysis of a k-out-of-n:G repairable system with single vacation. Appl Math Model 38(24):6075–6097
58. Wu CH, Ke JC (2014) Multi-server machine repair problems under a (V, R) synchronous single vacation policy. Appl Math Model 38(7–8):2180–2189
59. Zhang Y, Wu W, Tang Y (2017) Analysis of an k-out-of-n:G system with repairman's single vacation and shut off rule. Oper Res Perspect 4:29–38
60. Neuts MF (1981) Matrix-geometric solutions in stochastic models. Johns Hopkins University Press, Baltimore
61. Servi L, Finn S (2002) $M/M/1$ queue with working vacations $(M/M/1/WV)$. Perform Eval 50(1):41–52
62. Baba Y (2005) Analysis of a $GI/M/1$ queue with multiple working vacations. Oper Res Lett 33(2):201–209
63. Tian N, Ma Z, Liu M (2008) The discrete time Geom/Geom/1 queue with multiple working vacations. Appl Math Model 32(12):2941–2953
64. Selvaraju N, Goswami C (2013) Impatient customers in an $M/M/1$ queue with single and multiple working vacations. Comput Ind Eng 65(2):207–215
65. Guha D, Goswami V, Banik AD (2015) Equilibrium balking strategies in renewal input batch arrival queues with multiple and single working vacation. Perform Eval 94:1–24
66. Rajadurai P, Saravanarajan MC, Chandrasekaran VM (2018) A study on $M/G/1$ feedback retrial queue with subject to server breakdown and repair under multiple working vacation policy. Alex Eng J 57(2):947–962
67. Lin CH, Ke JC (2009) Multi-server system with single working vacation. Appl Math Model 33(7):2967–2977
68. Yang DY, Wang KH, Wu CH (2010) Optimization and sensitivity analysis of controlling arrivals in the queueing system with single working vacation. J Comput Appl Math 234(2):545–556
69. Li J, Tian N (2011) Performance analysis of a $GI/M/1$ queue with single working vacation. Appl Math Comput 217(19):4960–4971

70. Guha D, Banik AD (2013) On the renewal input batch-arrival queue under single and multiple working vacation policy with application to EPON. Inf Syst Oper Res 51(4):175–191
71. Kempa WM, Kobielnik M (2018) Transient solution for the queue-size distribution in a finite-buffer model with general independent input stream and single working vacation policy. Appl Math Model 59:614–628
72. Li J, Tian N (2007) The $M/M/1$ queue with working vacations and vacation interruption. J Syst Sci Syst Eng 16(1):121–127
73. Tao L, Liu Z, Wang Z (2011) The $GI/M/1$ queue with start-up period and single working vacation and Bernoulli vacation interruption. Appl Math Comput 218(8):4401–4413
74. Gao S, Liu Z (2013) An $M/G/1$ queue with single working vacation and vacation interruption under Bernoulli schedule. Appl Math Model 37(3):1564–1579
75. Tao L, Zhang L, Xu X, Gao S (2013) The $GI/Geo/1$ queue with Bernoulli-schedule-controlled vacation and vacation interruption. Comput Oper Res 40(7):1680–1692
76. Laxmi PV, Jyothsna K (2015) Impatient customer queue with Bernoulli schedule vacation interruption. Comput Oper Res 56:1–7
77. Rajadurai P, Chandrasekaran VM, Saravanarajan MC (2018) Analysis of an unreliable retrial G-queue with working vacations and vacation interruption under Bernoulli schedule. Ain Shams Eng J 9(4):567–580

Software Multi Up-Gradation Modeling Based on Different Scenarios

Adarsh Anand, Priyanka Gupta, Yoshinobu Tamura and Mangey Ram

Abstract Last decade has seen several multi up-gradation models and related software release policies. The unified modeling framework for developing multi-release software systems has gained tremendous lime light for prediction of software reliability. Various models have been developed to cater the diversified scenarios like those of imperfect debugging, stochasticity and testing effort related modeling. In the present work, using the unified modeling approach, it has been checked out which release performs best for a particular type of real life scenario. The data for Tandem computers has been utilized for validation purpose.

Keywords Imperfect debugging · Multi up-gradation · Stochastic behaviour · Testing effort · Weighted criteria approach

1 Introduction

Nowadays, software has become a "complexity sponge" because it has to accommodate evolving understanding, thereby, making it a facilitator of progress. With the surfacing of new technology, augmentation of dependencies over a software product is observed. In some areas like NASA, it is growing by a factor of ten in each decade. According to a study, it is estimated that Orion's primary flight security had exceeded 1Million lines of codes [1]. Therefore, more reliable software is demanded to qualify

A. Anand (✉) · P. Gupta
University of Delhi, Delhi, India
e-mail: adarsh.anand86@gmail.com

P. Gupta
e-mail: priyankag0794@gmail.com

Y. Tamura
Tokyo City University, Tokyo, Japan
e-mail: tamuray@tcu.ac.jp

M. Ram
Department of Mathematics, Graphic Era Deemed to be University, Dehradun, Uttarakhand, India
e-mail: mangeyram@gmail.com

M. Ram and H. Pham (eds.), *Advances in Reliability Analysis and its Applications*, Springer Series in Reliability Engineering,
https://doi.org/10.1007/978-3-030-31375-3_8

this escalating software reliant and safety critical systems. To cater this demand of high reliability, prolonged testing of the software is suggested before it is launched in the market. But, due to the presence of fierce competition and cost incurred in the testing process, the management ends up releasing the software early in the market. The end result is invariably a software product which has failed to deliver a standard of software products. A plethora of bugs are present in the released version of the software thereby generating difficulties for the management as well as for its users. Along with this issue, another major concern which the software's developers has to face is the continuous changes in the technology which forces its users to believe the fact that the software they are using is outdated and is declining in performance.

To address these kinds of issues, provide greater value to its customers and to increase its market presence, management comes up with the concept of multi up-gradation of the software. This multi up-gradation of the software can be treated as a boom or bust situation. The additional features of the upgraded version are treated as a boom by the users. But at the same time it is not viable to believe that any up-gradation done in the software will always be beneficial. Enormous changes in the code of the software are done for its up-gradation which increases the complexity of the code showing the way to the increase in the fault content. The presence of these faults in the software leads to the gaffes in the industry. For example: In 2018, December, Telco supplier Ericsson experienced an outage due to a bug which crippled the network of around 11 countries giving a major headache to more than 20 million Softbank and 15 million mobile customers in Japan and UK respectively, due to which company had paid millions of pounds as compensation [2]. Moreover, in 2015, F-35 Joint Strike Fighter plane was hit by a sporadic bug which corrupted all its sensors due to which it was not able to identify its targets [3]. Similarly, a lot of blunders have occurred in the past thereby creating the need of safe up-gradations in the software.

To cater this need of safe up-gradations, utilization of Software Reliability Growth Models (SRGM) is done. The models with the analogous objective are employed to predict the incidence of faults in the software. The information provided by the SRGM can be further utilised to gain the insights of the software behaviour and helps in planning the various strategies by the management. The literature contains enormous number of such type of SRGM in which this concept has been used to plan the multi up-grades of the software. Moreover, the target of researchers was always to develop such a SRGM which describes the behaviour of the software in a perfect manner. In order to achieve the target, presence of some scenarios such as imperfect debugging, testing effort and stochastic behaviour of fault correction rate were identified by them. Their detailed description is provided below.

In this era, the competition is increasing with brisk speed. Due to the presence of distributed computing system and enhancement in the functionality of the software, it is not viable to believe that testing team is able to remove a particular bug perfectly. In other words it can be said that despite the fact that the rigorous testing is performed, they are often starved of the time and resources required to do the job correctly which leads to the situation in which faults are removed in an inappropriate manner. The situation where a bug is not removed perfectly or increment in the fault content during the debugging process can be the two possible situations which may arise here. The above stated two situations are coined under one single term known as "Imperfect debugging". This may lead to consumer's disappointment, increased costs and a negative impact on the firm's status.

Apart from this, it is seen that each fault has its own peculiarities and they are subsequently assigned to different fixers for their resolution making it reasonable to assume that different amount of efforts are employed by the testing team for their removal. Therefore, the assumption of inclusion of constant efforts by the testing team during the testing phase of the software should be relaxed and the SRGM are developed incorporating the testing effort function.

Moreover, to sustain in the field of research, many mathematical models are developed by the researchers which are based on the assumption of discrete behaviour of the fault correction process. Technically, this assumption has failed to represent the real life scenarios of the software accurately. With the increase in the size of the software code, decrease in the number of faults corrected with respect to the original fault content is observed which causes randomness in the fault correction process. The root of this problem can typically be traced back to the presence of irregular fluctuations. This index (presence of irregular fluctuations) allows researchers to incorporate the stochastic behaviour of fault correction rate directly in the process.

Literature contains numerous number of such SRGM which deals with the above stated concepts. The reference [4] has showed the light on developing a mathematical framework describing the fault detection rate using the approach of hazard rate function. Many researchers have worked upon the different fault correction rates following exponential, S-shaped and Logistic distribution [4]. With the course of time, the need of inclusion of multi up-gradation was observed. The pioneer attempt in this direction was made by Kapur et al. [5] in which the concept of SRGM was combined with multi up-gradation of the software based on all previous releases. Authors [6] have come up with the model which is based on just previous release of the software. Advancement in the previous model was given by Singh et al. [7] in which two types of failure correction rate Logistic and Normal distribution were considered and the results were compared. Analysts like [8] and [9] have dealt with the faults in the operational phase of the software along with its multi release framework. All the above stated work was done considering the perfect nature of the debugging process. But soon this assumption was relaxed. Kapur et al. [10] have formulated a mathematical model combining the effect of different severity of faults along with the concept of multi up-gradation of the software. Singh et al. [11] had developed a model in which the fault severity of the software is modelled with the testing effort function. Many researchers like [12, 13] has worked in direction of formulating the stochastic model for successive software releases. Aggarwal et al. [14] have presented the concept of imperfect debugging along with that of multi release of the software. Whereas some authors like [15] have also been successful to incorporate the error generation form of imperfect debugging along with the stochastic behaviour of fault correction process. The authors [16] have shown the impact of presence of both users and testers in the debugging process of multi upgrades of the software with the help of convolution property. The authors [17] have innovatively taken care of logistic distribution for failure correction rate in the modelling of multi up-grades of software considering the impact of all previous releases.

From the above stated literature review, it can be culminated that a lot of researchers have worked on the concept of multi up-gradation of the software taking care of many diversified scenarios such as testing effort, imperfect debugging and stochastic behaviour taken one at a time. But the study in which all of these scenarios are compared parallelly under the generalized framework of multi up-gradation has gained very less attention. In this paper, the authors have uniquely identified all those unexplored areas and have tried to address the issue by comparing those four diversified scenarios under a single roof along with the concept of multi up-gradation of the software. Moreover, Logistic distribution has been employed to cater the flexible nature of failure correction rate.

Furthermore, in order to determine which scenario best describes a particular release, various goodness of fit criteria are compared. But this can also be the case where this comparison is not sufficient enough to give us the results. To overcome this problem, an approach of multi- criteria problem i.e. Weighted Criteria approach is used. Utilization of this technique has a very diverse literature background. Many authors like Anjum et al. [18] have used this concept in the area of software reliability in order to rank various software reliability models. Anand et al. [19] have ranked various innovation based diffusion models using this technique in the area of marketing as well. The field of vulnerability detection modeling has also being introduced with this concept and the ranking of various vulnerability detection models was done by Anand et al. [20].

The serial wise description of the presented research paper is described as follows: the notations used and the mathematical modeling for the study is presented in Sects. 2 and 3 respectively. Since the validation of the developed model is the eventual purpose of the research work, this is presented in the Sect. 4. Finally, the paper is concluded in the last section.

2 Notations

a_i	Total fault content generated in the software due to *i*th release, i = 1, 2, … n
t_i	Time of the *i*th release
$m_i(t)$	Total number of faults removed from the *i*th release of the software at time t
$H(t)$	Probability distribution function
b_i	Fault removal rate for *i*th release, i = 1, 2, … n
β_i	Learning parameter for *i*th release, i = 1, 2, 3, …n
σ_i	Magnitude for the irregular fluctuations for *i*th release
α_i	Rate at which faults are introduced in the *i*th release
ρ_i	Coefficient of perfect debugging for the *i*th release
$W(t)$	Cumulative testing effort in the interval $(0,\ t)$.

3 Methodology

There are substantial precedents which show the usage of unification approach [4] in determining the collective count of the faults eradicated from the software till time t. After following the same and all the basic assumptions of NHPP model, the same can be represented with the help of the following equation:

$$m(t) = a.H(t) \tag{1}$$

Now, the above Eq. (1) is framed in the context of multi up- gradation of the software and the resulting equations are represented below:

Release 1

The moment the coding team has done their part, the testing team comes into the picture. Handful amount of faults are removed by them before the software is made available for its users i.e. before its release. Mathematical representation of total number of faults removed prior to the first release is represented as:

$$m_1(t) = a_1.H_1(t) \tag{2}$$

where, $H_1(t)$ represents the rate at which the faults are removed at time t.

Release 2

The presence of cut-throat competition in the market has somehow forced the software firms to enrich their features by bringing upgrades. This enhancement in the features often leads to the augmentation in the fault content of the software. Therefore, meticulous efforts are applied by the testing team for their removal. But even after that, some of them remained reclined in the software. Therefore, recently generated faults of the newer version i.e.a_2 along with the remaining fault content which was not removed in the previous version of the software i.e. $a_1.(1 - H_1(t_1))$ are now to be removed with the newer fault correction rate $H_2(t)$. The mathematical representation of the same is provided as

$$m_2(t) = a_2.H_2(t - t_1) + a_1.(1 - H_1(t_1)).H_2(t - t_1), \quad t_1 \le t \le t_2 \tag{3}$$

Release 3

Due to the need of an hour, multiple software releases are done by the management which shows its way to the enhancement in the functionality of the software as well as in the overall fault content. In order to remove the peculiarities from the software, rigorous testing is continued in such a way that the leftover faults of the all previous versions coupled with the freshly generated ones, i.e. $a_2.(1 - H_2(t_2 - t_1))+$ $a_1.(1 - H_1(t_1)).(1 - H_2(t_2 - t_1))$ and a_3 respectively, are to be dealt with the newer correction rate $H_3(t)$. The mathematical representation of the above stated scenario can be expressed by the following equation

$$\begin{aligned} m_3(t) &= a_3.H_3(t-t_2) + a_2.(1-H_2(t_2-t_1))H_3(t-t_2) \\ &\quad + a_1.(1-H_1(t_1)).(1-H_2(t_2-t_1)).H_3(t-t_2) \end{aligned} \qquad t_2 \le t \le t_3 \tag{4}$$

The above equation number (4) can be generalised for the ith release of the software and resulted equation can be expressed as:

$$\begin{aligned} m_i(t) &= a_i.H_i(t-t_{i-1}) + a_{i-1}.(1-H_{i-1}(t_{i-1}-t_{i-2})).H_i(t-t_{i-1}) + \ldots.. \\ &\quad + a_1.(1-H_1(t_1)).(1-H_2(t_2-t_1)).(1-H_3(t_3-t_2))\ldots \\ &\quad \ldots(1-H_{i-1}(t_{i-1}-t_{i-2})).H_i((t-t_{i-1})) \end{aligned} \tag{5}$$

Along with the concept of generalized framework of multi up-gradation of the software the authors have assumed that the failure correction rate is following Logistic distribution. This assumption was considered to incorporate its flexible nature. The Logistic distribution is entitled in the following equation:

$$H(t) = \frac{1-\exp(-b.t)}{1+\beta.\exp(-b.t)} \tag{6}$$

For the inclusion of generality, the parameters of the Logistic distribution i.e. b and β are assumed to be different for each release. Along with this, the authors have studied the effect of four diversified scenarios parallelly in the above stated concept namely perfect debugging, imperfect debugging, testing effort and the presence of stochastic behaviour which are described as follows.

Case 1: Perfect debugging

This is the simplest scenario in which behaviour of the testing team is considered to be perfect in nature. All the faults perceived by the testing team are eventually removed. No alteration in the failure correction rate is seen in this case. Therefore, the eventual modeling for multi up-gradation of software under this scenario is presented as below:

$$\begin{aligned} m_1(t) &= a_1.H_1(t), && 0 \le t \le t_1 \\ m_2(t) &= a_2.H_2(t-t_1) + a_1.(1-H_1(t_1)).H_2(t-t_1), && t_1 \le t \le t_2 \\ m_3(t) &= a_3.H_3(t-t_2) + a_2.(1-H_2(t_2-t_1))H_3(t-t_2) && \\ &\quad + a_1.(1-H_1(t_1)).(1-H_2(t_2-t_1)).H_3(t-t_2) && t_2 \le t \le t_3 \end{aligned}$$

where, $H_i(t) = \frac{1-\exp(-b_i.t)}{1+\beta_i.\exp(-b_i.t)}$, $i = 1, 2, 3, \ldots$ Similarly, the formulation can be made for the ith release of the software.

Case 2: Imperfect debugging

This case deals with the scenario in which eradication of faults follows slightly imperfect nature. Both types of imperfect debugging i.e. introduction of faults and the partial removal of faults are considered. The final representation of multi up-gradation modeling under this scenario is presented below:

$$\begin{aligned} m_1(t) &= \tfrac{a_1}{1-\alpha_1}.H(t) && 0 \le t \le t_1 \\ m_2(t) &= \tfrac{a_2}{1-\alpha_2}.H_2(t-t_1) + \tfrac{a_1}{1-\alpha_1}.(1-H_1(t_1)).H_2(t-t_1) && t_1 \le t \le t_2 \\ m_3(t) &= \tfrac{a_3}{1-\alpha_3}.H_3(t-t_2) + \tfrac{a_2}{1-\alpha_2}.(1-H_2(t_2-t_1))H_3(t-t_2)+ && \\ &+\tfrac{a_1}{1-\alpha_1}.(1-H_1(t_1)).(1-H_2(t_2-t_1)).H_3(t-t_2) && t_2 \le t \le t_3 \end{aligned}$$

where, $H_i(t) = \left[1 - \left(\frac{(1+\beta_i)e^{(-b_i.t)}}{1+\beta_i.e^{(-b_i.t)}}\right)^{\rho_i(1-\alpha_i)}\right]$, $i = 1, 2, 3, \ldots$ represents the eventual failure correction rate. Similar formulation can be made for the ith release of the software.

Case 3: Testing effort

The need for the incorporation of testing effort while forming the SRGM's cannot be ignored. Therefore, the authors have employed this concept along with the multi up-gradation of software which is being presented in the following equation:

$$\begin{aligned} m_1(W(t)) &= a_1.H_1(W(t)) && 0 \le t \le t_1 \\ m_2(W(t)) &= a_2.H_2(W(t-t_1)) + a_1.(1-H_1(W(t_1))).H_2(W(t-t_1)) && t_1 \le t \le t_2 \\ m_3(W(t)) &= a_3.H_3(W(t-t_2)) + a_2.(1-H_2(W(t_2-t_1))).H_3(W(t-t_2)) && \\ &+a_1.(1-H_1(W(t_1))).(1-H_2(W(t_2-t_1))).H_3(W(t-t_2)) && t_2 \le t \le t_3 \end{aligned}$$

where, $H(t) = \frac{1-\exp(-b.W(t))}{1+\beta.\exp(-b.W(t))}$, $i = 1, 2, 3, \ldots$ represents the improved failure correction rate in which efforts applied by the testers i.e. $W(t)$ is assumed to follow Logistic Pattern which is represented as: $W(t) = \frac{1-\exp(-b.t)}{1+\beta.\exp(-b.t)}$ and $m_i(W(t))$ represents faults removed in the ith release of the software after applying $W(t)$ effort at time t. The similar formulation can be made for the ith release of the software.

Case 4: Stochastic behaviour

The inclusion of stochastic behaviour of fault correction process has resulted in the slight modulation in the fault correction rate. The formulated multi up-gradation and the resulting failure correction rate are expressed by the following equations:

$$\begin{aligned} m_1(t) &= a_1.H_1(t) \quad 0 \le t \le t_1 \\ m_2(t) &= a_2.H_2(t-t_1) + a_1.(1-H_1(t_1)).H_2(t-t_1) \quad t_1 \le t \le t_2 \\ m_3(t) &= a_3.H_3(t-t_2) + a_2.(1-H_2(t_2-t_1))H_3(t-t_2) \\ &+a_1.(1-H_1(t_1)).(1-H_2(t_2-t_1)).H_3(t-t_2) \quad t_2 \le t \le t_3 \end{aligned}$$

where, $H(t) = \left[1 - \left(\frac{1+\beta}{1+\beta.e^{(-b.t)}}\right).e^{\left(-b.t+\frac{\sigma^2.t}{2}\right)}\right]$ is the rate by which failures are rectified. Similar formulations can also be made for the ith release of the software.

4 Numerical Analysis

In this section the outlook for the validation of the above explained model is presented. The parameters of the 4 discussed cases are estimated on the 4 release data set of Tandem Computers [21] using the software package SAS. The results are presented in the Tables 1, 2, 3 and 4.

The graphical representation of the latent number of faults removed at any time for each release is represented by Figs. 1, 2, 3, and 4.

To check for the predictive capability of the presented model, various comparison criteria are chosen. The values of all these measures for the 4 releases under the various scenarios is presented in the Tables 5, 6, 7 and 8.

Table 1 Parameter values for perfect debugging for each release

Parameter	Release 1	Release 2	Release 3	Release 4
a	110.83	125.00	69.00	45
b	0.172	0.198	0.362	0.198
β	1.2505	2.378	6.160	3.475

Table 2 Parameter values for imperfect debugging for each release

Parameter	Release 1	Release 2	Release 3	Release 4
a	104.385	123.856	65.95189	43.9558
b	0.192771	0.32809	0.617768	0.772632
β	1.003258	2.935176	4.1591	4.68397
ρ	0.862	0.591	0.357	0.134672
α	0.0701	0.0284	0.065	0.09

Table 3 Parameter values for testing effort for each release

Parameter	Release 1	Release 2	Release 3	Release 4
a	108.2664	120.2126	63.70673	42.025
b	0.0342	0.034044	0.048575	0.012455
β	1.387842	2.39801	1.12411	0.2041

Table 4 Parameter values for each release under stochastic behaviour

Parameter	Release 1	Release 2	Release 3	Release 4
a	104.0402	127.9851	63.75601	44.3214
b	0.2562	0.2978	0.574428	0.297672
β	1.644134	0.748623	0.654187	0.381055
σ	0.2701	0.519616	0.859	0.624

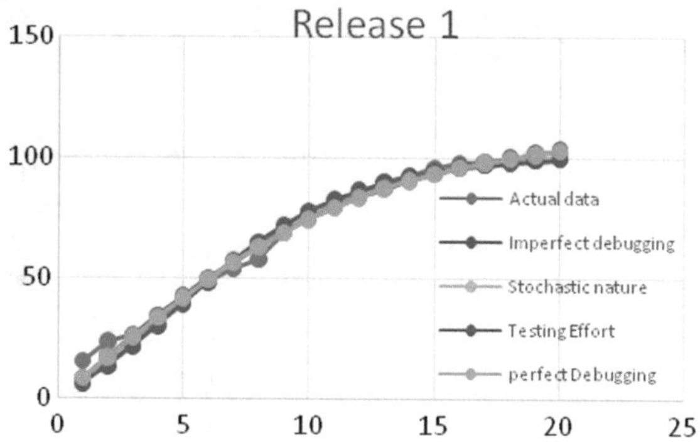

Fig. 1 Number of faults removed for Release 1

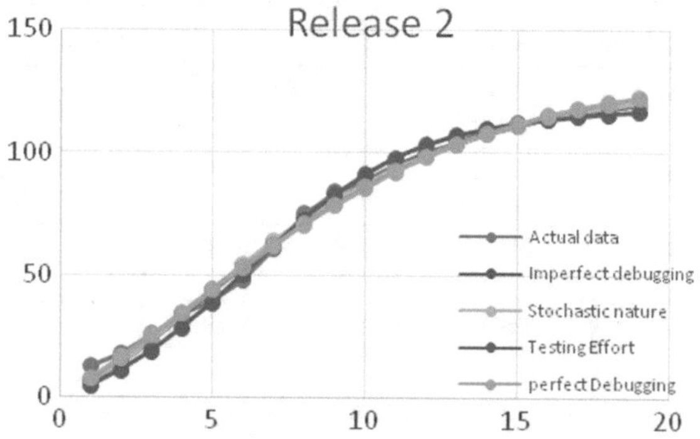

Fig. 2 Number of faults removed for Release 2

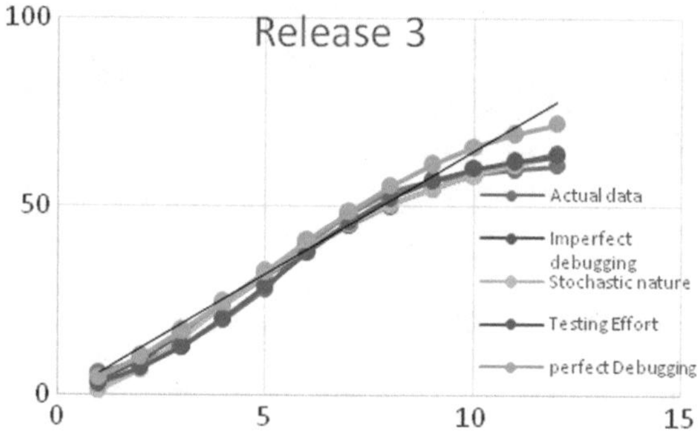

Fig. 3 Number of faults removed for Release 3

Moreover, to determine which scenario best describes the given releases, Weighted criteria approach has been used. The final results for each release are presented in the Tables 9, 10, 11 and 12. Table 13 represents the best fit scenarios for each release.

Furthermore, numerous interpretations can also be made from Table 13, which can be used to study the behaviour of the software. For release 1, the scenario with perfect debugging exhibit the best fit among all the considered cases. It can be said that the utmost utilization of resources was done by software testing team during

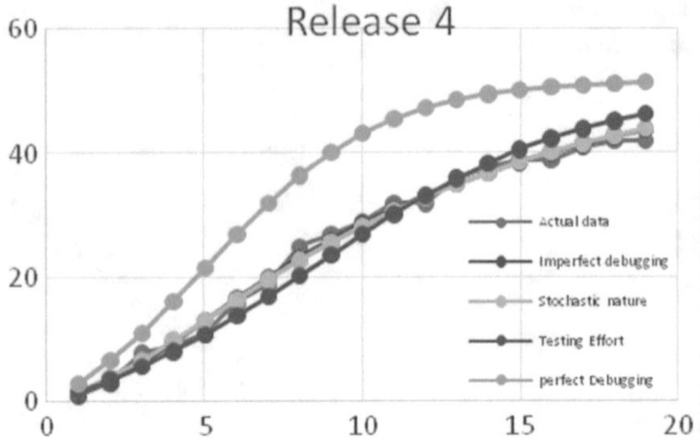

Fig. 4 Number of faults removed for Release 4

Table 5 Values of different comparison criteria for release 1

Release 1		SSE	MSE	RMSE	R-Square	Bias	Variance
Model-1	Perfect debugging	152.49	8.97	3.07	0.989	−0.824	3.432
Model-2	Imperfect debugging	184.3	10.84	3.29	0.9887	−0.424	3.204
Model-3	Stochastic behaviour	267.2	14.85	3.85	0.9836	−0.733	3.970
Model-4	Testing effort	325.4	18.08	4.25	0.98	−0.946	4.467

Table 6 Values of different comparison criteria for release 2

Release 2		SSE	MSE	RMSE	R-Square	Bias	Variance
Model-1	Perfect debugging	135.2	8.45	2.98	0.994	−0.188	2.998
Model-2	Imperfect debugging	128.3	8.01	2.83	0.994	−0.373	2.751
Model-3	Stochastic behaviour	166.3	10.39	3.22	0.993	−0.389	3.117
Model-4	Testing effort	250.3	14.72	3.83	0.99	−1.146	4.249

Table 7 Values of different comparison criteria for release 3

Release 3		SSE	MSE	RMSE	R-Square	Bias	Variance
Model-1	Perfect debugging	71.4	7.14	0.79	0.983	4.026	9.194
Model-2	Imperfect debugging	79.72	7.97	2.82	0.984	0.129	2.702
Model-3	Stochastic behaviour	92.06	10.22	3.19	0.981	−0.268	2.933
Model-4	Testing effort	38.33	3.48	1.86	0.992	−0.270	1.929

Table 8 Values of different comparison criteria for release 4

Release 4		SSE	MSE	RMSE	R-Square	Bias	Variance
Model-1	Perfect debugging	26.4	1.65	0.79	0.991	10.221	21.338
Model-2	Imperfect debugging	23.52	1.47	1.21	0.993	0.0314	1.144
Model-3	Stochastic behaviour	28.24	1.66	1.28	0.992	0.109	1.267
Model-4	Testing effort	109.5	6.08	2.46	0.969	−0.112	2.474

Table 9 Ranking of different models for Release 1

Model	Sum of weight	Sum of weighted value	Model value	Model rank
Perfect debugging	0.414	0.425	1.026	1
Imperfect debugging	1.611	36.365	22.565	2
Stochastic behaviour	3.585	192.091	53.579	3
Testing effort	5	353.179	70.635	4

Table 10 Ranking of different models for Release 2

Model	Sum of weight	Sum of weighted Value	Model value	Model rank
Perfect debugging	1.616	9.117	5.639	2
Imperfect debugging	0.806	−0.301	−0.373	1
Stochastic behaviour	2.396	57.493	23.994	3
Testing effort	5	274.098	54.819	4

Table 11 Ranking of different models for Release 3

Model	Sum of weight	Sum of weighted Value	Model value	Model rank
Perfect debugging	4.053	61.910	15.273	2
Imperfect debugging	3.252	70.151	21.569	3
Stochastic behaviour	4.138	106.879	25.824	4
Testing effort	0.445	0.831	1.866	1

Table 12 Ranking of different models for Release 4

Model	Sum of weight	Sum of weighted value	Model value	Model rank
Perfect debugging	2.171	32.603	15.017	3
Imperfect debugging	0.262	0.302	1.150	1
Stochastic behaviour	0.476	2.064	4.336	2
Testing effort	4.065	119.180	29.319	4

Table 13 Best fir scenario for each release

Releases	Release 1	Release 2	Release 3	Release 4
Best scenarios	Perfect debugging	Imperfect debugging	Testing effort	Imperfect debugging

this phase. But as soon as first version is released in the market, all the allocated resources are assumed to be bifurcated among two teams- one who continues testing for the released version of the software and the other who are working on the newly developing code for the next version of the software. This decrease in the amount of resources has led to the removal of faults in the imperfect manner. Moreover, the dominance of testing effort was observed in the third release followed by the imperfect debugging in the fourth release of the software.

5 Conclusion

The intent of this paper is to consider the increasingly ambitious requirements of the customers and the benefits of situating new features in the software. For this, generalized framework of multi up-gradation of the software has been studied. Moreover, this study was tasked to examine the impact of various scenarios like perfect and imperfect debugging, testing effort and stochastic nature on the developed model. The flexibility of the presented modeling is maintained by using the Logistic Distribution for the failure correction rate. The goodness of fit analysis has been done of the data set of Tandem Computers which consists of data of 4 releases followed by the ranking of each scenario for each release using weighted criteria approach.

Acknowledgements The research work presented in this chapter is supported by grants to the first author from DST, via DST PURSE Phase II, India.

References

1. https://www.nasa.gov/pdf/418878main_FSWC_Final_Report.pdf. Accessed date 22 Oct 2018
2. https://www.tricentis.com/blog/software-fail-awards-2018/. Accessed date 04 Feb 2019
3. https://www.computerworlduk.com/galleries/infrastructure/top-software-failures-recent-history-3599618/. Accessed date 21 Oct 2018
4. Kapur PK, Pham H, Gupta A, Jha PC (2011) Software reliability assessment with OR applications. Springer, London
5. Kapur PK, Tandon A, Kaur G (2010) Multi up-gradation software reliability model. In: 2010 2nd international conference on reliability, safety and hazard (ICRESH). IEEE, pp 468–474
6. Kapur PK, Anand A, Singh O (2011) Modeling successive software up-gradations with faults of different severity. In: Proceedings of the 5th national conference on computing for Nation development. ISSN 0973-7529; ISBN 978-93-80544-00-7

7. Singh O, Kapur PK, Khatri SK, Singh JNP (2012) Software reliability growth modeling for successive releases. In: Proceeding of 4th international conference on quality. Reliability and infocom technology (ICQRIT), pp 77–87
8. Anand A, Singh A, Kapur PK, Das S (2014) Modeling conjoint effect of faults testified from operational phase for successive software releases. In: Proceedings of the 5th international conference on life cycle engineering and management (ICDQM), pp 83–94
9. Anand A, Singh O, Das S (2015) Fault severity based multi up-gradation modeling considering testing and operational profile. Int J Comput Appl 124(4)
10. Kapur PK, Anand A, Singh O, Hoda MN (2011) Modeling successive software up-gradations with faults of different severity. In: Proceedings of the 5th national conference, INDIACom, pp 351–356
11. Singh O, Aggrawal D, Anand A, Kapur PK (2015) Fault severity based multi-release SRGM with testing resources. Int J Syst Assur Eng Manage 6(1):36–43
12. Singh O, Kapur PK, Anand A, Singh J (2009) Stochastic differential equation based modeling for multiple generations of software. In: Proceedings of fourth international conference on quality, reliability and infocom technology (ICQRIT), trends and future directions, Narosa Publications, pp 122–131
13. Singh O, Kapur PK, Anand A (2011) A stochastic formulation of successive software releases with faults severity. In: 2011 IEEE international conference on industrial engineering and engineering management (IEEM). IEEE, pp 136–140
14. Aggarwal AG, Kapur PK, Garmabaki AS (2011) Imperfect debugging software reliability growth model for multiple releases. In: Proceedings of the 5th national conference on computing for nation development-INDIACOM, New Delhi
15. Anand A, Deepika, Verma AK, Ram M (2018) Revisiting error generation and stochastic differential equation-based software reliability growth models. In: System reliability management (pp 65–78). CRC Press, Boca Raton
16. Anand A, Das S, Aggrawal D, Kapur PK (2018) Reliability analysis for upgraded software with updates. In: Quality, IT and business operations (pp 323–333). Springer, Singapore
17. Das S, Aggrawal D, Anand A (2019) An alternative approach for reliability growth modeling of a multi-upgraded software system. In: Recent advancements in software reliability assurance (pp 93–105). CRC Press, Boca Raton
18. Anjum M, Haque MA, Ahmad N (2013) Analysis and ranking of software reliability models based on weighted criteria value. IJ Inf Technol Comput Sci 2:1–14
19. Anand A, Kapur PK, Agarwal M, Aggrawal D (2014) Generalized innovation diffusion modeling & weighted criteria based ranking. In: Proceedings of 3rd international conference on reliability, infocom technologies and optimization. IEEE, pp 1–6
20. Anand A, Bhatt N (2016) Vulnerability discovery modeling and weighted criteria based ranking. J Ind Soc Probab Stat 17(1):1–10
21. Wood A (1996) Predicting software reliability. Computer 29(11):69–77

A Hidden Markov Model for a Day-Ahead Prediction of Half-Hourly Energy Demand in Romanian Electricity Market

Anatoli Paul Ulmeanu

Abstract In this chapter we introduce an original approach to predict a day-ahead energy demand, based on machine learning and pattern recognition. Our Hidden Markov Model (HMM) is a simple and explainable model that uses integer sequences to define emission probability distributions attached to states. We develop a Mathematica code which relies on the maximum likelihood principle in HMM environment. Based on these exploratory results, we conclude that the HMM approach is an efficient way in modeling short-term/day-ahead energy demand prediction, especially during peak period(s) and in accounting for the inherent stochastic nature of demand conditions. The model can be easily extended to predict energy demand values for more than one day in the future. However, the accuracy of such predictions would decrease as we expect.

Keywords Forecasting · Hidden Markov Model · Energy demand · Pattern recognition

1 Introduction

The forecast of the national electrical energy demand is a highly active research area, given the interest of Electricity Transmission System Operators (ETSO), Natural Gas Transmission Operators (NGTO) and Storage System Operators (SSO), in the fields of energy markets.

Several machine learning techniques have been used in the field of forecasting energy demand data, a number of them proving successful. These techniques include the Fuzzy Logic, the Neural Networks and the Support Vector Machines.

Hidden Markov Models (HMM) have been notably applied to the given problem in the past, which is unforeseen given the time-dependent nature of the energy demand.

A. P. Ulmeanu (✉)
Department of Power Generation and Use, Polytechnic University of Bucharest, 060042 Bucharest, Romania
e-mail: Paul.Ulmeanu@upb.ro

M. Ram and H. Pham (eds.), *Advances in Reliability Analysis and its Applications*, Springer Series in Reliability Engineering,
https://doi.org/10.1007/978-3-030-31375-3_9

HMMs have been successfully applied in the areas of Speech Recognition, DNA sequencing, and ECG analysis (see [4] for a full review).

Accurate short-term prediction of electrical energy demand has recently become iimportant because of its vital role in the management functions and decision making processes in European Electricity Market. In Romania, Transelectrica S.A. is the only operator providing the services of electricity transmission, operational technical management of the Romanian Power System (RPS) and electricity market administration (by means of its legal personality subsidiary OPCOM S.A.), such domains being considered as a natural monopoly under the law. This service consists mainly in providing the transmission of active electricity between two or more points of the transmission grid belonging to the RPS, while observing the continuity, safety, metering management system and quality norms. Given the dynamic and stochastic nature of energy demand, this chapter proposes a stochastic approach based on Hidden Markov Model, for prediction of the half-hourly national energy consumption, for a day ahead. The data used in the study was gathered from real-time electricity demand in Romania, over last ten years. The dynamic changes of conditions are addressed with state transition probabilities. For a sequence of energy metering data / observations, HMMs estimate the most likely sequence of states. Therefore, the chapter concludes that the HMM approach was successful in modeling short-term / day-ahead energy demand prediction, especially during peak period(s) and in accounting for the inherent stochastic nature of demand conditions.

2 HMM Strategy

2.1 Assumptions and Preliminaries

A discrete-time Hidden Markov Model is a stochastic process generated by two interrelated probabilistic mechanisms (cf., e.g. [2, 4]).

These are: (a) an underlying unobserved Markov chain (MC) and (b) a set of random functions, each of them associated with its respective state.

At discrete instants of time, the process is assumed to be in a certain state and an observation is generated by the random function corresponding to the current state. The underlying MC then changes its state according to its transition matrix. An observer sees only the output of the random functions associated with each state and cannot directly observe the states of the underlying MC. Consequently, the MC is said to be hidden in the observations. The objective is to estimate the states of the chain, given the observations.

To be more specific, a HMM is described by:

(a) An unobserved Markov chain $J = (J_n)_{n \in \mathbb{N}}$ of state space $\mathbf{E} = \{1, 2, \ldots, M\}$, where $\mathbb{N}$ is the set of nonnegative integers; the behavior of the Markov chain is described by the probability transition matrix $\mathbf{P} = (p_{ij})_{i,j \in \mathbb{E}}$

where $p_{ij} = \mathbb{P}(J_{n+1} = j | J_n = i), n \in \mathbb{N}$, and by its initial distribution denoted by $\boldsymbol{\alpha} = (\alpha(i))_{i \in \mathbb{E}}$, where $\alpha(i) = \mathbb{P}(J_0 = i)$;

(b) An observed sequence of random variables $Y = (Y_n)_{n \in \mathbb{N}}$ with state space $\mathbf{A} = \{1, 2, \ldots, Q\}$. The random variables $(Y_n)_{n \in \mathbb{N}}$ are assumed to be conditionally independent, given $(J_n)_{n \in \mathbb{N}}$. That is to say the following conditional independence holds true.

$$\mathbb{P}(Y_n = a | Y_{n-1} = .; ..; Y_0 = .; J_n = i; \ldots J_0 = .) = \mathbb{P}(Y_n = a | J_n = i) = R_{i;a}$$

For the given state spaces $\mathbf{E}$ and $\mathbf{A}$, let us denote by $\mathbb{M}_{E \times A}$ the set of nonnegative 2-dimensional matrices on $\mathbf{E} \times \mathbf{A}$. So, $\mathbf{R} = \left(R_{i;a}; i \in \mathbf{E}, a \in \mathbf{A}\right) \in \mathbb{M}_{\mathbf{E} \times \mathbf{A}}$ is the conditional distribution of the chain Y, given the unobserved underlying Markov chain J. This is called the emission probability matrix.

Let us now specify the state spaces of the processes $(Y_n)_{n \in \mathbb{N}}$ and $(J_n)_{n \in \mathbb{N}}$ in our case.

Throughout this chapter, we will assume that:

- The HMM is homogeneous with respect to time, that is to say that its transition probabilities do not depend on the time index n and also, the conditional distribution of Y_n, given J_n, does not depend on n as well;
- The Markov chain $(J_n)_{n \in \mathbb{N}}$ is ergodic;
- The underlying Markov chain is stationary; that means that the initial distribution is the stationary one. In other words, the process under study is assumed to be in equilibrium.

Note that a non-stationary HMM is completely determined by its parameters $\theta = (\boldsymbol{\alpha}, \mathbf{P}, \mathbf{R})$, while a stationary HMM is completely determined by its parameters $\theta = (\mathbf{P}, \mathbf{R})$. Also note that the time parameter is discrete and the state spaces of these two probabilistic mechanisms are both finite and discrete.

There are three statistical problems associated with HMMs that are of particular interest to us:

(a) Given an observation sequence $O = (o_1, \ldots, o_T)$ and a space of possible models, how do we adjust the parameters so as to find the model θ that maximizes $\mathbb{P}(O|\theta)$?

(b) Given a model $\theta = (\boldsymbol{\alpha}, \mathbf{P}, \mathbf{R})$ and an observation sequence O, what is the underlying state sequence $S = (s_1, \ldots, s_T)$ that best "explains" the observations?

(c) Given a model $\theta = (\boldsymbol{\alpha}, \mathbf{P}, \mathbf{R})$ and an underlying state sequence $S = (s_1, \ldots, s_T)$, find the most probably underlying path $(s_{T+1}, \ldots, s_{2T})$ and the most probably prediction track $(o_{T+1}, \ldots, o_{2T})$ generated by the HMM model.

.

Here, we briefly describe these three problems, for more details the reader is referred to [1, 3] and [5] for an overview of statistical and information-theoretic aspects of HMMs.

The first problem is of the most significant interest because it deals with the training of the model. In the times series analysis case, it is about training the model with the past historical data. Only after training can we use the model to perform further predictions. The second problem is about finding out the hidden path.

(a) Training problem.
Given the observations and the set of possible states, our objective is to estimate the parameters set θ. In a non-parametric context, the objective is to find the most likely set of transitions and emission probabilities. The estimation is achieved while the parameters set θ is the most probable set that generate the observations. The likelihood function $\mathcal{L}$ of the observation sequence $o_0, \ldots, o_N$ for a given parameters set θ is

$$\mathbb{P}(Y_0 = o_0, Y_1 = o_1, \ldots Y_N = o_N|\theta)$$
$$= \sum_{j_0 \in \mathbf{E}} \cdots \sum_{j_N \in \mathbf{E}} \alpha(j_0) R_{j_0;o_0} \prod_{l=1}^{N} p_{j_{l-1};j_l} R_{j_l;o_l}$$

The problem doesn't have a global solution in the sense that the likelihood function cannot be globally maximized. However, we can choose θ which locally maximizes the likelihood function by means of an iterative procedure, such as the Baum–Welch algorithm, see [1]. The Baum–Welch algorithm is the most efficient method to obtain the maximum likelihood estimate defined by

$$\hat{\theta}_{ML} = arg \ \max_{\theta} \mathbb{P}(Y_0 = o_0, Y_1 = o_1, \ldots Y_N = o_N|\theta)$$

It is worth mentioning, however, that the direct computation of the likelihood values causes the combinatorial explosion problem. Hence, the most relevant characteristic of the Baum–Welch algorithm is that can provide feasible computational cost that overcome the direct calculation of the likelihood function.

(b) Finding the hidden path.
For an HMM with M hidden states and an observation sequence of T observations, there are M^T possible hidden sequences. For real tasks, where M and T are both large, M^T is a very large number, so we cannot compute the total observation likelihood by computing a separate observation likelihood for each hidden state sequence and then summing them. Instead of using such an exponential algorithm, we use an efficient $O(M^{2T})$ algorithm called the forward algorithm. The forward algorithm is based on dynamic programming, that is, an algorithm that uses a table to store intermediate values as it builds up the probability of the observation sequence. The forward algorithm computes the observation probability by summing over the probabilities of all possible hidden state paths that could generate the observation sequence, but it does so efficiently by implicitly folding each of these paths into a single forward trellis, see [4].

(c) Forecasting.
Using the trained HMM, the likelihood value for current day's dataset $\mathcal{D}$ is calculated. For example, say the current day is $\mathcal{M}onday$ and the likelihood value for this day is $\mathcal{L}$, then from the past dataset using the HMM we locate all

the $\mathcal{M}ondays$ having those instances that would produce the same $\mathcal{L}$ or nearest to the $\mathcal{L}$ likelihood value. That is we locate the past $\mathcal{M}ondays$ where the energy demand behaviour is similar to that of the current day's set $\mathcal{D}$. Let's say $\mathcal{M}$ is a such day. Assuming that the next day's energy demand $\mathcal{D}+1$ should follow about the same past data pattern such as $\mathcal{M}+1$, we use its underlying path $S_{\mathcal{M}+1}$ of $\mathcal{M}+1$ to generate, via the HMM, the forecast path of $\mathcal{D}+1$, i.e. $O_{\mathcal{D}+1}$, in terms of the most probable observed energy demand.

2.2 Approach

Training of the above HMM from given sequences of observations is done using the Baum-Welch algorithm which uses Expectation-Maximization (EM) to arrive at the optimal parameters for the HMM, see [3].

Our approach is presented in full, as implemented in Mathematica.

In our model the records are the half-hourly mean electricity demand data in the form of the 2-dimensional vector: (Day/Month/Year HH:MM, Half-hourly Mean Power Demand), as shown in Fig. 1. As shown in Fig. 2b, we train a HMM for 24 hidden states, following the Gaussian model, for $NPaths = 1090$ daily energy demand paths, i.e. 52320 data points. A data point Δ_i is the difference between the current observed energy demand data O_i and the previously observed energy demad data O_{i-1}, half-hour ago. Once the model has been trained, we have the best estimations in terms of means and standard values for each Gaussian distribution, as shown in Table 1. These values are given in MW.

We are now in position to decode a daily observed path, in terms of relative values of energy demand (variations) $ToDayR$, as shown in Figs. 3 and 4. The list of hidden Markov states are obtained in Mathematica via Viterbi algorithm decoding, following the call of the procedure *FindHiddenMarkovStates[ToDayR, hmm,"ViterbiDecoding"]*.

As our objective was to predict the next day's energy demand for a specific market using aforementioned HMM, the next step is to decode all the observed daily paths, in have a benchmark of patterns. This has been achieved via the Mathematica code shown in Fig. 6. An example revealing the first three records in this benchmark is shown in Fig. 7. Using the HMM, the likelihood value for current day's dataset x is calculated via Mathematica function called $LogLikelihood[DiscreteMarkovProcess[hmm], x]]$.

As mentioned already, we locate into the benchmark of patterns where the energy demand behaviour is similar to that of the current day's set x, given the current day name day. This algorithm searching the hidden patterns having the same or almost the same likelihood as x and the same day name as day is presented in Fig. 8. A such example searching in $WZGALL$ a similar hidden path as shown in Fig. 5 in presented in Fig. 9. Hence, the most probable energy demand for the next day, i.e. 6 May, is now achieved via the mean values μ given in the Table 1:

DD/MM/YYYY HH:MM	MeanPower (MW)
01/01/2015 00:00	7275.7
01/01/2015 00:30	7090.3
01/01/2015 01:00	6919.0
01/01/2015 01:30	6748.0
01/01/2015 02:00	6647.7
01/01/2015 02:30	6500.7
01/01/2015 03:00	6452.3
01/01/2015 03:30	6331.7
01/01/2015 04:00	6232.5
01/01/2015 04:30	6197.3

Fig. 1 Example of half-hourly recorded data metering the electricity demand

```
ClearAll["Global`*"];

DIR = "C:\\Users\\Administrator\\Desktop\\SEN\\Date\\";

FileDATA = DIR <> "HalfSEN-2011--2018.xlsx";

<< JLink`;
InstallJava[];
ReinstallJava[JVMArguments → "-Xmx5g"]

T =.; data =.;

T = Import[FileDATA];

Δ = Dimensions[T][[2]];

DateTime =.; HalfHourlyMeanPower =.; ΔHalfHourlyMeanPower =.;

HalfHourlyMeanPower=T[[1,1;;Δ]][[1;;Δ,2]];

data=Table[0.,{i,Δ},{j,2}];
```

(a) Importing the records of data

```
ED = Range[1, 48];

ΔHalfHourlyMeanPower = Table[0, {i, 1, Δ}];

ΔHalfHourlyMeanPower[[1]] = 0;
For[i = 2, i < Δ, i++,
  ΔHalfHourlyMeanPower[[i]] = HalfHourlyMeanPower[[i]] - HalfHourlyMeanPower[[i - 1]]];

ZHourlyMeanPowerC = Partition[ΔHalfHourlyMeanPower, 48];
(* parsing each day in 48 half-hours *)

TS = TemporalData[ZHourlyMeanPowerC, {ED}]

TemporalData[ Time: 1 to 48 Data points: 52 320 Paths: 1090 ]

hmm =.;

hmm = EstimatedProcess[TS, HiddenMarkovProcess[24, "Gaussian"]]
```

(b) Creating the Temporal Data

Fig. 2 Training the HMM for 24 hidden states

Table 1 hmm[[3]]

Hidden state symbol	(μ, σ)	Hidden state symbol	(μ, σ)
1	(267.79, 96.6921)	2	(−184.757, 61.696)
3	(−27.8328, 48.7138)	4	(−32.4731, 48.1861)
5	(−5.70205, 40.8809)	6	(457.57, 178.405)
7	(−111.855, 51.2164)	8	(15.3543, 34.4825)
9	(−82.433, 45.5238)	10	(−34.5825, 56.0938)
11	(−30.8202, 50.8992)	12	(−2.56807, 26.0771)
13	(−12.9156, 53.8194)	14	(−359.008, 232.463)
15	(59.1676, 43.9878)	16	(55.9772, 70.4242)
17	(−40.6597, 59.9462)	18	(−249.178, 76.4531)
19	(−37.2164, 46.8883)	20	(4.09228, 39.6486)
21	(−35.5523, 53.7459)	22	(132.844, 59.8831)
23	(38.0422, 51.5838)	24	(−53.6988, 52.5562)

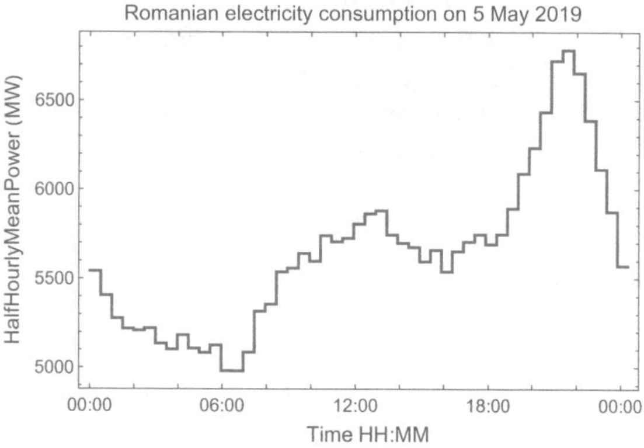

Fig. 3 A path of daily observed data

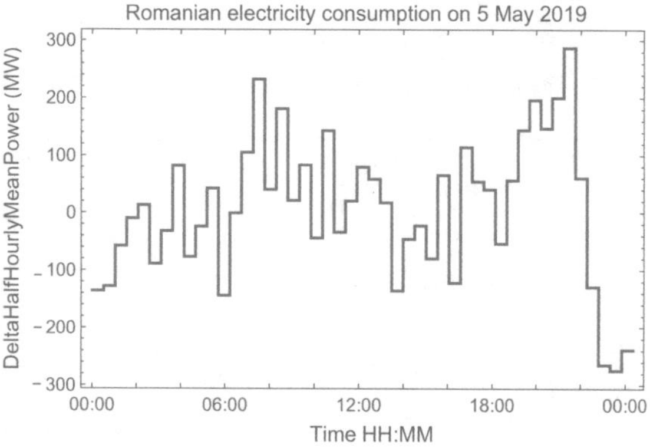

Fig. 4 Half hourly variation of the energy demand on 5 May 2019

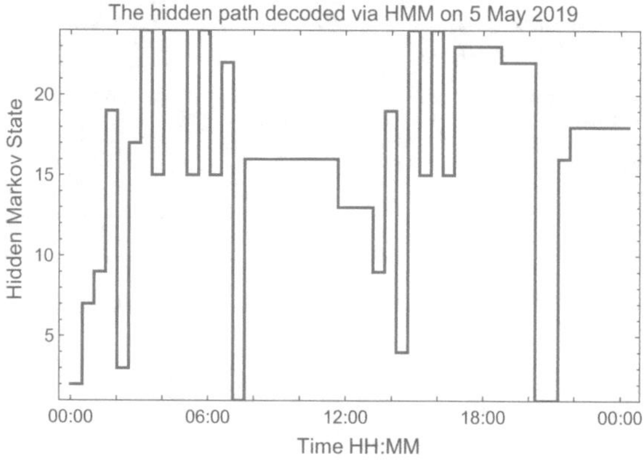

Fig. 5 The hidden path obtained via the HMM decoding on 5 May 2019

```
NPaths=TS["PathCount"];

aposteriorstates=.;WZGALL=.;WZGALL={};
For[i =1,i<NPaths,i++,
  aposteriorstates=
   FindHiddenMarkovStates[ZHourlyMeanPowerC[[i]],hmm,"ViterbiDecoding"];
  t =T[[i]][[1]];
  Y={DateValue[t,"DayName"],aposteriorstates};
   WZGALL=Append[WZGALL,Y]];

Export[DIR<>"DatePatterns--ALL"<>".xls",List@WZGALL//Normal,"XLS"];
```

Fig. 6 The hidden path obtained via the HMM decoding on 5 May 2019

```
WZGALL[[1 ;; 3]]
```

```
{{Thursday, {14, 2, 2, 2, 7, 7, 9, 9, 9, 19, 3, 20, 24, 11, 24, 11, 24, 20, 8, 8, 8, 24, 5,
24, 5, 24, 5, 8, 8, 23, 23, 23, 22, 1, 1, 1, 16, 16, 13, 13, 13, 13, 7, 9, 9, 18, 18, 18}},
{Friday, {2, 7, 7, 9, 9, 4, 11, 24, 15, 23, 23, 23, 23, 23, 23, 22, 1, 16, 16, 16, 16, 13, 13,
13, 13, 9, 19, 3, 17, 10, 24, 15, 23, 22, 1, 1, 1, 16, 13, 13, 13, 13, 9, 18, 9, 18, 18, 18}},
{Sunday, {14, 2, 7, 9, 19, 3, 24, 15, 23, 23, 22, 22, 1, 1, 1, 1, 1, 16, 16, 13, 13, 13, 9, 9,
24, 15, 24, 24, 15, 24, 15, 24, 15, 24, 15, 23, 23, 23, 22, 1, 6, 1, 18, 18, 18, 18, 18, 18}}}
```

Fig. 7 The first three hidden paths decoded via the HMM

```
DatetoDigit[day_] := Module[{z = 0}, If[day == Monday, z = 1];
   If[day == Tuesday, z = 2];
   If[day == Wednesday, z = 3];
   If[day == Thursday, z = 4];
   If[day == Friday, z = 5];
   If[day == Saturday, z = 6];
   If[day == Sunday, z = 7];
   Return[z]];
```

```
DatetoDigit[Sunday]
```

```
7
```

```
FindLogLikehood[day_, x_, y_] := Module[{i = 1, z = Dimensions[y][[1]], dif = 0.5, LogLMin = 0,
LogL = Table[10^9, {i, NPaths}]},

  While[i <= z , If[day == WZGALL[[i]][[1]] && (DatetoDigit[WZGALL[[i + 1]][[1]]] -
DatetoDigit[day] == 1 || DatetoDigit[WZGALL[[i + 1]][[1]]] - DatetoDigit[day] == -6), LogL[[i]] =
Abs[ LogLikelihood[DiscreteMarkovProcess[hmm], y[[i]][[2]]] -
LogLikelihood[DiscreteMarkovProcess[hmm], x]]; i++]; LogLMin = Ordering[LogL, 2];
  If[LogL[[LogLMin[[1]]]] > dif && LogL[[LogLMin[[2]]]] > dif && (DatetoDigit[WZGALL[[i +
1]][[1]]] - DatetoDigit[day] == 1 || DatetoDigit[WZGALL[[i + 1]][[1]]] - DatetoDigit[day] == -6),
Return[-1], Return[{LogLMin[[1]], WZGALL[[LogLMin[[1]] + 1]][[2]]}]]]
```

Fig. 8 Find in list y a similar path as x, given the current *day* name

```
HP5May = {2, 7, 9, 19, 3, 17, 24, 15, 24, 24, 15, 24, 15, 22, 1, 16, 16, 16, 16, 16, 16, 16, 16,
13, 13, 13, 9, 19, 4, 24, 15, 24, 15, 23, 23, 23, 23, 22, 22, 22, 1, 1, 16, 18, 18, 18, 18, 18};
```

```
HPNextDay = FindLogLikehood[Sunday, HP5May, WZGALL][[2]]
```

```
{14, 2, 7, 9, 9, 4, 17, 20, 24, 15, 23, 22, 22, 22, 1, 1, 1, 1, 1, 1, 16, 16, 16, 16, 16,
 16, 16, 13, 13, 9, 19, 13, 9, 5, 5, 24, 15, 23, 23, 22, 1, 1, 16, 18, 18, 18, 18, 18}
```

Fig. 9 Forecasting the next day energy demand in terms of hidden path

3 Exploratory Results

The suggested algorithm was tested on electricity demand data published by the Romanian Electricity Transmission Company, namely Transelectrica S.A. (www.transelectrica.ro).

The data was collected over a time period of ten years (2009–2018). The data were split into the training, validation and test datasets, as show in Figs. 10, 11, 12 and 13.

Based on these exploratory results, we conclude that the HMM approach is an efficient way in modeling short-term / day-ahead energy demand prediction, especially during peak period(s) and in accounting for the inherent stochastic nature of demand conditions.

The model can be easily extended to predict energy demand values for more than one day in the future. However, the accuracy of such predictions would decrease as we expect.

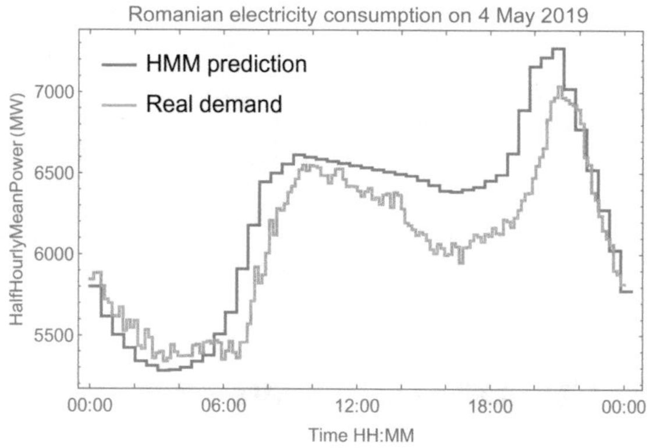

Fig. 10 Forecasting the energy demand for Saturday 04th of May 2019

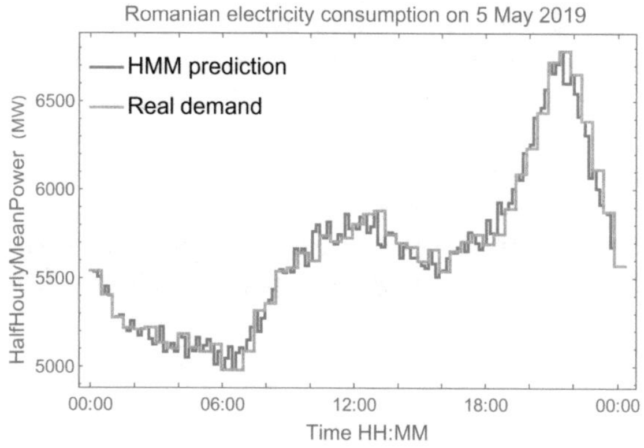

Fig. 11 Forecasting the energy demand for Sunday 5th of May 2019

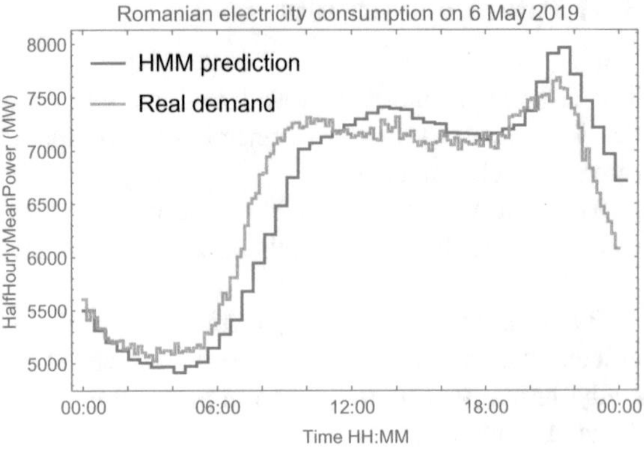

Fig. 12 Forecasting the energy demand for Monday 06th of May 2019

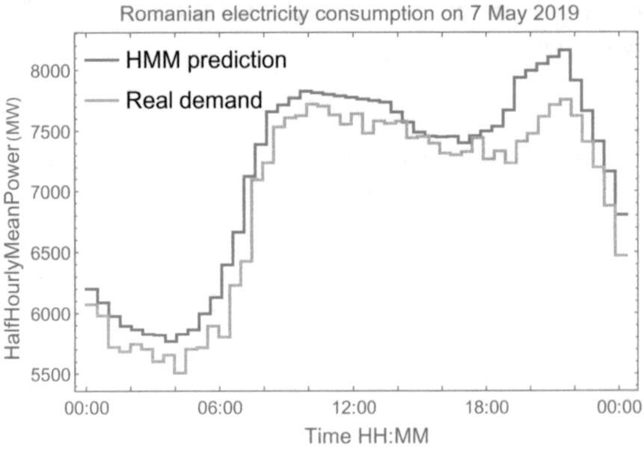

Fig. 13 Forecasting the energy demand for Tuesday 07th of May 2019

4 Conclusions and Future Work

In this chapter, we have presented a hidden Markov model-based algorithm to predict a day-ahead energy demand, based on machine learning and pattern recognition. We have used real data collected through meters in Transelectrica S.A., the electricity transmission system operator in Romania. We have selected HMM for prediction due to its inherent benefits over other AI-based algorithms. HMM models can also perform anomaly detection in electricity consumption data. In future work, to further establish and validate prediction results of our proposed model, we will perform temporal granularity analysis. For this purpose, we will evaluate the prediction accuracy for every quarter of hour, half-hourly and hourly data aggregation.

References

1. Baum LE, Petrie T, Soules G, Weiss N (1970) A Maximization Technique Occurring in the Statistical Analysis of Probabilistic Functions of Markov Chains. Ann Math Stat 41(1):164–171. https://doi.org/10.2307/2239727
2. Ephraim Y, Merhav N (2002) Hidden Markov processes. IEEE Trans Inf Theory 48(6):1518–1569. https://doi.org/10.1109/TIT.2002.1003838
3. Olivier Cappé TR, Moulines Eric (2005) Inference in Hidden Markov Models. Springer Series in Statistics, Springer-Verlag, New York. https://doi.org/10.1007/0-387-28982-8
4. Rabiner LR (1989) A tutorial on hidden Markov models and selected applications in speech recognition. Proc. IEEE 77(2):257–286. https://doi.org/10.1109/5.18626
5. Viterbi A (1967) Error bounds for convolutional codes and an asymptotically optimum decoding algorithm. IEEE Trans Inf Theory 13(2):260–269. https://doi.org/10.1109/TIT.1967.1054010

A General (Universal) Form of Multivariate Survival Functions in Theoretical and Modeling Aspect of Multicomponent System Reliability Analysis

Jerzy K. Filus and Lidia Z. Filus

Abstract One of important classical reliability problems can be formulated as follows. Given a k-component system (k = 2, 3, …) with series reliability structure where the life-times X_1, …, X_k of the components are nonnegative stochastically dependent random variables. In order to determine the system's (as a whole) reliability function as well as for some system maintenance analysis, one needs to find or construct a proper joint probability distribution of the random vector (X_1, …,X_k) which may be expressed in terms of the joint reliability (survival) function. Numerous particular solutions for this problem are present in the literature, to mention only [Freund in J Am Stat Assoc 56:971–77, 1961 1, Gumbel in J Am Stat Assoc 55:698–707, 1960 2, Marshall and Olkin in J Appl Probab 4:291–303, 1967 3]. Many other, not directly associated with reliability, k-variate probability distributions were invented [Kotz et al. in Continuous Multivariate Distributions, Wiley, New York, 2000 4]. Some of them later turned out to be applicable to the above considered reliability problem. However, the need for proper models still highly exceeds the existing supply. In this chapter we present not only particular bivariate and k-variate new models, but, first of all, a general method for their construction competitive to the copula methodology [Sklar A in Fonctions de repartition a n dimensions et leurs marges, Publications de l'Institut de Statistique de l'Universite de Paris, pp. 229–231, 1959 5]. The method follows the invented universal representation of any bivariate and k-variate survival function different from the corresponding copula representation. A comparison of our representation with the one given by copulas is provided. Also, some new bivariate models for 2-component series systems are presented. Possible applications of our models and methods may go far beyond the reliability context, especially toward bio-medical and econometric areas.

J. K. Filus
Department of Mathematics and Computer Science, Oakton Community College, Des Plaines, USA
e-mail: jkfilus98@gmail.com

L. Z. Filus (✉)
Department of Mathematics, Northeastern Illinois University, Chicago, USA
e-mail: L-Filus@neiu.edu

M. Ram and H. Pham (eds.), *Advances in Reliability Analysis and its Applications*, Springer Series in Reliability Engineering,
https://doi.org/10.1007/978-3-030-31375-3_10

Keywords Reliability · Multivariate probability distributions · Universal representations · Copula methods comparison · Marginal and baseline forms of survival functions

1 Introduction

This chapter contains revisions and extensions of our previous paper [6] associated with some universal representation of bivariate probability distributions in the form of joint survival functions. First of all, the constructions method, which in [6] follows the Aalen version [7] of the Cox model [8], here became a more general theory formally independent from the Aalen origin. Between others, in the present formulation the existence of hazard rates of the marginal distributions (which lies in the foundation of the Aalen and the Cox based approach) is not necessary and therefore discrete distributions cases as well as variety of mixed cases we cover here too.

Moreover, missing in [6] is the second natural representation of bivariate survival functions. The first, also considered in [6], relies on factoring out from the expression on any bivariate survival model the two marginal univariates. In the second representation of the same bivariate function two baseline survival functions, in general different than the marginals, are factored out too. This approach, considered in Sect. 4 of this chapter, is especially relevant in various modeling situations where one considers preexisting "laboratory conditions" in which the units under the investigation (system components, in particular) are tested in physical separation of each other. The so obtained nonnegative random quantities, say T_1, T_2, (life-times, for example, in general different than X_1, X_2) are by nature independent.

In Sect. 4 we investigate the relationship between the baseline factors and the corresponding marginal factors in the proposed two bivariate model representations. In [6] we emphasized the case where both representations are identical. This, unfortunately, is not always the case. We found that there are two, in general different, representations of the same bivariate survival (reliability) function. The simple and nice relationship between the two is expressed by Theorem 1 in Sect. 4 of this work. One representation (mathematically more typical: marginal) we described in Sect. 2 and the other (more typical and intuitive in a huge number of applications) is the main topic of Sect. 4. Both the representations stand for analytical representations of the same bivariate model and, possibly, each model possesses both representations, which in some cases are identical.

At least the marginal factors representation is universal as based on the simple arithmetic identity following definitions (1) and (2) in next section.

This representation of any bivariate survival function S(x, y) (in the product form $S_1(x)\ S_2(y)\ J(x, y)$, where $S_1(x)$, $S_2(y)$ are the marginals) has not only theoretical value but also allows for many effective constructions, specially applicable for two component series system reliability. The constructions are then reduced to finding the (dependence) function J(x, y), given in advance two marginal survival functions

$S_1(x)$, $S_2(y)$. The analytical criteria for the function J(x, y) to be proper for "connecting" the given marginals $S_1(x)$, $S_2(y)$ into the legitimate bivariate models are given in Sect. 2.

In the cases most important from the reliability point of view, the criteria for J(x, y) reduce to being equivalent to a solution of the derived integral equations and inequalities. Thus, a part of the theory reduces to analyzing and solving those integral equations and inequalities. An important part of them are linear equations. Also, some of their solutions (so the corresponding survival functions) are immediately obtainable and the corresponding bivariate models have good properties also from the reliability (and of any application) view point.

In Sect. 5 we extended our considerations to k-dimensional cases ($k \geq 3$). Most of the multivariate analysis, however, were performed for the case $k = 3$. Nevertheless, the general result for an arbitrary k was provided. That was in association with the obtained 'recurrence formula' for the transition from ($k - 1$)-dimensional marginal survival functions to the k-dimensional one. Therefore, theoretically, the method of the construction works for any $k = 3, 4, \ldots$ Unfortunately, from application view point, it is limited because of the growing fast complexity of the analytical expressions. In our opinion, starting from the case $k = 4$ and, for sure, for all the cases $k \geq 5$ the computer supplies and technics are indispensable.

On the other hand, the case $k = 2$ can readily be handled in purely analytical ways and can, efficiently, be extended in other direction, namely toward the construction of bivariate stochastic processes, given two univariate processes. This possibility was signalized as the Remark at end of Sect. 4 and developed more in [9].

As for the mentioned theoretical value, especially of that 'marginal factors representation', the presented results and methods may be considered as competitive [10] to the methods associated with the notion of copulas [5]. Some comparison of the two methodologies was given in Sect. 3 as more associated with the marginal representation. The Sklar's theorem was shortly compared with our version of this theorem formulated instead for survival functions. Especially, structures of underlying two functions: the copula and the marginal factors representation exhibit interesting differences. Efficiency of the two methods were compared too.

2 Marginal Factors Representation

2.1 Suppose, we face the following typical problem as, in particular, often is met in reliability of two component system. Given two arbitrary marginal survival (reliability) functions $S_1(x)$, $S_2(y)$ of random variables, say, X, Y that in reliability settings are interpreted as two system component life-times. Suppose X and Y are stochastically dependent and the task is to determine their joint probability distribution in the form of a joint survival function S(x, y). The latter is defined by the usual formula:

$$S(x, y) = P(X \geq x, Y \geq y).$$

The theory, described below, is a general one with possible applications to various areas of the "real world". In this work, however we specially stress the reliability settings and applications.

Generally speaking the main problem we encounter is the old one:

Given two (or more) univariate survival functions one seeks bivariate (or k-variate, $k \geq 3$) survival functions for which they remain the marginal. The usual way to solve such problem is to find a proper copula [5] that turns two or more corresponding cdfs into the joint cdf. This method has, however, some drawbacks. First of all, finding a proper copula is, in general, a hard task. Secondly, for most of the copulas an underlying physical content of the dependence as encountered during the modeling process is pretty unclear. Almost always the copulas remain rather purely mathematical formulas without sufficient reference to the modeled reality. Even if statistical analysis of an available data shows some fit to the model given by the underlying copula, an understanding of that fit, mostly, is lost.

Here, we propose an alternative method for finding a joint survival function (instead of the joint distribution function [cdf]) with no use of copulas. This, rather new, method seems to be more natural, easier and closer related to modeled realities.

Namely, we claim that the common formula valid for any bivariate survival function can be expressed in the factored form:

$$S(x, y) = S_1(x)S_2(y)\, J(x, y) \tag{1}$$

where $S_1(x)$, $S_2(y)$ are the marginal survival functions of arbitrary random variables X, Y.

[Basically, the univariate survival functions of X, Y may be taken from any two arbitrary (in general different) classes of probability distributions. As mentioned, the X, Y may be interpreted as the life-times.]

The, here introduced, "dependence function" J(x, y), that we propose to call the "**Aalen factor**" or "**joiner**", is uniquely determined by S(x, y) by means of the obvious formula:

$$J(x, y) = S(x, y)/S_1(x)S_2(y). \tag{2}$$

If, for some x or y, we have $S_1(x) = 0$ or $S_2(y) = 0$ then, for these x, y, the joiner is undefined and in such a case we set instead of (1): $S(x, y) = 0$.

Realize, there always exists the "trivial joiner": $J(x, y) = 1$ for all x, y, which corresponds to the case when the random variables X, Y are independent.

Recall, that both the functions $S_1(x)$, $S_2(y)$ are the marginals of S(x, y) and therefore are also uniquely determined by the joint survival function S(x, y) upon substituting

$$S_1(x) = S(x, 0) \text{ and } S_2(y) = S(0, y).$$

Thus, every fixed survival function S(x, y) uniquely and effectively determines its joiner J(x, y) while, given two fixed marginals $S_1(x)$, $S_2(y)$, a single proper joiner uniquely determines the corresponding survival function. However, the joiner is not unique in the sense that, given a fixed pair $(S_1(x), S_2(y))$, different joiners determine different models S(x, y) with the same marginals. In other words, given fixed two marginals there is one to one relationship between all the possible joiners J(x, y) and the bivariate survival functions S(x, y) [all having the same common marginals].

2.2 At this point two questions arise.

First, given any fixed pair of marginals $S_1(x)$, $S_2(y)$, is a class of nontrivial joiners J(x, y), that "connect" these marginals into a bivariate survival function, nonempty?

2.2.A A The second question (see, [10]) that is about the whole class of such joiners can be stated as follows:

Given any two marginal survival functions $S_1(x)$ and $S_2(y)$, for what functions J(x, y) will the product $S_1(x)\,S_2(y)\,J(x, y)$ be a valid survival function?

The first question can immediately be positively answered at least in the case when both the marginals have the probability densities and therefore are represented by the corresponding hazard (failure) rates, say, $\lambda_1(x)$, $\lambda_2(x)$ [the situation typical in reliability settings].

As it will be shown later the following nontrivial elements belonging to this class are

$$J(x,y)=\exp\left[-a\int_0^x\int_0^y \lambda_1(t)\lambda_2(u)\,dt\,du\right] \tag{3}$$

where $0 \le a \le 1$, and, therefore, the corresponding joint survival functions can be expressed as follows:

$$\begin{aligned}S(x,y)&=S_1(x)S_2(y)J(x,y)\\&=\exp\left[-\int_0^x \lambda_1(t)dt-a\int_0^x\int_0^y \lambda_1(t)\lambda_2(u)\,dt\,du-\int_0^y \lambda_2(u)du\right]\end{aligned} \tag{4}$$

To answer the second question first we find some necessary conditions for any real measurable function $J^*(x, y)$ defined for nonnegative real x, y that, given two survival functions $S_1(x)$, $S_2(y)$, the product

$S_1(x)\,S_2(y)\,J^*(x, y) = S^*(x, y)$ is a valid survival function. If so, then according to the adopted notation, we will have that $J^*(x, y) = J(x, y)$ and $S^*(x, y) = S(x, y)$. Later, we will seek for some necessary and sufficient conditions for that.

Now, from formula (2) one obtains:

$$S_2(y)J(x,y)=S(x,y)/S_1(x)=S_2(y|x)=P(Y\ge y|X\ge x).$$

Thus, to find the required necessary conditions for the fact $J^*(x, y) = J(x, y)$ taking place, realize that for each value of x, the product $S_2(y)\, J(x, y)$ must be a valid (univariate) survival function in y. For any x we must then have $S_2(0)\, J(x, 0) = 1$, and therefore we obtain $J(x, 0) = 1$.

Moreover, the product $S_2(y)\, J(x, y)$ must be nonnegative [and so is $J(x, y)$] and a nonincreasing function of y. There is obviously no necessity that $J(x, y)$ alone is always (given a fixed x) nonincreasing while the product $S_2(y)\, J(x, y)$ always is. There is also no necessity that $J(x, y) \to 0$ as $y \to \infty$.

In particular, for the independence case, we have $J(x, y) = 1$, for each x, y. Thus, necessary conditions for $J(x, y)$, given the fixed survival functions $S_1(x)$, $S_2(y)$, that the product $S_1(x)\, S_2(y)\, J(x, y)$ is a valid bivariate survival function, are:

(1) $J(x, y)$ is a nonnegative real function and, for each x, $J(x, 0) = 1$,
(2) $J(x, y)$ is continuous from the right with respect to each of its variables,
(3) For each x, the product $S_2(y)\, J(x, y)$ is a nonincreasing function of y and it approaches zero as $y \to \infty$. Notice, here we always assume that, for every considered Y and for every x, $P(Y = \infty \mid X \geq x) = 0$, otherwise of course, the considered product need not to approach zero.

When considering the product $S_1(x)\, J(x, y) = P(X \geq x \mid Y \geq y)$ one obtains two additional conditions "dual" to (1) and (3) above. Namely, $J(x, y)$ must also satisfy the following:

(1*) For each y, $J(0, y) = 1$,
(3*) For each y, the product $S_1(x)\, J(x, y)$ is a nonincreasing function of x and it approaches zero as $x \to \infty$ [with the similar assumption as the one associated with condition (3)].

The logical conjunction of all the above five conditions is still only the necessary condition for the product $S_1(x)\, S_2(y)\, J(x, y)$ to be a legitimate bivariate survival function.

The following key condition when added to the above five conditions (1)–(3*) allows to form the six conditions whose logical conjunction stands for a necessary and sufficient condition characterizing the class of all joiners $J(x, y)$ "fitting" to a given pair of the marginal survival functions $S_1(x)$, $S_2(y)$.

Let $J(x, y)$ be an arbitrary function satisfying the five conditions (1)–(3*).

Any product $S_1(x)\, S_2(y)\, J(x, y)$ now will shortly be denoted by $S(x, y)$. It is not yet necessarily a legitimate bivariate survival function unless it satisfies some conditions specified below.

Let us start to analyze these (sufficient) conditions.

For example, if the second mixed derivatives $\partial^2/\partial x\partial y\, S(x, y)$ and $\partial^2/\partial y\partial x\, S(x, y)$ exist and are equal to each other, the mentioned key condition for $S(x, y)$ to be a survival function is the uniform nonnegativity condition:

$$\partial^2/\partial x\partial y\, S(x, y) \geq 0 \tag{5}$$

for all nonnegative x, y.

Under this [as well as under the conditions (1)–(3*)] the derivative is the joint probability density function and S(x, y) is a survival function for the random vector (X, Y).

However, as considered in the light of general theory, the second derivative $\partial^2/\partial x\partial y$ S(x, y) may not exist, while S(x, y) still may be a valid survival function. This happens, for example, in the case when at least one of the two random variables X, Y is of discrete type. The condition (5) must then be replaced by a more general condition which, in the special case when the second mixed derivative $\partial^2/\partial x\partial y$ S(x, y) exists, essentially reduces to (5).

For the general analysis realize that the following expression

$$\{[S(\mathbf{x}+h,\mathbf{y}+k)-S(\mathbf{x},\mathbf{y}+k)]-[S(\mathbf{x}+h,\mathbf{y})-S(\mathbf{x},\mathbf{y})]\}/hk \tag{6}$$

is always defined, at least for some of the values $\mathbf{x} \geq 0$, $\mathbf{y} \geq 0$ together with some $h > 0$ and $k > 0$.

Expression (6) has exactly <u>the form</u> of the differential quotient for $\partial^2/\partial x\partial y$ S(x, y) whenever the latter second derivative exists. If it does not exist, (6) still has the same form that exists on its own.

It's numerator equals to a "Stjelties-like measure" determined by S(x, y) (this is thought off as an 'extended measure', possibly taking on all the [not necessarily non-negative] real values) of any rectangular in the first quadrant of the cartesian plane with the vertices (x, y), (**x** + h, **y**$_)$, (**x**, **y** + k), (**x** + h, **y** + k), (h > 0, k > 0) on which the function S(x, y) is defined.

The specific condition (5) we replace by the general condition

$$0 \leq [S(x+h,y+k)-S(x,y+k)]-[S(x+h,y)-S(x_0,y)] \leq 1 \tag{7}$$

for all the, mentioned above rectangulars. The second inequality in (7) as the consequence of the previous conditions (1), (3), (1*) and (3*) is automatically satisfied.

Thus, formula (7), when satisfied, brings to the "Stjelties-like measure" of each the considered rectangulars an interpretation of 'probability measure'.

If the fraction (6) is well defined for all the values x, y, h, k and if the limit of this fraction for (h, k) $\to$ (0, 0) exists then condition (5) is the direct consequence of the first inequality in (7).

Other form of (7), possibly easier for finding solutions, is the following sufficient condition for S(x, y) = $\mathbf{S_1(x)\,S_2(y)\,J(x,y)}$ to be a legitimate survival function:

$$S(x+h,y)+S(x,y+k) \geq S(x,y)+S(x+h,y+k), \tag{7*}$$

of course, together with the five conditions (1)–(3*).

Recall that, in practical investigations, all the time, the symbols S(,) should be replaced by the expressions $S_1(.)\,S_2(.)\,J(,)$ with the proper arguments as present in (7) and (7*).

Finding specific solutions S(x, y) of inequality (7*) sometimes may be problematic.

However, this will be an easier task if we replace the conditions (3) and (3*) by a stronger condition stating that not only the products $S_2(y)$ J(x, y) and $S_1(x)$ J(x, y) must be nonincreasing functions of y and x, respectively, but that the function J(x, y) itself is a nonincreasing function of both x and y. Under this assumption many joiners are immediately available. Realize that every other bivariate survival function, say R(x, y) is itself a joiner (i.e., J(x,) = R(x, y) is a sufficient condition for the considered product $S_1(x)$ $S_2(y)$ J(x, y) to be a survival function, see [10]) for any pair ($S_1(x)$, $S_2(y)$).

Moreover, since the arithmetic product of any two survival functions is again a valid survival function, the semigroup structure for this subset of the class of the joiners is given, where the sub-semigroup of actual survival functions, that we are seeking for, is generated by the product $S_1(x)$ $S_2(y)$ and therefore has no unity. The unity for the semigroup of the, now considered, type of the joiners (only) can be created by simply adding to it the trivial joiner $J_0(x,y) = 1$ which itself is not a survival function for the random variables that take on finite values only.

Notice, by the way, that all the elements of the considered semigroup with the unity $J_0(x, y)$ are irreversible.

As one can see, the class of the joiners proper for a given pair ($S_1(x)$, $S_2(y)$) is quite wide.

Anyway, given any pair ($S_1(x)$, $S_2(y)$) (not necessarily from the same class of the distributions) many bivariate models S(x, y), having $S_1(x)$, $S_2(y)$ as their common marginals can be constructed, although not every function J*(x, y), satisfying (1)–(3*), is a proper joiner for a given fixed pair ($S_1(x)$, $S_2(y)$).

Until now, our attention was concentrated on the problem of determining a class of joiners that "fit" to a given fixed pair ($S_1(x)$, $S_2(y)$) of univariate survival (reliability) functions. To this subject, in the case when the derivative $\partial^2/\partial x \partial y$ $S_1(x)$ $S_2(y)$ J(x, y) equal to $\partial^2/\partial y \partial x$ $S_1(x)$ $S_2(y)$ J(x, y) exists, we return soon in next subsection.

Now, the question above stated is reversed.

Namely, suppose there is given a function (a candidate for the joiner) J*(x, y). One may ask:

For which pairs ($S_1(x)$, $S_2(y)$) of survival functions the product $S_1(x)$ $S_2(y)$ J*(x, y) is a valid survival function. By the way, this problem's solutions can be based on the same inequalities (7) or, whenever appropriate, (5) even if one would seek different functions (i.e., $S_1(x)$ and $S_2(y)$) as possible solutions.

Other version (see, [10]) of the last problem is the following:

Find the joiners (if they exist) J(x, y) that would be proper for any pair ($S_1(x)$, $S_2(y)$) of the univariate survival functions, provided the variables x, y of all the underlying functions run through the same domain.

As for the answer, first realize that the trivial joiner J(x, y) = 1 satisfies the requirement of the problem.

At the moment, we can not find the whole class of such joiners. Nevertheless, as a partial answer, we may state that all the joiners J(x, y) that are themselves bivariate survival functions are proper joiners for all the pairs ($S_1(x)$, $S_2(y)$) such that in the product $S_1(x)$ $S_2(y)$ J(x, y) the variables x, y run through the same domain. This follows from the fact that the product of any two survival functions (defined on the same set of arguments) is a survival function.

It is yet unclear to us if this class of the joiners (together with the trivial joiner) is the maximal one.

Weaker version of the last question can be stated as follows. Find nontrivial joiners J(x, y) such that there (always) exist survival functions $S_1(x)$, $S_2(y)$ so that the product $S_1(x)\,S_2(y)\,J(x, y)$ is the survival function.

As for the answer to the last question, realize that the class of all the joiners (3) has this property, where the hazard rates of the corresponding $S_1(x)$, $S_2(y)$ exist and are $\lambda_1(t)$, $\lambda_2(y)$.

2.2.B From now on, we will continue trying to answer our original question on determining a class of the joiners J(x, y) that "fit" to a given fixed pair $(S_1(x), S_2(y))$.

Let us now concentrate on the important (also for reliability problems) case when the derivatives $\partial^2/\partial x\partial y\, S(x,y) = \partial^2/\partial y\partial x\, S(x,y)$ does exist.

Then, one can replace the general condition (7) or (7*) by nonnegativity of this derivative.

Formula (5) together with the condition $S(0, 0) = 1$, means that the above derivative is a joint probability density of the considered random vector (X, Y) and thus the involved joiner is the proper one.

Since the considered second derivative of the product $S_1(x)\,S_2(y)\,J(x, y)$ exists, the hazard (failure) rates $\lambda_1(x)$ and $\lambda_2(y)$ corresponding to the survival functions $S_1(x)$, $S_2(y)$, respectively, do exist.

Moreover, any candidate J(x, y) for a proper joiner satisfying the five conditions (1)–(3*) can be represented by a continuous function (x, y) (bounded from below by a properly chosen nonpositive number) so that

$$J(x, y) = \exp\left[-\int_0^x\int_0^y \psi(t, u)\,du\,dt\right]. \tag{8}$$

Taking all together under consideration one realizes that the candidate for the survival function can be represented as follows:

$$S(x, y) = \exp\left[-\int_0^x \lambda_1(t)dt - \int_0^x\int_0^y \psi(t, u)du\,dt - \int_0^y \lambda_2(u)du\right]. \tag{9}$$

In particular, for the simplicity, one may consider, as a first approach, the separated variables functions:

$$\psi(t, u) = \psi_1(t)\psi_2(u).$$

Given the marginals $S_1(x)$, $S_2(y)$ as represented by the hazard rates $\lambda_1(x)$, $\lambda_2(y)$, respectively, our problem of finding proper joiners, say, in the form (8), reduces to finding the corresponding proper functions $\psi(x, y)$ such that the derivative $\partial^2/\partial x\partial y$

taken from the right hand side of (9) is nonnegative. The latter means that the probability density corresponding to (9) does exist and therefore the function $\psi(x, y)$ "fits" to the hazard rates $\lambda_1(x)$, $\lambda_2(y)$. To find, the proper function $\psi(x, y)$, given the functions $\lambda_1(x)$, $\lambda_2(y)$ we calculate the derivative $\partial^2/\partial x\partial y$ from the right hand side of (9) and set it nonnegative. As a result of some simplification we obtain the following integral inequality with respect to $\psi(x, y)$:

$$\left[\lambda_1(x) + \int_0^y \psi(x, u)du\right]^x \left[\lambda_2(y) + \int_0^x \psi(t, y)dt\right] \geq \psi(x, y). \quad (10)$$

Now, this part of the creating theory reduces to solving integral inequality (10), given the marginal distributions as represented by the hazard (failure) rates $\lambda_1(x)$, $\lambda_2(y)$.

When assuming $\psi(x, y) \geq 0$ (10) can be simplified to the following stronger inequality:

$$\lambda_1(x)\lambda_2(y) \geq \psi(x, y) \quad (10^*)$$

So that, if $\psi(x, y) \geq 0$ is satisfied then every solution of (10*) is a solution of (10) [obviously the reverse statement is false].

Thus, some solutions of (10) can immediately be found using (10*).

For example, one can easily check that $\psi(x, y) = a\lambda_1(x)\,\lambda_2(y)$ $(0 \leq a \leq 1)$ is a solution of (10).

As the result one obtains the bivariate survival function given by (4).

At this point realize that other more general solution one obtains if the coefficient 'a' in (4) will be replaced by any continuous function a(x, y) such that $0 \leq a(x, y) \leq 1$.

In particular, we propose the function $a(x, y) = c\exp[-b_1x - b_2y]$,
where $0 \leq c \leq 1$, while b_1 and b_2 are nonnegative real.

Explicitly, the new model, which is a generalization of (4), has the following form:

$$S(x, y) = \exp\left\{-\int_0^x \lambda_1(t)dt - \int_0^y \lambda_2(u)du - c\exp[-b_1x - b_2y]\int_0^x\int_0^y \lambda_1(t)\,\lambda_2(u)dt\,du\right\} \quad (11)$$

Other generalization of (4) one obtains upon substituting in (4)

$$a = a(x, y) = c\exp[-bxy] \quad (12)$$

with a nonnegative b.

Realize that the solution (4) or its generalizations such as (11) and the one given by (12) are well chosen for any pair of the marginals represented by (variable) pairs $(\lambda_1(x), \lambda_2(y))$.

Here, the condition $\psi(x, y) \geq 0$ follows automatically.

Now, suppose there is given any fixed pair $(\lambda_1(x), \lambda_2(y))$ of the marginals. Consider the following integral equation in $\psi(x, y)$:

$$\lambda_1(x)\lambda_2(y) + \lambda_1(x)\int_0^x \psi(t, y)dt + \lambda_2(y)\int_0^y \psi(x, u)du = \psi(x, y). \tag{13}$$

Realize that upon the same as before assumption that $\psi(x, y) \geq 0$, one can see that all solutions of integral Eq. (13) are at the same time the solutions (in $\psi(x, y)$) of inequality (10).

In this (continuous) case the initially stated problem: "given the marginals find the [bivariate] joint probability distribution", reduces to solving integral Eq. (13) which is (nonhomogeneous) linear with respect to $\psi(x, y)$. This equation may be reduced to the following homogeneous linear:

$$\lambda_1(x)\int_0^x \psi(t, y)dt + \lambda_2(y)\int_0^y \psi(x, u)\, du = \psi(x, y). \tag{14}$$

Realize, all solutions of (14) also are solutions of the inequality (10) although [as in the case of (13)] not all those solutions.

Equations (13) and (14) can be dramatically simplified if to consider only constant hazard rates λ_1, λ_2.

Obviously, in such a case, the marginals are reduced to the exponential only.

Solving the Eqs. (13) or (14) for the more general Weibullian case $\lambda_1(x) = \lambda_1\alpha\, x^{\alpha-1}$ and $\lambda_2(y) = \lambda_2\beta y^{\beta-1}$ is not much more difficult from the integral equations theory viewpoint.

Anyway, the theory and methods for solving the classes of integral Eqs. (13) and (14) in general seems to be treatable also in purely analytical ways.

3 Universality of Joiner Representation of Bivariariate Survival Functions

3.1 Actually, the **universality** of the joiner representation (1) of any bivariate survival function, given any marginals $S_1(x)$, $S_2(y)$ is a consequence of the simple (always true) arithmetic identity [for $S_1(x) \neq 0$ and $S_2(y) \neq 0$ with $S(x, y) = 0$ otherwise] obtained by substituting (2) to (1). Realize that, at least in the case the, corresponding to $S_1(x)$, $S_2(y)$, marginal hazard rates $\lambda_1(x)$, $\lambda_2(y)$ exist a nonempty class of nontrivial joiners (3) exists. This class produces the model (4) which seems to be important both theoretically and with respect to possible applications. Numerous modifications of this model such as (11) and (12) are at hand too. So, there is no problem to find some bivariate distribution "fitting" the initial marginals of the continuous type. More solutions one can find solving integral Eqs. (13) or (14) or finally the integral inequality (10*) or the most general inequality (10). In the case there is no both hazard rates corresponding to $S_1(x)$, $S_2(y)$ one can try to apply inequality (7*) for any joiner's candidate, say $J^*(x, y)$. By the way, as mentioned above, also any bivariate survival (reliability) function $R(x, y)$ is a joiner proper for all possible pairs $\{S_1(x), S_2(y)\}$ of the marginals. Shortly, there is no problem in finding some bivariate models for a given marginals. The problem only may lie in a proper choice among them (using statistical methods) or in finding more possibilities.

The situation is then similar to that when, given two marginal distribution functions, one seeks for proper copulas. On one side, every known copula produces a joint cdf for which the given in advance univariate cdfs are the marginals. On the other, a number of such copulas is limited and, usually, a big problem occurs when one wants to find a copula that fits to a given situation in applications. Unlike with the model (4), for example, in most of the copula cases its hard to see a relationship between a given practical situation in the real world and (pure) mathematical formula that the given copula is based on.

At this point, compare our approach and the following Sklar's [5] theorem for the bivariate case:

"Every bivariate distribution function $F(x_1, x_2)$ can be expressed in terms of its marginal cdf's $F_1(x_1), F_2(x_2)$ and a copula $C(,)$, so that $F(x_1, x_2) = C(F_1(x_1), F_2(x_2))$."

In the case we consider, the corresponding version of above theorem can be formulated as follows:

"Every bivariate survival function $S(x,y)$ can be expressed in terms of its marginal survival functions $S_1(x)$, $S_2(y)$ and a function $D(S_1(x), S_2(y)) = S_1(x)\, S_2(y)\, J(x, y)$ for some other function $J(x, y)$".

[Both above theorems have their natural extension to an arbitrary dimension $k = 3, 4, \ldots$, see Sect. 5 of this chapter].

The essential difference between the theorems lies in the form of the functions $C(F_1(x_1), F_2(x_2))$ and $D(S_1(x), S_2(y))$ which are (different) universal representations of the same bivariate distributions. First, realize the copula $C(F_1(x_1), F_2(x_2))$ only depends on the distributions $F_1(x_1)$, $F_2(x_2)$ and not on the separate values of x_1 or x_2. On the other hand, for the expression $D(S_1(x), S_2(y))$ the univariate distributions, as characterized by $S_1(x)$, $S_2(y)$, are both factored out and are not anymore present in $J(x, y)$. The "rest" $J(x, y)$ of the expression $D(,)$ only depends on x and y alone and describe the underlying stochastic dependence between the random variables X, Y (and not between the marginal distributions). The marginal and the dependence structures are analytically separated which makes it easier to analyze both.

Besides, in general, for practical purposes it's easier to find (during a modeling process) a proper joiner, given a pair $S_1(x)$, $S_2(y)$, than a proper copula, given $F_1(x_1)$, $F_2(x_2)$.

Sklar's theorem, at the moment, may seem to be more general. However, the methods here applied for survival functions, as restricted to nonnegative real values only, may readily be extended to all the distribution functions.

3.2 When talking about ***universality*** of the "joiner representation" of bivariate survival functions, the question that comes up is the relationship between that representation and the representation by the method of **parameter dependence** [11]. Recall that the conditional survival function, as defined by the parameter dependence method, is given by formula $S_2(y \mid x) = P(Y \geq y \mid X \geq x) = S_2(y; Q_2(x))$ [where Q_2 alone is a scalar or vector parameter in $S_2(y; Q_2) = P(Y \geq y)$, independent of x]. Since we always have $S_2(y \mid x) / S_2(y) = J(x,y)$, we get

$$J(x, y) = S_2(y; Q_2(x))/S_2(y; Q_2). \tag{15}$$

Finding the joiner while using the method of parameter dependence reduces to applying the last quotient. If instead, we start that method with $S_1(x \mid y) = S_1(x \mid Q_1(y))$ we obtain the same joiner in alternative form:

$$J(x, y) = S_1(x; Q_1(y))/S_1(x; Q_1). \tag{16}$$

Thus, in light of formulas (15) and (16), the parameter dependence representation of bivariate survival functions may be thought of as a method of finding the corresponding joiners.

As clearly follows from formulas (15) and (16), the method of construction by the joiner is more general than the method of parameter dependence [11]. Also it is more general than the method of finding the conditional survival functions by the use [6] of either the Cox [8] or the Aalen [7] model. Actually, the "joiner method" contains all three methods as special cases.

4 Baseline Factors Representation

There is an essential difference between the marginal distributions given by $S_1(x)$, $S_2(y)$ and the corresponding baseline distributions, say, $B_1(x)$, $B_2(y)$ when they exist. The baseline distributions, we started to consider in [6], are usually given in the "beginning" of the modeling process as the distributions of independent random quantities, say T_1, T_2 (in general different than the "in system" random variables X, Y considered above), "before" any "physical" interactions between the objects happen.

{In the frameworks we consider by "physical" we mean not necessarily strictly physical interactions but any "real world" nonmathematical interactions such as social, psychological, financial, etc.}

In **reliability** modeling this preexisting "beginning" can, for example, be the same as the "laboratory conditions" where two system components life times are tested separately from each other before these statistically same components are put to work into a system. Once the physical interactions between the components occur, the original independent random quantities T_1, T_2 (each having a given baseline survival function) turn into the (above considered) dependent random variables, say X, Y, whose joint distribution is given by the joint survival function S(x, y) . As a classical example let us follow Freund's [1] bivariate exponential model: two airplane engines work together on the same wing of the plane. Before they are put to work within a real plane, each of them is tested separately (in idealized laboratory conditions) under an artificial load equal to the constant load within the airplane when the work conditions are normal, undisturbed. The load does not change its value during the test so the times to failure in such "ideal" conditions are independent random variables, say T_1, T_2 . In Freund's model their probability distributions are considered to be exponential with the failure rates λ_1, λ_2 respectively. When the statistically the same (but physically other) engines are put to work within the real plane the situation is different. Namely, when during the flight one of the engines fails the other has to stand more load [this never happens in the laboratory conditions], and therefore its failure rate increases from the value λ_i (i = 1, 2) to a higher value, say, λ_i*. With this higher value of the load, the remaining engine works until it, possibly, fails too. Now, times to failure are different and are dependent random variables X, Y having the joint survival function S(x, y) which in the preceding example is the famous Freund [1] model. The marginal distributions of X and Y are, in this case, different from the (baseline) distributions of T_1 and T_2 respectively. Such procedure involving the obtained (when the objects are tested in physical separation) baseline distributions is pretty common in various applications. The difference between the baseline and the marginal distributions parallels the difference between the objects physical separation and work within a system where "physical" interactions take place. In most (if not in all) cases the (same) joint survival function S(x, y) can also be factored as

$$S(x, y) = B_1(x)B_2(y)K(x, y). \tag{17}$$

In some cases, such as the first Gumbel bivariate exponential survival function [2], baseline distributions are the same as the marginal. We then have $B_1(x) = S_1(x)$, $B_2(y) = S_2(y)$, and, consequently, K(x, y) = J(x, y) for all x, y.

This, however, is not the case for Freund [1] for Marshall and Olkin [3] and many other (see [4]) models.

Our next goal is to consider the analytical relationship between representations (1) and (17) of (possibly) any survival function S(x, y). First, realize that since S(x, 0) = $S_1(x)$ and S(0, y) = $S_2(y)$ it follows from representation (1) that J(x, 0) = J(0, y) = 1 for each x and.

The latter relationship does not always hold for the "**system function**" K(x, y).

Now we will prove the following theorem that relates representations (1) and (17) of any joint survival function S(x, y).

Theorem 1 Let S(x, y) be any bivariate survival function. Suppose both baseline survival functions $B_1(x)$, $B_2(y)$ for S(x, y) exist. The following relationships between representations (1) and (17) of the considered bivariate survival function hold:

$S_1(x) = B_1(x)K(x, 0)$,

$S_2(y) = B_2(y)K(0, y)$.

Moreover, we have that

$$J(x, y) = K(x, y)/K(x, 0)\,K(0, y).$$

Proof As it follows from (17), upon the conditions K(x, 0) = K(0, y) = 1, the representations (1) and (17) are identical, i.e., both baseline distributions are marginals and so J(x, y) = K(x, y). This happens in the case of the first Gumbel bivariate exponential survival function [2].

More generally, this is the case when

$$J(x, y) = K(x, y) = \exp\left[-\int_0^x\int_0^y \psi(t; u)dudt\right] \text{and}\, \psi(t; u) \ge 0.$$

However, it is not the case for the Freund [1] bivariate exponential and also not for the classical Marshall and Olkin [3] bivariate model. (The reason is that in these two models the integral $\int_0^x \int_0^y \psi(t; u)du\,dt$ is differently calculated for $x \le y$ than for $x > y$.)

Now, suppose the two representations are not identical (but still are equivalent). Consider representation (17): $S(x, y) = B_1(x)B_2(y)K(x, y)$.

From that we can find the marginals

$$S(x, 0) = S_1(x) = B_1(x)\,K(x, 0),$$
$$S(0, y) = S_2(y) = B_2(y)\,K(0, y),$$

and the first part of theorem is proven.

Substitute the above result into representation (1). We obtain:

$$\begin{aligned} S(x, y) &= S_1(x)S_2(y)J(x, y) \\ &= \{B_1(x)\,K(x, 0)\}\{B_2(y)K(0, y)\}J(x, y) \\ &= B_1(x)B_2(y)K(x, 0)K(0, y)\,J(x, y), \end{aligned}$$

The latter implies that K(x, y) = K(x, 0) K(0, y) J(x, y) which concludes the proof. □

Thus, every proper function K(x, y) uniquely determines the joiner J(x, y) but the equation K(x, y) = K(x, 0) K(0, y) J(x, y) suggests that for a given J(x, y) more than one solution for K(x, y) may exist depending on the initial values K(x, 0), K(0, y) . However, at least one ("trivial") solution always exists upon the initial conditions K(x, 0) = K(0, y) = 1 . Then, of course, K(x, y) = J(x, y) for each x and y.

It may be helpful to realize that every function K(x, y) must satisfy the condition **K(0, 0) = 1.**

To see this consider the identity $S(0, 0) = B_1(0)*B_2(0)*K(0, 0)$, i.e., 1 = 1*1*K(0, 0), where the symbol '*' denotes arithmetic multiplication.

The next question that arises is: given two arbitrary baseline survival functions $B_1(x)$, $B_2(y)$. What conditions must satisfy the class of functions K(x, y) in order that the products $B_1(x)\ B_2(y)\ K(x, y)$ are valid bivariate survival functions, say S(x, y)? This problem we are leaving open.

Finally notice that for a given pair $B_1(x)$, $B_2(y)$ the corresponding proper function K(x, y) can be considered as an analytical description of stochasticity of the "physical" system the, originally independent, objects are put into. Usually, the arising stochastic interactions are described by one or more additional parameters that when equal to zero (or to some other particular value(s)) the interactions cease to exist, and the function K(x, y) becomes 1. This is the case for Freund model [1] when the increment of failure rate caused by the failure of the other engine is zero. Similar situation is encountered in the Marshall and Olkin model [3] when the failure rate of multiple failure becomes zero. Other examples can be given too [4].

Remark All the above considerations can also be extended to the following situation, see [9]. Given two, totally described, stochastic processes, say $\{X_t\}$, $\{Y_t\}$ with time 't' running for both through the same set T. Thus, as we assume, for every $t \in T$, the marginal survival functions $S_{1t}(x_t)$, $S_{2t}(y_t)$ of the random variables X_t, Y_t are given. They may be treated as functions of (additional variable) time. According to the above considerations, for each time epoch t, we can construct a class of bivariate survival functions

$$S_t(x_t, y_t) = S_{1t}(x_t)S_{2t}(y_t)J_t(x_t, y_t).$$

Then, under some assumptions on the time t dependence, we will obtain a class of bivariate stochastic processes, see [9]. In the case of continuous time $t \in T$, a continuity of the initial functions $S_{1t}(x_t)$, $S_{2t}(y_t)$ as well as of the constructed function $S_t(x_t, y_t)$ with respect to time t may be essential.

Possibly, some more conditions for the process' joiner $J_t(x_t, y_t)$ (as a function of time t) must be satisfied in order that the survival functions $S_t(x_t, y_t)$ form the full description of the so created bivariate stochastic process $\{X_t, Y_t\}$. This question is the subject of further investigations.

In practical applications it may be useful if the initial functions $S_{1t}(x_t)$, $S_{2t}(y_t)$ are considered to be baseline rather than marginal. Thus, it seems that the theory of bivariate random vectors (X, Y) may be extended to the theory of bivariate stochastic processes $\{X_t, Y_t\}$. We expect applications of such a theory of bivariate stochastic processes in construction of new models for systems maintenance, econometrics and biomedical phenomena. □

5 k-Variate Survival Functions Universal Representation

5.1 First we consider the k = 3 case.

Given a tri-variate random vector (X_1, X_2, X_3) such that $P(X_1 \geq 0, X_2 \geq 0, X_3 \geq 0) = 1$.

Our task is to determine survival (reliability) functions:

$$R(x_1, x_2, x_3) = P(X_1 \geq x_1, X_2 \geq x_2, X_3 \geq x_3) \quad (18)$$

given the univariate marginals $R_1(x_1)$, $R_2(x_2)$, $R_3(x_3)$ i.e., the survival functions of the dependent random variables X_1, X_2, X_3 respectively.

As the first step of this procedure we must find, by the methods described in sections above, all the three marginal bivariate survival functions $R_{12}(x_1, x_2)$, $R_{13}(x_1, x_3)$, $R_{23}(x_2, x_3)$.

Recall, the latter task reduces to finding (or choosing) the corresponding joiners:

$$J_{12}(x_1, x_2), J_{13}(x_1, x_3), J_{23}(x_2, x_3).$$

At the second step realize that any trivariate survival function (18) of the random vector (X_1, X_2, X_3) can always be represented by the **universal** form:

$$R(x_1, x_2, x_3) = \{R(x_1, x_2, x_3)/R_{12}(x_1, x_2)R_{13}(x_1, x_3)R_{23}(x_2, x_3)\} \cdot R_{12}(x_1, x_2)R_{13}(x_1, x_3)R_{23}(x_2, x_3) \quad (19)$$

where

$$R(x_1, x_2, x_3)/R_{12}(x_1, x_2)\, R_{13}(x_1, x_3)\, R_{23}(x_2, x_3) = J^*_{123}(x_1, x_2, x_3), \quad (20)$$

while all the factors in the denominator of (20) are assumed to be nonzero.

Otherwise, for those arguments x_1, x_2, x_3 for which at least one of the factors is zero we set $R(x_1, x_2, x_3) = 0$ and proclaim the 'joiner' $J^*_{123}(x_1, x_2, x_3)$ undefined.

Thus, one obtains the following first universal representation of any trivariate survival function:

$$R(x_1, x_2, x_3) = J^*_{123}(x_1, x_2, x_3) \cdot R_{12}(x_1, x_2)R_{13}(x_1, x_3)R_{23}(x_2, x_3), \quad (21)$$

where $J^*_{123}(x_1, x_2, x_3)$ is a properly chosen [dependence] function of the three variables.

This form is universal since it follows the (always true) arithmetic identity (19).

So, after the first step is done, the problem reduces to determining the class of "first tri-joiners" $J^*_{123}(x_1, x_2, x_3)$ i.e., the class of (possibly) all functions defined on the set $K = \{(x_1, x_2, x_3) \in R^3 \mid x_1 \geq 0, x_2 \geq 0, x_3 \geq 0\}$ such that, given the three bivariate survival functions $R_{12}(x_1, x_2)$, $R_{13}(x_1, x_3)$, $R_{23}(x_2, x_3)$, the product $J^*_{123}(x_1, x_2, x_3) . R_{12}(x_1, x_2)\, R_{13}(x_1, x_3)\, R_{23}(x_2, x_3)$ is a legitimate tri-variate survival function defined on K (see, [10]).

Remark Actually, all the 3D problems (k = 3) of both the construction and the characterization [of the $J^*_{123}(x_1, x_2, x_3)$'s class] reduces to that formulated above. The accompanying effort to solve it (i.e., to find the general characterization or a specific model) is, in the essence, not much different than that in the bivariate case. The only difference relies on higher complexity of underlying analytical expressions. □

One can expand (20) as follows:

$$\begin{aligned} R(x_1, x_2, x_3) &= J^*_{123}(x_1, x_2, x_3) \cdot R_{12}(x_1, x_2) R_{13}(x_1, x_3) R_{23}(x_2, x_3) \\ &= J^*_{123}(x_1, x_2, x_3)[J_{12}(x_1, x_2) R_1(x_1) R_2(x_2)] \\ &\quad [J_{13}(x_1, x_3) R_1(x_1) R_3(x_3)][J_{23}(x_2, x_3) R_2(x_2) R_3(x_3)] \\ &= J^*_{123}(x_1, x_2, x_3).J_{12}(x_1, x_2) J_{13}(x_1, x_3) J_{23}(x_2, x_3) \\ &\quad [R_1(x_1)]^2 [R_2(x_2)]^2 [R_3(x_3)]^2. \end{aligned} \tag{22}$$

Formula (22) is the <u>second universal representation</u> of <u>**any**</u> trivariate survival function $R(x_1, x_2, x_3)$.

At this point, restrict our attention to the second universal representation of $R(x_1, x_2, x_3)$ for the case when the third derivative $\partial^3 R(x_1, x_2, x_3)/\partial x_1 \partial x_2 \partial x_3$ exists and is continuous.

In this case the hazard rates $\lambda_1(x_1)$, $\lambda_2(x_2)$, $\lambda_3(x_3)$ for the marginals $R_1(x_1)\, R_2(x_2)$, $R_3(x_3)$ exist.

Besides, assume that we have at disposal the following representations of bivariate and trivariate joiners:

$$J_{12}(x_1, x_2) = \exp[-\int_0^{x1}\int_0^{x2} \psi_{12}(t; u)du\, dt],$$

$$J_{13}(x_1, x_3) = \exp[-\int_0^{x1}\int_0^{x3} \psi_{13}(t; u)du\, dt],$$

$$J_{23}(x_2, x_3) = \exp[-\int_0^{x2}\int_0^{x3} \psi_{23}(t; u)du\, dt]$$

$$J*_{123}(x_1, x_2, x_3) = \exp[-\int_0^{x1}\int_0^{x2}\int_0^{x3} \psi_{123}(t, u, v)dv\, du\, dt].$$

Now, in this "continuous case", we have the following expression for the (hipothetical) joint survival function (22) in the exponential form:

$$R(x_1, x_2, x_3) = \exp\Bigg[-2\int_0^{x1} \lambda_1(t)dt - 2\int_0^{x2} \lambda_2(u)du - 2\int_0^{x3} \lambda_3(v)dv$$
$$-\int_0^{x1}\int_0^{x2} \psi_{12}(t; u)du\, dt - \int_0^{x1}\int_0^{x3} \psi_{13}(t; u)du\, dt$$
$$-\int_0^{x2}\int_0^{x3} \psi_{23}(t; u)du\, dt - \int_0^{x1}\int_0^{x2}\int_0^{x3} \psi_{123}(t, u, v)dv\, du\, dt\Bigg] \quad (23)$$

From now on, the task relies on verifying when (23) represents a legitimate 3-variate survival function. In other words, given the three univariate marginals as represented by the hazard rates $\lambda_1(x_1)$, $\lambda_2(x_2)$, $\lambda_3(x_3)$ the question is ([10]), for which 'joiner equivalent functions'

$$\psi_{12}(x_1; x_2),\ \psi_{13}(x_1; x_3),\ \psi_{23}(x_2; x_3) \text{ and } \psi_{123}(x_1, x_2, x_3)$$

is (23) a valid survival function ?

The answer is proceeded as follows:

Suppose the functions $\psi_{12}(x_1; x_2)$, $\psi_{13}(x_1; x_3)$, $\psi_{23}(x_2; x_3)$ and $\psi_{123}(x_1, x_2, x_3)$ are continuous.

Then, there exist the continuous third mixed derivative: $\partial^3 R(x_1, x_2, x_3)/\partial x_1 \partial x_2 \partial x_3$ which, in order the joint density exists, must be uniformly non-negative. Thus, given the nonnegative functions $\lambda_1(x_1)$, $\lambda_2(x_2)$, $\lambda_3(x_3)$ we need to solve the following inequality with respect to the (possible) dependence functions $\psi_{12}(x_1; x_2)$, $\psi_{13}(x_1; x_2)$, $\psi_{23}(x_1; x_2)$, $\psi_{123}(x_1, x_2, x_3)$

$$\partial^3 R(x_1, x_2, x_3)/\partial x_1 \partial x_2 \partial x_3 \geq 0, \quad (24)$$

where $R(x_1, x_2, x_3)$ is given by (23).

After the calculations are performed the inequality (24) reduces to the following inequality:

$$\left\{2\lambda_2(x_2) + \int_0^{x1} \psi_{12}(t; x_2)dt + \int_0^{x3} \psi_{23}(x_2; u)du + \int_0^{x1}\int_0^{x3} \psi_{123}(t, x_2, v)dv\, dt\right\}$$

$$\left\{2\lambda_3(x_3) + \int_0^{x1} \psi_{13}(t, x_3)dt + \int_0^{x2} \psi_{23}(t; x_3)dt + \int_0^{x1}\int_0^{x2} \psi_{123}(t, u, x_3)du\, dt\right\}$$

$$\geq \psi_{23}(x_2; x_3) + \int_0^{x1} \psi_{123}(t, x_2, x_3)dt; \tag{25}$$

Above inequality is equivalent to (24) if:

$$2\lambda_1(x_1) + \int_0^{x2} \psi_{12}(x_1; u)du + \int_0^{x3} \psi_{13}(x_1; u)du + \int_0^{x2}\int_0^{x3} \psi_{123}(x_1; u; v)dv\, du \neq 0, \tag{25*}$$

The inequalities (24) and (25), together with (25*), must be satisfied uniformly.

Special Cases:

If to restrict the considerations to the case ψ_{123} $(x_1, x_2, x_3) = 0$ [the 3-independence], then the general inequality (25) reduces to the following:

$$\left\{2\lambda_2(x_2) + \int_0^{x1} \psi_{12}(t; x_2)dt + \int_0^{x3} \psi_{23}(x_2; u)du\right\}$$
$$\left\{2\lambda_3(x_3) + \int_0^{x1} \psi_{13}(t, x_3)dt + \int_0^{x2} \psi_{23}(t; x_3)dt\right\} \geq \psi_{23}(x_2; x_3). \tag{26}$$

Also, if in the above we set all ψ_{ij} $(x_i; x_j) \geq 0$ then we obtain a very simple version of (26):

$$4\lambda_2(x_2)\lambda_3(x_3) \geq \psi_{23}(x_2; x_3). \tag{27}$$

From the later we immediately obtain a specific solution:

$$\psi_{23}(x_2; x_3) = 4a\, \lambda_2(x_2)\lambda_3(x_3),$$

where $0 \leq a \leq 1$.

Other $\psi_{\mathbf{ij}}$ $(x_\mathbf{i}; x_\mathbf{j})$ are nonnegative and obviously, $\psi_{\mathbf{123}}$ $(x_\mathbf{1}, x_\mathbf{2}, x_\mathbf{3}) = 0$.

In particular, we also may chose:

$$\psi_{\mathbf{12}}(x_\mathbf{1}; x_\mathbf{2}) = 4b\, \lambda_1(x_1)\lambda_2(x_2) \text{ and } \psi_{13}(x_1, x_3) = 4c\, \lambda_1(x_1)\lambda_3(x_3)$$

with $0 \leq b, c \leq 1$.

This very particular (the simplest but possibly the most important?) case of the trivariate survival function (with the 3-independence) can be written in the following explicit form:

$$R(x_1, x_2, x_3) = \exp\Bigg[-2\int_0^{x1} \lambda_1(t)dt - 2\int_0^{x2} \lambda_2(u)du - 2\int_0^{x3} \lambda_3(v)dv - 4b\int_0^{x1}\int_0^{x2} \lambda_1(t)\lambda_2(u)du\,dt - 4c\int_0^{x1}\int_0^{x3} \lambda_1(t)\lambda_3(u)dudt_2 - 4a\int_0^{x2}\int_0^{x3} \lambda_2(t)\lambda_3(u)du\,dt\Bigg]. \quad (28)$$

Given then the hazard rates $\lambda_1(x_1)$, $\lambda_2(x_2)$, $\lambda_3(x_3)$, the only statistical problem is to estimate the parameters a, b, c and then to verify [statistically] the whole model (28) to what extend does it fit to a given data.

Remark As we (initially) have assumed all the four functions ψ_{12} $(x_1; x_2)$, ψ_{13} $(x_1; x_3)$, ψ_{23} $(x_2; x_3)$, ψ_{123} (x_1, x_2, x_3) present in inequality (25) were unknown so that we faced the process of solving (25) with respect to the four unknown functions. This assumption is, however, not necessary. The three functions ψ_{12} $(x_1; x_2)$, ψ_{13} $(x_1; x_3)$, ψ_{23} $(x_2; x_3)$ (or some of them) may be known in advance as either just assumed or as solutions of three separate bivariate problems described in Sect. 2. In such a case the inequality (25) together with (25*) contains only one unknown function ψ_{123} (x_1, x_2, x_3).

This simplifies the procedure of finding solutions (with respect to ψ_{123} (x_1, x_2, x_3) alone).

What, in this case, comes first to mind is verification of the statistical hypothesis:

$\psi_{123}(x_1, x_2, x_3) = 0$ [the 3-independence] versus the alternative
$\psi_{123}(x_1, x_2, x_3) \neq 0$

given particular data.

This, of course, may bring a solution ψ_{123} $(x_1, x_2, x_3) = 0$ for (25).

Various other procedures in analysing solutions of (25) apparently are involved. The theory may be developed from this point on. However, that is out of scope of this, relatively short, work. □

5.2 As for the general k-variate case (k = 3, 4, ….) we need to proceed in the way similar as in the case of k = 3.

Refer to formulae (19) and (20) for the case k = 3. So, by analogy in the general case k = 3, 4, …, we need to assume that we have, in advance (from all previous steps of the procedure of the construction), all the k − 1 dimensional marginal survival functions of the random vector $(X_1, X_2, \ldots, X_k)$.

The task is to procced the recurrence transition from all the k − 1 variate involved cases to the k-variate case. For that we apply the following formula analogic to formula (19):

$$R(x_1, x_2, \ldots, x_k) = \{R(x_1, x_2, \ldots, x_k)/R_{12\ldots(k-1)}(x_1, x_2, \ldots, x_{k-1})R_{12\ldots(k-2)k}(x_1, x_2, \ldots, x_{k-2}, x_k) \ldots R_{23\ldots k}(x_2, x_3, \ldots, x_k)\}R_{12\ldots(k-1)}(x_1, x_2, \ldots, x_{k-1}) R_{12\ldots(k-2)k}(x_1, x_2, \ldots, x_{k-2}, x_k) \ldots R_{23\ldots k}(x_2, x_3, \ldots, x_k), \quad (19^*)$$

where

$$R(x_1, x_2, \ldots, x_k)/R_{12\ldots(k-1)}(x_1, x_2, \ldots, x_{k-1})R_{12\ldots(k-2)k}(x_1, x_2, \ldots, x_{k-2}, x_k) \ldots R_{23\ldots k}(x_2, x_3, \ldots, x_k) = J^*_{12} \cdots {}_k (x_1, x_2, \ldots, x_k).$$

Now, the most general formula for the k-variate survival function is:

$$R(x_1, x_2, \ldots, x_k) = J^*_{12} \cdots {}_k (x_1, x_2, \ldots, x_k) R_{12\ldots(k-1)}(x_1, x_2, \ldots, x_{k-1})R_{12\ldots(k-2)k}(x_1, x_2, \ldots, x_{k-2}, x_k) \ldots R_{23\ldots k}(x_2, x_3, \ldots, x_k) \quad (29)$$

Equation (29) is the First Universal Form of any k-variate survival function for k = 2, 3, … since it follows the arithmetic identity (19*) [for the nonzero product:

$$R_{12\ldots(k-1)}(x_1, x_2, \ldots, x_{k-1})R_{12\ldots(k-2)k}(x_1, x_2, \ldots, x_{k-2}, x_k) \ldots R_{23\ldots k}(x_2, x_3, \ldots, x_k)].$$

In the particular case (of "k-independence") we may have (assume):

$$J^*_{12\ldots k}(x_1, x_2, \ldots, x_k) = 1.$$

Then, obviously, we have:

$$R(x_1, x_2, \ldots, x_k) = R_{12\ldots(k-1)}(x_1, x_2, \ldots, x_{k-1}) R_{12\ldots(k-2)k}(x_1, x_2, \ldots, x_{k-2}, x_k) \ldots R_{23\ldots k}(x_2, x_3, \ldots, x_k) \quad (30)$$

6 Conclusions

This work and our preceding works [6] and [9] evolve from the original Cox [8] and Aalen [7] paradigms for stochastic dependence. However, a quite new general theory, independent of the Aalen origin, emerges. This is a theory on the structure common to all (!) k-variate (k = 2, 3, …) survival functions. Basically, this structure may be considered as described by the recurrence formula (29), as well as, in the case k = 3, by (22).

Thus, the theory relates any k-variate probability distribution to all its marginals of degrees m with $1 \leq m \leq k - 1$. This relation follows the creating theory. On the other hand, the theory provides methods for the construction of new models. As for the theory's structure and the associated constructions, one initially assumes knowledge of the parameters and other properties of all univariate probability distributions. That is the theory's level zero. This initial information plays in this theory a role similar to the role of axioms in formalized mathematical theories. At the next step we consider the factored form (1) of all possible bivariate distributions as level one of the theory and of the underlying constructions. Based on this (bivariate) level we investigate, as level two, the theory of the tri-variate distributions which uses information obtained at the two previous levels. Then we go to the next level $k = 4$ and so on. Thus, each level of analysis and constructions assumes knowledge of all previous ones.

The transition from level $(k - 1)$ $(k = 2, 3, \ldots)$ to level k obeys formula (29). The set of problems that remains in association with any such transition is to determine methods of finding proper k-joiners $J_{12 \ldots k}(x_1, x_2, \ldots, x_k)$. Thus, the theory contains the structure and methods of construction for any finite dimension $k = 2, 3, \ldots$. As it follows from formulas (22), (23) [this $k = 3$ case can obviously be extended to any dimension], all the constructions rely on finding or choosing proper joiners which in many important cases can be found by solving proper integral equations such as (13) or (14) or, first of all, integral inequality (10) when $k = 2$. For higher k the integral equations and inequalities will be more complex. Anyway, an important part of the theory contains analysis and solving some (possibly linear) integral equations. Application of the created theory to **reliability** is obvious since the methods of construction based on finding proper joiners create a huge number of new k-variate models, many of them applicable to multicomponent system reliability description.

As mentioned in the text, the theory (we propose to name it the "joiner theory") is in a good sense competitive to the copula methodology. In many cases our theory seems to produce many more solutions so, possibly, more particular models can be constructed. When the hazard (failure) rates exist (as it is the case for most of reliability and other important applications) the Aalen paradigm works, and therefore the relation between the stochastic model and the modeled "physical" reality may be quite well understood, while typically, it is not the case with the copula modeling.

Besides, the extension of our theory from random vectors with nonnegative univariate marginals to any marginals is actually straightforward. Namely, having any distribution function, say $F(x_i)$, one can define the "survival function" $S(x_i)$, simply as $1 - F(x_i)$. Now instead of $S(0)$ we often will consider $S(-\infty) = 1$ or $S(b) = 1$ for some real b.

Such extension of our theory to all k-variate distributions (not only based on nonnegative marginals) looks straightforward. That would provide the full generality. Notice, however, that the nonnegative random variables actually fulfill the needs of reliability theory at least whenever life-times are of concern.

Our theory requires much further development especially in its statistical part which was omitted in this chapter.

Since for one practical problem one can find, in general, many models as solutions it is vital to investigate or even create statistical methods of discrimination to single out the best possibility with respect to fit to a given data.

References

1. Freund JE (1961) A bivariate extension of the exponential distribution. J Amer Stat Assoc 56:971–77
2. Gumbel EJ (1960) Bivariate exponential distributions. J Am Stat Assoc 55:698–707
3. Marshall AW, Olkin I (1967) A generalized bivariate exponential distribution. J Appl Probab 4:291–303
4. Kotz S, Balakrishnan N, Johnson NL (2000) Continuous multivariate distributions, vol 1, 2nd ed. Wiley, New York
5. Sklar A (1959) Fonctions de repartition a n dimensions et leurs marges, vol 8. Publications de l'Institut de Statistique de l'Universite de Paris, pp 229–231
6. Filus JK, Filus LZ (2017) The Cox-Aalen models as framework for construction of bivariate probability distributions, universal representation. J Stat Sci Appl 5:56–63
7. Aalen OO (1989) A linear regression model for the analysis of the life times. Stat Med 8:907–925
8. Cox DR (1972) Regression models and life tables (with discussion). J R Stat Soc B 74:187–220
9. Filus JK, Filus LZ (2017) General method for construction of bivariate stochastic processes given two marginal processes. Presentation at 7th international conference on risk analysis, ICRA 7 May 4 2017, Northeastern Illinois University, Chicago, IL
10. Arnold B (2017) Private communication, December 2017
11. Filus JK, Filus LZ (2013) A method for multivariate probability distributions construction via parameter dependence. In: Communications in statistics—theory and methods, vol 42, no 4, pp 716–721

An Exact Method for Solving a Least-Cost Attack on Networks

Asma Ben Yaghlane, Mehdi Mrad, Anis Gharbi and M. Naceur Azaiez

Abstract The chapter focuses on treating a particular problem of intelligent threats. We consider a least cost attack in a network-interdiction framework with a single source and a single destination. We suppose that a network is disabled if all links of a full cut set fail. That is, no flow can reach its destination node. We propose an exact solution under a budget constraint accounting for a threshold of success probability of the attack on targeted cut sets. We investigate the efficiency of the suggested solution through some illustrations. We extend the results to the case of networks with multiple sources/destinations.

Keywords Attack on networks · Cut set · Integer programming · Reliability

1 Introduction

Intelligent threats represent today a real challenge for critical infrastructure protection. An important sub-class of problems deals with attacks on networks. These include areas of investigation referred to as cyber-attacks, adversarial network analysis, interdiction networks, and recently network survivability problems.

A. B. Yaghlane (✉) · A. Gharbi · M. Naceur Azaiez
Tunis Business School, BADEM Lab, Université de Tunis, 2074 Al-Mourouj, Tunis, Tunisia
e-mail: asma_benyaghlane@yahoo.fr

A. Gharbi
e-mail: gharbianis2001@yahoo.com

M. Naceur Azaiez
e-mail: naceur.azaiez@tbs.rnu.tn

A. B. Yaghlane
Université Internationale de Tunis, 48, Rue Des Minéraux, 2035 Charguia I, Tunis, Tunisia

M. Mrad
Industrial Engineering Department, King Saud University, P. O. Box 800, Riyadh 11421, Saudi Arabia
e-mail: mradmehdiisg@yahoo.fr

M. Ram and H. Pham (eds.), *Advances in Reliability Analysis and its Applications*, Springer Series in Reliability Engineering,
https://doi.org/10.1007/978-3-030-31375-3_11

From practical point of view, one may consider preserving the functionality of critical infrastructure such as electricity distribution networks or oil pipeline networks from potential terrorist attacks. It is crucial for governments to protect related vital services from potential intelligent threats given the heavy consequences of corresponding attacks. If it is not feasible to totally prevent the attack, it should be possible to identify strategies to considerably reduce its damage.

While interdiction networks place important focus on rerouting flow upon failure of some nodes/arcs, network survivability (Ben Yaghlane et al. [2]) exclusively focus on attacks that completely prevent flow from reaching its destinations. In particular, from the attacker perspective, the concern is to fully disable a cut set of the network. Ben Yaghlane et al. [2] provide some tools to approach the problem of determining the cut set with the highest probability of failure. The related techniques are mainly centered on the min-cut problem and the minimal spanning tree problem. In Ben Yaghlane et al. [2] however, the reliability versus the vulnerability of the network is the main concern of the defender versus the attacker. The cost of attacking resources is not accounted for. In the present paper, we attempt to identify, for the attacker problem, the optimal cut set to target that yields the minimum expected attacking cost. The same problem has been treated by Gharbi et al. [5] attempting to identify some suitable heuristics. In this chapter, not only we provide an exact solution, but also we show that our method by far is more efficient than that of Gharbi et al. [5].

2 Literature Review

Defense/attack strategies have been extensively modeled through various techniques to approach intelligent threats and critical infrastructure protection. A review paper by Hausken and Levitin [6] classifies contributions in the area according to the system structure, defense measures, and attack tactics and circumstances. When considering attacks on networks, most of the investigations deal with interdiction networks. Royset and Wood [17] consider a bi-objective model to approach a maximum-flow network-interdiction problem. Zenklusen [19] considers attacks on arcs and nodes of a flow-max network problem with a single source and a single destination. The approach concerns using pseudo-polynomial algorithms that enable determining the minimum attack budget needed to remove enough links and nodes in a way to ensure demand dissatisfaction of flow at the destination node. This represents one of few papers that explicitly account for resource availability.

The *k-Most Vital Arcs Problem* represents a challenging stream in this area (e.g., Malik et al. [9], Ball et al. [1], Corley and Sha [4]). Ramirez-Marquez et al. [15] consider an optimal defensive resource allocation strategy for a source-destination flow network where attack resources are evenly distributed among links. Ramirez-Marquez et al. [13] consider defensive resource allocation to each link against a potential interdiction strategy and offer an evolutionary algorithm to identify the optimal allocation strategies.

Some extensions of network interdiction problems include the stochastic context. One may refer to Pan and Morton [12], Ramirez-Marquez and Rocco [14], and Morton [10] as examples. Interdiction networks in transportation problems receive some particular attention. For instance, Reilly et al. [16] consider the problem of targeting carriers of hazardous materials by terrorists. The government may determine carriers' movement restrictions in time and space to avoid terrorist attacks. Bier and Hausken [3] explore the interdiction network problem in a transportation system having two arcs with congestion. They consider three types of strategic actors, the government, the attacker, and the drivers. Hausken and Zhuang [7] consider congestion management in transportation problems including the presence of adversary players such as attackers and defenders. In rail-truck intermodal transportation, Sarhadi et al. [18] develop an analytical framework using a decomposition-based heuristic technique to minimize the impact of the worst-case attack scenario. Cyber-attacks represent another important area of investigation for the class of attacks on computer networks. One may refer to Obert et al. [11] and Hu et al. [8] for more details.

The problem under investigation in this chapter has been attempted by Gharbi et al. [5]. The suggested solution methodology is promising but fails to provide effective solutions particularly when the size gets large. We will prove that our method is by far much more efficient. Further, it yields an exact solution.

3 Problem Statement

We consider a network that transfers some flow from source nodes to some destinations. The network is targeted by an attacker that seeks to disconnect it by fully disabling a cut set as opposed to the general setting of interdiction networks; in which the attackers may content by disturbing the flow circulation. A cut set is a set of links whose failure will warrant the failure of the entire network in the sense that it becomes unable to send flow to the destinations. An equivalent mathematical definition will be offered in Sect. 4.2.1. The attacker attempts to identify the least attack cost while ensuring a minimum probability of a successful attack suggesting a chance constraint in the model. In fact, it is assumed that the probability of disconnecting one link upon attack is known. The attack occurs one at a time on the various links of the network. A link may be attempted at most once. An attack on a cut set is a sequence of attacks on its corresponding links. A pre-defined attack cost for each link is given. Also, the attacker has some limited budget; so that the cost of any attack on any cut set (i.e., any sequence of attacks on all links of the cut set) must be within the budget constraint. The problem is to determine which cut set to target to minimize the attack cost while satisfying the chance constraint on the success probability of the selected attack. The problem can initially be treated in the special case of a single source and a single destination. Then, the solution method may extend to account for multiple sources and destinations.

4 Solution Method

4.1 Notations

Let $G(V, E)$ be a graph

V	set of nodes
E	set of arcs
c_{ij}	attacking cost of arc (i, j) $\forall$ $(i, j) \in E$
p_{ij}	a predefined survival probability upon attack of arc (i, j) $\forall$ $(i, j) \in E$
B	the available budget of the attack
P	the threshold value of the success probability of the attack
s	the source node
t	the destination node
$\lvert.\rvert$	the cardinality of the given set
Z(.)	the expected cost of a given cut set.

4.2 Solution Procedure

The proposed algorithm relies on the enumeration of all feasible cut sets between the source node s and the destination node t. Obviously, the number of cut sets between s and t is huge and there is no polynomial algorithm that can enumerate all the cut sets between two nodes in a graph. However, for the attacker problem, not all the cut sets are feasible since we have a budget constraint and a success probability constraint. Consequently, these two last constraints will dramatically reduce the number of feasible cut sets. Hence, an explicit enumeration is eventually possible. For that aim, we first solve the bi-criteria cut problem, where we minimize the budget under success probability constraint. If the value of the obtained solution is larger than B, then the attacker problem is not feasible. Otherwise, the obtained cut set is a feasible solution. In order to generate further feasible cut sets, we add constraints that prevent any previously generated cut set to appear and solve the bi-criteria cut problem again until the space of solutions will be empty.

At this level, the expected attack cost function for a generated cut set S will be assessed using the following formula:

$$Z(S) = c_{i_1,j_1} + \sum_{k=2}^{|S|}\left(\prod_{h=1}^{k-1}(1 - p_{i_h,j_h})\right)c_{i_k,j_k}$$

The cut set with the least expected attack cost is certainly optimal.

4.2.1 Cut Set Definition

A cut set between the source s and the destination t in the graph $G(V, E)$ is a set of arcs that splits the set of nodes V into two disjoint subsets R_1 and R_2 such that:

$R_1 \cup R_2 = V$

$R_1 \cap R_2 = \emptyset$

$s \in R_1$

$t \in R_2$

A min-cut is a cut set whose attacking cost is minimum.

4.2.2 Formulation of the Bi-Criteria Cut

Decision variables

$$x_{ij} = \begin{cases} 1 & if\, edge\{i, j\}\, is\, considered\, in\, the \min - \text{cut} \\ 0 & otherwise \end{cases} \forall (i, j) \in E$$

$$y_i = \begin{cases} 1 & if\, node\, i \in R_2 \\ 0 & if\, node\, i \in R_1 \end{cases} \forall i \in V$$

The corresponding Integer Linear Program is given by

$$\textbf{Min} \sum_{(i,j)\in E} c_{ij} x_{ij} \tag{1}$$

$$x_{ij} \geq y_j - y_i \forall (i \cdot j) \in E \tag{2}$$

$$y_t \geq y_s + 1 \tag{3}$$

$$\sum_{(i,j)\in E} \ln(1 - p_{ij}) \mathrm{x_{ij}} \geq Ln(P) \tag{4}$$

$$y_i \in \{0, 1\} \forall i \in V \tag{5}$$

$$x_{ij} \in \{0, 1\} \forall (i.j) \in E \tag{6}$$

The objective function (1) attempts to minimize the total cost of the attack. If this cost of the optimal solution is larger than *B*, then the attacker problem does not have a feasible solution. Constraints (2) and (3) force the set of selected arcs in the solution to be a cut set. Constraint (4) is a chance constraint and therefore ensures that the success probability of the attack is larger than the predefined threshold value *P*. Constraints (5) and (6) are the integrality constraints of the decision variables.

It is worth noting that this integer linear program will only generate non-redundant or minimal cut sets. That is, it would not consider cut sets containing arcs that, when deleted, the new set is still a cut set. In fact, the objective function will avoid adding useless arcs with additional costs. Denote by ILP (1–6) the ILP having the six constraints specified above.

In order to generate a new cut set that is different from the old generated one, we suggest adding a new ILP which consists of the old one augmented by a new constraint (7) as follows:

$$\sum_{\{i,j\}\in S} x_{ij} \leq |S| - 1 \tag{7}$$

Constraint (7) ensures that the arcs of any cut set S cannot appear together in the optimal solution of the ILP. Thus, at each iteration of our algorithm, if a new cut set is generated, an additional constraint related to this cut set will be appended. The algorithm will sequentially generate a number of feasible cut sets satisfying the budget constraint as well as the chance constraint related to the success probability of attack and will identify the best among all identified cut sets.

Algorithm

Step *0*

$k = 0$;

$ILP_k = ILP\ (1\text{–}6)$

Step 1

Solve ILP_k

If the objective function is larger than B or the space of solutions is empty (no feasible solution exists), then go to step (3), else let X_k be the obtained solution and go to step (2) at which an attempt of a new feasible cut set will be generated.

Step 2

Add a constraint (7) ! That avoids the solution X_k to appear again, to ILP_k.

$k = k + 1$;

Go to step (1)

Step 3

If $k = 0$, then the problem is infeasible. Else, assess all the obtained feasible solutions using the expected attack function and consider the best one as the optimal solution of the attacker problem.

Given that the network has only a finite number of cut sets, the algorithm will stop after finitely many iterations.

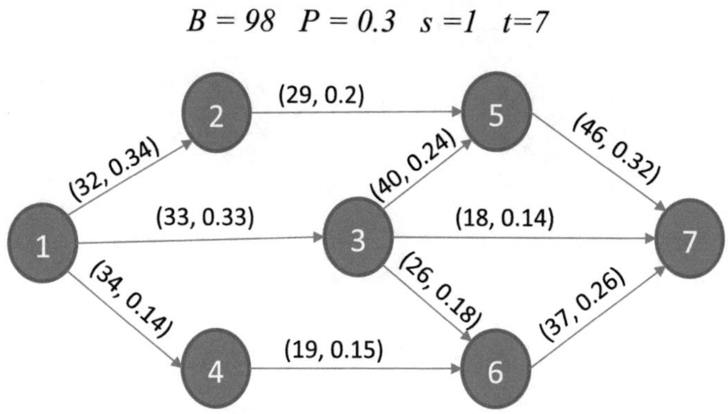

Fig. 1 Network of the illustration

4.3 Illustration

Consider the following network:

$$B = 98 \quad P = 0.3 \quad s = 1 \quad t = 7$$

See Fig. 1.

ILP (0)

$$\begin{aligned}
\text{Min } & 32x_{12} + 29x_{25} + 46x_{57} + 40x_{35} + 33x_{13} + 18x_{37} + 26x_{36} \\
& + 34x_{14} + 19x_{46} + 37x_{67} \\
\text{s.t. } & x_{12} \geq y_2 - y_1 \\
& x_{25} \geq y_5 - y_2 \\
& x_{57} \geq y_7 - y_5 \\
& x_{35} \geq y_5 - y_3 \\
& x_{13} \geq y_3 - y_1 \\
& x_{37} \geq y_7 - y_3 \\
& x_{36} \geq y_6 - y_3 \\
& x_{14} \geq y_4 - y_1 \\
& x_{46} \geq y_6 - y_4 \\
& x_{67} \geq y_7 - y_6 \\
& y_7 \geq y_1 + 1
\end{aligned}$$

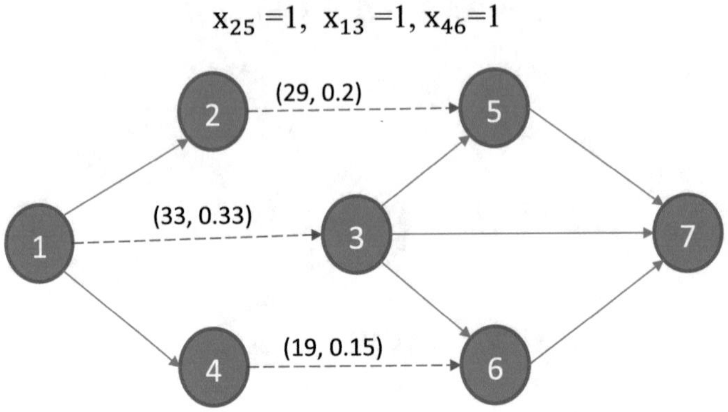

Fig. 2 Generated cut set at first iteration

$$-0.416x_{12} - 0.223x_{25} - 0.386x_{57} - 0.274x_{35} - 0.400x_{13} - 0.151\,18x_{37}$$
$$-0.198x_{36} - 0151x_{14} - 0.163x_{46} - 0.301x_{67} \geq -1.050$$

$$y_i \in \{0, 1\} \forall i \in V$$
$$x_{ij} \in \{0, 1\} \forall (i.j) \in E$$

Solution of ILP (0)

$$x_{25} = 1, \quad x_{13} = 1, \quad x_{46} = 1$$

See Fig. 2.

Attack cost = 29 + 33 + 19 = 81 < 98
Probability of success = (1 − 0.2) * (1 − 033) * (1 − 0.15) = 0.4556 > 0.35
Expected attack cost = 33 + 19 * (0.67) + 29 * (0.67) * 0.85 = 62.2455

ILP (1)

$$\textbf{Min}\ 32x_{12} + 29x_{25} + 46x_{57} + 40x_{35} + 33x_{13} + 18x_{37} + 26x_{36} + 34x_{14} + 19x_{46} + 37x_{67}$$

$$\textbf{s.t.}\ x_{12} \geq y_2 - y_1$$
$$x_{25} \geq y_5 - y_2$$
$$x_{57} \geq y_7 - y_5$$
$$x_{35} \geq y_5 - y_3$$
$$x_{13} \geq y_3 - y_1$$
$$x_{37} \geq y_7 - y_3$$
$$x_{36} \geq y_6 - y_3$$
$$x_{14} \geq y_4 - y_1$$
$$x_{46} \geq y_6 - y_4$$
$$x_{67} \geq y_7 - y_6$$
$$y_7 \geq y_1 + 1$$
$$-0.416x_{12} - 0.223x_{25} - 0.386x_{57} - 0.274x_{35} - 0.400x_{13} - 0.151\ 18x_{37} - 0.198x_{36} - 0151x_{14} - 0.163x_{46} - 0.301x_{67} \geq -1.050$$

$$x_{25} + x_{13} + x_{46} \leq 2$$
$$y_i \in \{0, 1\} \forall i \in V$$
$$x_{ij} \in \{0, 1\} \forall (i.j) \in E$$

Solution of ILP (1)

See Fig. 3.

$$x_{12} = 1, \quad x_{13} = 1, \quad x_{46} = 1$$

Attack cost = $32 + 33 + 19 = 84 < 98$
Probability of success = $(1 - 0.34) * (1 - 033) * (1 - 0.15) = 0.36448 > 0.35$
Expected attack cost = $32 + 33 * (0.66) + 19 * (0.66) * 0.67 = 62.1818$

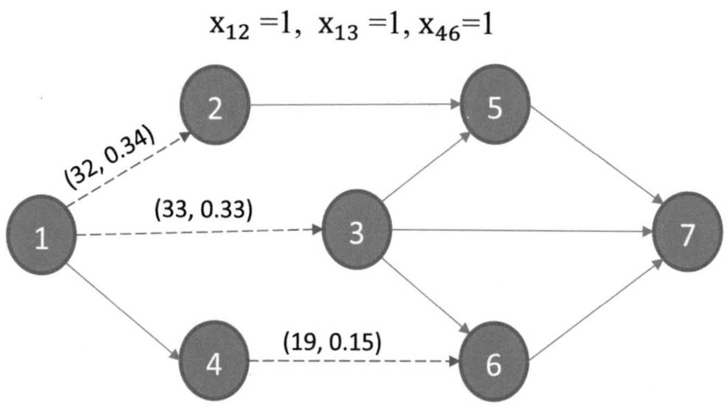

Fig. 3 Generated cut set at second iteration

ILP (2)

$$
\begin{aligned}
&Min\, 32x_{12} + 29x_{25} + 46x_{57} + 40x_{35} + 33x_{13} + 18x_{37} + 26x_{36} \\
&\quad + 34x_{14} + 19x_{46} + 37x_{67} \\
&s.t.\, x_{12} \geq y_2 - y_1 \\
&x_{25} \geq y_5 - y_2 \\
&x_{57} \geq y_7 - y_5 \\
&x_{35} \geq y_5 - y_3 \\
&x_{13} \geq y_3 - y_1 \\
&x_{37} \geq y_7 - y_3 \\
&x_{36} \geq y_6 - y_3 \\
&x_{14} \geq y_4 - y_1 \\
&x_{46} \geq y_6 - y_4 \\
&x_{67} \geq y_7 - y_6 \\
&y_7 \geq y_1 + 1
\end{aligned}
$$

$$
\begin{aligned}
&-0.416x_{12} - 0.223x_{25} - 0.386x_{57} - 0.274x_{35} - 0.400x_{13} - 0.151\,18x_{37} - 0.198x_{36} - 0151x_{14} \\
&\quad -0.163x_{46} - 0.301x_{67} \geq -1.050
\end{aligned}
$$

$$
\begin{aligned}
&x_{25} + x_{13} + x_{46} \leq 2, \quad x_{12} + x_{13} + x_{46} \leq 2 \\
&y_i \in \{0, 1\} \forall i \in V \\
&x_{ij} \in \{0, 1\} \forall (i.j) \in E
\end{aligned}
$$

Solution of ILP (2)

$$
\mathrm{x}_{25} = 1, \mathrm{x}_{13} = 1, \mathrm{x}_{14} = 1
$$

See Fig. 4.

Attack cost = 33 + 29 + 34 = 96 < 98
Probability of success = (1 – 0.33) * (1 – 0.20) * (1 – 0.14) = 0.46096 > 0.35
Expected attack cost = 33 + 29 * (0.67) + 34 * (0.67) * 0.8 = 70.654

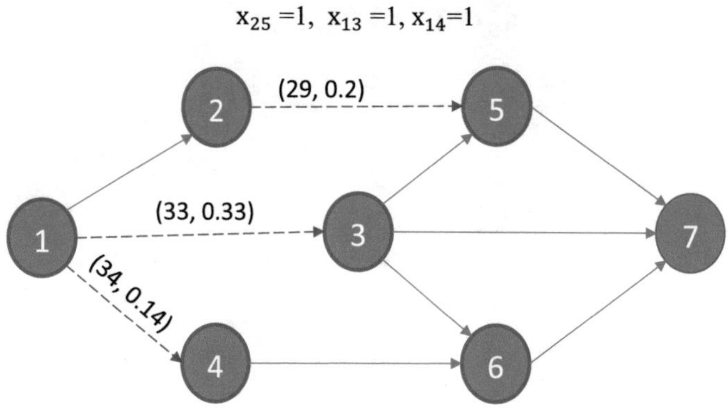

Fig. 4 Generated cut set at third iteration

ILP (3)

$$Min\, 32x_{12} + 29x_{25} + 46x_{57} + 40x_{35} + 33x_{13} + 18x_{37} + 26x_{36} + 34x_{14} + 19x_{46} + 37x_{67}$$

$$S.t.\, x_{12} \geq y_2 - y_1$$
$$x_{25} \geq y_5 - y_2$$
$$x_{57} \geq y_7 - y_5$$
$$x_{35} \geq y_5 - y_3$$
$$x_{13} \geq y_3 - y_1$$
$$x_{37} \geq y_7 - y_3$$
$$x_{36} \geq y_6 - y_3$$
$$x_{14} \geq y_4 - y_1$$
$$x_{46} \geq y_6 - y_4$$
$$x_{67} \geq y_7 - y_6$$
$$y_7 \geq y_1 + 1$$
$$-0.416x_{12} - 0.223x_{25} - 0.386x_{57} - 0.274x_{35} - 0.400x_{13} - 0.151\,18x_{37} - 0.198x_{36} - 0151x_{14} - 0.163x_{46} - 0.301x_{67} \geq -1.050$$

$$x_{25} + x_{13} + x_{46} \leq 2, \quad x_{12} + x_{13} + x_{46} \leq 2, \quad x_{25} + x_{13} + x_{14} \leq 2$$
$$y_i \in \{0, 1\} \forall i \in V$$
$$x_{ij} \in \{0, 1\} \forall (i.j) \in E$$

Solution of ILP (3)

$$x_{57} = 1, x_{13} = 1, x_{46} = 1$$

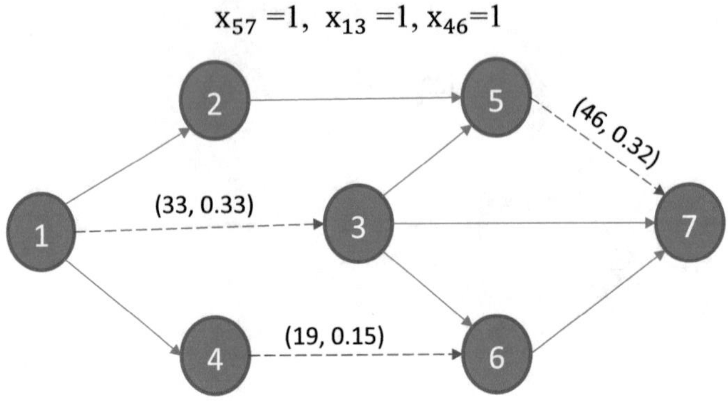

Fig. 5 Generated cut set at fourth iteration

See Fig. 5.

Attack cost = 33 + 19 + 46 = 98
Probability of success = (1 − 0.33) * (1 − 0.15) * (1 − 0.32) = 0.38726 > 0.35
Expected attack cost = 33 + 19 * (0.67) + 46 * (0.67) * 0.85 = 71.927

ILP (4)

$$
\begin{gathered}
Min\,32x_{12} + 29x_{25} + 46x_{57} + 40x_{35} + 33x_{13} + 18x_{37} + 26x_{36} \\
+ 34x_{14} + 19x_{46} + 37x_{67} \\
\boldsymbol{s.t.}\, x_{12} \geq y_2 - y_1 \\
x_{25} \geq y_5 - y_2 \\
x_{57} \geq y_7 - y_5 \\
x_{35} \geq y_5 - y_3 \\
x_{13} \geq y_3 - y_1 \\
x_{37} \geq y_7 - y_3 \\
x_{36} \geq y_6 - y_3 \\
x_{14} \geq y_4 - y_1 \\
x_{46} \geq y_6 - y_4 \\
x_{67} \geq y_7 - y_6 \\
y_7 \geq y_1 + 1
\end{gathered}
$$

$$
\begin{gathered}
-0.416x_{12} - 0.223x_{25} - 0.386x_{57} - 0.274x_{35} - 0.400x_{13} - 0.151\,18x_{37} \\
-0.198x_{36} - 0151x_{14} - 0.163x_{46} - 0.301x_{67} \geq -1.050 \\
x_{25} + x_{13} + x_{46} \leq 2, \quad x_{12} + x_{13} + x_{46} \leq 2, x_{25} + x_{13} + x_{14} \leq 2 \quad x_{57} + x_{13} + x_{46} \leq 2 \\
y_i \in \{0, 1\} \forall i \in V \\
x_{ij} \in \{0, 1\} \forall (i.j) \in E
\end{gathered}
$$

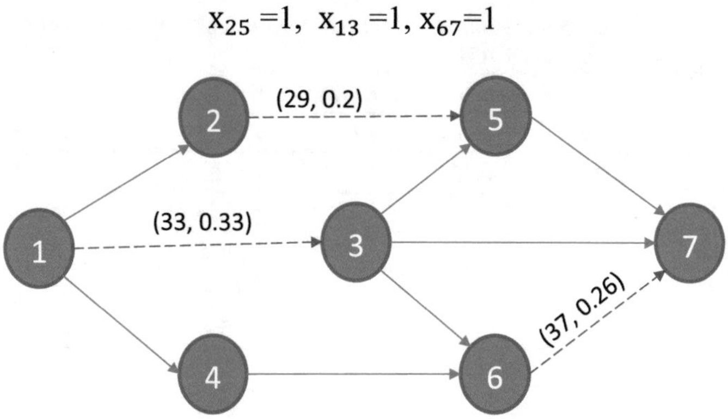

Fig. 6 Generated cut set at fifth iteration

Solution of ILP (4)

$$x_{25} = 1, \ x_{13} = 1, \ x_{67} = 1$$

See Fig. 6.

Attack cost = 33 + 37 + 29 = 99 > 98

This cut is not feasible since it violates the budget constraint. Hence, the algorithm stops by comparing the expected costs of the previously generated cut sets. It follows that the optimal cut set to target is the one represented in Fig. 3 with an expected cost of *62.1818*.

5 Experimental Study and Discussion

An experimental study is conducted to investigate the efficiency of the suggested solution approach. It entirely relies on the instances generated by Gharbi et al. [5] for comparative purposes. The first part of the study is devoted to compare our suggested method with the one by Gharbi et al. [5]. In the second part, we investigate the ability of our approach to tackle large scale networks.

In the comparative study, we consider networks with sizes ranging between *10* and *30* nodes. We identify the number of unsolved instances by each method as well as the efficiency of each method using the ratio consisting of the solution time of Gharbi et al. [5] over that of the current solution approach. We display below the following notations.

Us: number of unsolved instances within 1800s

Table 1 Comparison between the current cut sets generation approach and the branch and bound algorithm proposed by Gharbi et al. [5]

		Current approach		Gharbi et al. [5]		
		Us	*Time*	*Us*	*Time*	*Time_ratio*
n	10	0	0.08	0	28.39	355.096936
	15	0	1.21	4	253.12	209.62752
	20	0	0.51	14	975.29	1927.06975
	25	0	0.46	25	1366.46	2997.11575
	30	0	0.52	32	1523.05	2924.86437

Time: average computational time of solved instances
Time_ratio: Gharbi et al. [5]_time/current approach_time

The instances used in Table 1 are described in Gharbi et al. [5] as follows:

> The attacking costs are drawn from the discrete uniform distribution on [20, 60] for s and t related links, and on [1, 40] for the remaining ones. The survival probabilities upon attack are generated between 0.1 and 0.35 for s and t related links, and between 0.05 and 0.3 for the remaining links. The number of nodes n is taken equal to 10, 15, 20, 25, and 30. The degree of each node is randomly generated between 2 and dmax where dmax $\in \{4, 5, 6, 7\}$. For each (n, dmax) combination, 10 instances were randomly generated.

It should be noted that the two algorithms are not tested on the same computer. In fact, Gharbi et al. [5] use a Quad 2.8 GHz Personal Computer with 4 GB RAM while the current experimentation uses an Intel®Core™ i7-4720 HQ processor with 2.6 GHz and 16 Gb of available memory. The two computers are of comparable performance. It should be clear however that the time_ratio is extremely high proving that the current approach by far outperforms the other one. Further, in Gharbi et al. [5], many instances were unsolved practically for all sizes of the network. Of course, our exact method allows solving all instances.

The second part of the experiment exclusively concerns the current approach to test its ability to solve large networks. The used instances are generated similarly to Gharbi et al. [5] except that the number of nodes n is taken equal to *50*, *100*, *150*, and *200*. The degree of each node is randomly generated between 2 and *dmax* where *dmax* is either *10* or *20*. For each (*n, dmax*) combination, *10* instances are randomly generated. All the computational experiments are conducted on an Intel®Core ™ i7-4720 HQ processor with 2.6 GHz and 16 Gb of available memory. We display below the relevant notations:

N	number of nodes
d	maximum degree of each node
Av_m	average number of arcs
Us	number of unsolved instances (among the *10* tested instances)
Av_t	average computational time
Max_T	maximum computational time
Min_T	minimum computational time

Av_cuts average number of generated cut sets
Max_cuts maximum number of generated cut sets
Min_cuts minimum number of generated cut sets (Table 2).

The maximum solution time over all instances is within *300* s. The average computational time is always within *60* s and most often below *five* seconds. The minimum time never reaches *one* second. This proves that our method is very efficient even for large scale networks. Note that the first *ten* instances related to a size of *50* nodes witness an unfeasible solution. All remaining instances provide optimal solutions.

The problem can easily be extended to account for networks with multiple sources/destinations. In fact, it suffices to add a virtual source/destination node with attack cost equals infinity of all arcs linking the original sources/destinations to the virtual one. Clearly, this extension will not affect the efficiency of the suggested solution method. In fact, it turns out to be a special case where some (virtual) nodes have infinite costs and hence cannot be attacked.

6 Conclusions and Perspectives

The problem considered in this paper attempts to identify the optimal attack strategy on a network that completely prevents the flow from reaching its destination. We assume that the network has a single source and a single destination. Hence, the attack is considered as successful if a full cut set is disabled. Therefore, the problem has to do with identifying the least-cost cut set to target. It is assumed that the cost of attacking any given link is known. All candidate cut sets must satisfy a minimum level of success probability to be considered as feasible. Further, a budget constraint is to be applied so that any cut set requiring a cost that exceeds this budget is to be excluded. While some attempts of attacks on a particular cut set may be stopped before trying all related links, the attacker will estimate the cost of attacking this cut set by considering the sum of the costs of its individual links (including perhaps those that may not be attempted). The cost measure to be optimized however is the expected cost of attack on a given cut set.

The proposed methodology consists on formulating a sequence of ILP to identify distinct cut sets satisfying both the budget and chance constraints. The best expected cost among all generated cut sets will be therefore the optimal solution for the problem. An illustration is provided to explain all the steps of the solution method. Further, the solution is tested against an existing approach and proves to be by far more effective and efficient. Moreover, it shows that it can adequately handle large scale networks. The method is naturally extended to account for multiple sources and destinations.

Among possible avenues for future research, one may attempt to identify the best k cut sets ($k > 1$) to target in a scope of increasing the chance of totally disconnecting the network. Another area of future work would be to identify optimal defensive strategies. For instance, the defender may consider strengthening the optimal targeted

Table 2 Results of the cut sets generation approach on large instances

n	d	Av_m	US	Av_T	Max_T	Min_T	Av_Cuts	Max_Cuts	Min_Cuts
50	10	316	0	0.3316	1.187	0.031	7.8	29	0
50	20	517.1	0	1.9755	16.933	0.038	13.9	100	1
100	10	617.6	0	15.7351	139.808	0.047	53.8	404	1
100	20	1049.4	0	0.456	1.341	0.069	4.1	10	1
150	10	925.4	0	4.5121	33.249	0.085	20	118	1
150	20	1590.5	0	4.8398	33.877	0.116	13.1	84	1
200	10	1290.6	0	47.5064	272.605	0.116	76.8	330	1
200	20	2137.2	0	1.2925	4.76	0.136	5	14	1

cut sets or alternatively to strengthen some path sets that may ensure the arrival of the flow to the destinations. A third interesting extension of this work may consider modeling the problem in a game-theoretic setting where the players are respectively the attacker and the defender of the network.

References

1. Ball MO, Golden BL, Vohra RV (1989) Finding the most vital arcs in a network. Oper Res Lett 8(2):73–76
2. Ben Yaghlane A, Azaiez MN, Mrad M (2019) System survivability in the context of interdiction networks. Reliab Eng Syst Saf 185:362–371
3. Bier VM, Hausken K (2013) Defending and attacking a network of two arcs subject to traffic congestion. Reliab Eng Syst Saf 112:214–224
4. Corley HW, Sha DY (1982) Most vital links and nodes in weighted networks. Oper Res Lett 1(4):157–160
5. Gharbi A, Azaiez MN, Kharbeche M (2010) Minimizing expected attacking cost in networks. Electr Notes Discr Math 36:947–954
6. Hausken K, Levitin G (2012) Review of systems defense and attack models. Int J Perform Eng 8(4):355–366
7. Hausken K, Zhuang J (2015) (eds) Game theoretic analysis of congestion, safety and security: networks, air traffic and emergency departments. Springer, New York, ISSN 1614-7839, ISSN 2196-999X (electronic). https://doi.org/10.1007/978-3-319-13009-5
8. Hu F, Lua Y, Vasilakos VA, Haoc Q, Maa R, Patil Y, Zhanga T, Lua J, Li X, Xiong NN (2016) Robust cyber–physical systems: concept, models, and implementation. Future Gen Comput Syst 56:449–475
9. Malik K, Mittal AK, Gupta SK (1989) The k most vital arcs in the shortest path problem. Oper Res Lett 8(4):223–227
10. Morton DP (2011) Stochastic network interdiction. Encyclopedia of Operations Research and Management Science. Wiley, Cochran
11. Obert J, Pivkina I, Huang H, Cao H (2016) Proactively applied encryption in multipath networks. Comput Secur 58:106–124
12. Pan F, Morton DP (2008) Minimizing a stochastic maximum-reliability path. Networks 52(3):111–119
13. Ramirez-Marquez JE, Rocco C, Levitin G (2011) Optimal network protection against diverse interdictor strategies. Reliab Eng Syst Saf 96(3):374–382
14. Ramirez-Marquez JE, Rocco C (2009) Stochastic network interdiction optimization via capacitated network reliability modeling and probabilistic solution discovery. Reliab Eng Syst Saf 94(5):913–921
15. Ramirez-Marquez JE, Rocco C, Levitin G (2009) Optimal protection of general source-sink networks via evolutionary techniques. Reliab Eng Syst Saf 94(10):1676–1684
16. Reilly A, Nozick L, Xu N, Jones D (2012) Game theory-based identification of facility use restrictions for the movement of hazardous materials under terrorist threat. Transp Res Part E 48:115–131
17. Royset JO, Wood RK (2007) Solving the bi-objective maximum-flow network- interdiction problem. INFORMS J Comput 19(2):175–184
18. Sarhadi H, Tulett DM, Verma M (2017) An analytical approach to the protection planning of a rail intermodal terminal network. Eur J Oper Res 257(2):511–525
19. Zenklusen R (2010) Network flow interdiction on planar graphs. Discr Appl Math 158:1441–1455

Reliability Analysis of Complex Repairable System in Thermal Power Plant

Dilbagh Panchal, Mohit Tyagi, Anish Sachdeva and R. K. Garg

Abstract The aim of this work is to study the performance issues of cooling tower unit of a coal fired thermal power plant. The complex series- parallel arrangement of repairable system has been modeled with Petri-Net (PN) approach. Reliability parameters of the considered system have been tabulated using fuzzy Lambda-Tau approach. Triangular Membership Function (TFN) has been used for considering the vagueness in the collected operational failure rate and repair time data. Failure dynamics of the considered system was studied on the basis of increasing and decreasing trend of reliability parameters. The analysis result has been supplied to the maintenance engineer of the plant to fix the optimum maintenance interval for the considered unit.

Keywords Thermal power plant · Reliability · Failure rate · Repair time · TFN and maintenance interval

1 Introduction

Minimizing or elimination of sudden breakdown from a plant operation is one of the great challenge for maintenance engineer as failure is a common observable fact associated with plant operation. An industrial system is complex arrangement of various system/subsystems/components and due to this complexity failure prediction during plant operation is a tough ask from the maintenance engineer [1]. With this

---

D. Panchal (✉) · M. Tyagi · A. Sachdeva · R. K. Garg
Department of Industrial and Production Engineering, Dr. B.R Ambedkar National Institute of Technology, Jalandhar, Punjab 144011, India
e-mail: panchald@nitj.ac.in

M. Tyagi
e-mail: tyagim@nitj.ac.in

A. Sachdeva
e-mail: asachdeva@nitj.ac.in

R. K. Garg
e-mail: gargrk@nitj.ac.in

M. Ram and H. Pham (eds.), *Advances in Reliability Analysis and its Applications*, Springer Series in Reliability Engineering,
https://doi.org/10.1007/978-3-030-31375-3_12

various other factors such as human error and availability of incomplete plant operational data make the job of maintenance engineer/system analyst more complex. A minor failure in the subsystem/component of a system results in sudden breakdown in the plant operation and due to this the system become unavailable which directly affects the plant profitability [2–4]. Sudden failure in the plant operation not only contributes to production cost of the plant but also a great threat to its operational safety. Therefore, for minimizing or eliminating such losses it is essential to maintain the availability and maintainability aspects of the system. For maintaining the availability and maintainability aspects of the system it is essential to study the failure behavior of the system which helps plant engineer to develop an optimal maintenance strategy for the considered system. Hence, for the failure behavior study of a system its reliability analysis is very useful which is presented in this work.

2 Research Background

In the past various authors have studied the performance issues of complex repairable systems. In order to study the failure behavior of the system different mathematical concepts have been used for developing the modelling equations. In the past Markov modelling was used by various researchers for performing the reliability analysis of the different complex industrial system. Arora and Kumar [5] developed Markov modelling for carrying the reliability, availability and maintainability analysis of steam and power generation systems in a coal fired thermal power. With the time it has been implemented by many other researchers for studying the performance issues of different industrial system [6–8]. The main limitation of this Markov modelling based approach was that it makes use of crisp failure and repair time information for analysis. Uncertainty or vagueness is always there in the collected data and this uncertainty was not considered in the previous studies due to limitation of Markov approach. Therefore, the results obtained with the implementation of Markov approach may be highly uncertain/vague. To overcome this limitation fuzzy set theory based model were developed by the researchers and were presented with their application in different field. Sharma et al. [9] implemented fuzzy lambda tau approach for reliability analysis of a system in a paper production plant. Sharma and Sharma [2] again expounded the application of fuzzy lambda tau approach for press unit of the paper industry. Panchal and Kumar [10] presented the application of fuzzy methodology based reliability approach for studying the failure behavior of compressor house unit of coal fired thermal power industry. Garg [11] analyzed the performance of repairable system with the implementation of vague Lambda-Tau approach. Panchal and Kumar [12] presented an integrated framework for studying the failure behavior of water treatment plant in a thermal power industry. Various reliability parameters were computed under fuzzy environment and failure dynamics for the system were analyzed. Panchal et al. [3] developed a novel framework for studying the failure behavior of chlorine gas plant of a chemical industry. The reliability based study of the plant proves to be very useful for taking maintenance decision for the risky component of the plant. The proposed model was very useful for developing Reliability Based Maintenance (RBM) for the sub-system/component of the considered industry.

For the above studies it has been noted that cooling tower plant of the coal fired thermal power plant has not yet been investigated for studying its failure dynamics for taking maintenance decision. Therefore, in the current work fuzzy lambda-Tau approach has been applied to tabulate the various reliability parameters of the considered system.

3 Proposed Framework

For the reliability analysis of the considered system its series- parallel arrangement has been studied and on the basis of this study PN model with AND/OR gate has been generated. Failure rate and repair time data has been collected from plant expert and maintenance log book records etc. and its fuzzification was done using Triangular Membership Function (TMF). Mathematical modelling was done according to plant's general arrangement and various reliability parameters were tabulated at different spread. Increasing/decreasing trend of various reliability parameters has been studied and fixing of maintenance intervals was done. The flow chart for the proposed work has been shown in Fig. 1.

4 Fuzzy Concept and Reliability Approach

4.1 Fuzzy Set Theory Basics

The basic fuzzy set concepts used for considering the vagueness in the collected failure and repair time data used for reliability analysis of the considered system are discussed below [13–17].

4.1.1 **Crisp and Fuzzy Set**: A crisp set is represented by equation as [18]:

$$M_A(x) = \begin{cases} 1, \ if \ x \in A \\ 0, \ if \ x \notin A \end{cases} \tag{1}$$

Unlike the crisp set, a fuzzy set can accommodate various degrees of membership on the real-continuous interval [0, 1], where the endpoints of 0 and 1 confirm to no membership and full membership respectively.

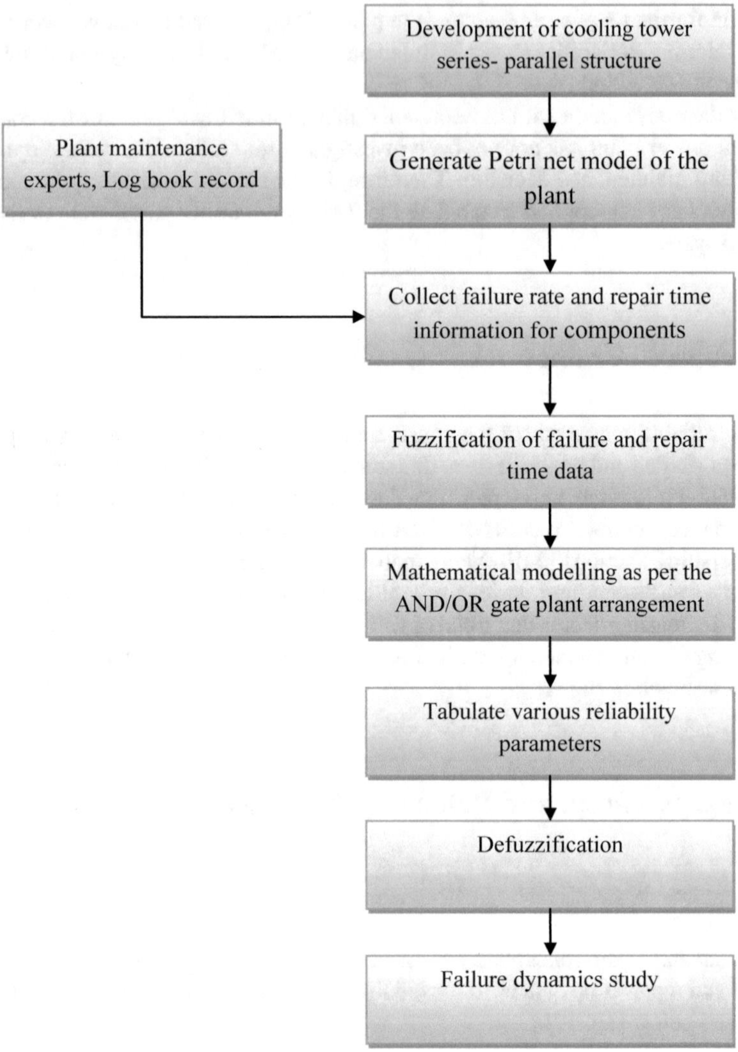

Fig. 1 Flow chart for reliability analysis

A fuzzy set can be represented by a functional mapping as:

$$\mu_{\tilde{A}}(x) \epsilon \tilde{A} \quad (2)$$

where, $\mu_{\tilde{A}}(x)$: degree of membership of element x in fuzzy set $\tilde{A}$.

4.1.2 **Membership Function**: A Triangular Membership function is represented by equation as [19]:

$$s \quad (3)$$

where, P, Q, $R \rightarrow$ the upper, mean and lower bound respectively.

4.1.3 α- cut values for triangular fuzzy set, is given by Eq. 4 as:

$$\tilde{M}^{\alpha} = \left[(Q - P^{\alpha})\alpha + P^{\alpha}, -\left(R^{(\alpha)} - Q\right)\alpha + R^{(\alpha)}\right] \quad (4)$$

4.2 Fuzzy Lambda-Tau Approach

Lambda-Tau approach is a quantitative tool for performing the reliability analysis of the complex industrial system under uncertain environment [20]. This approach makes use of failure rate and repair time data of various component of the system for behavior study of the complex operating system. Consideration of vagueness in the failure and repair time makes this approach more powerful than the Markov approach. Recently it has been implanted by various authors in different area namely paper mill, thermal power industry, chemical industry and sugar mill etc. [2, 3]. The various steps of fuzzy Lambda-Tau approach are as follows:

Step-1 Collect failure rate and repair data related to various component of the considered system.

Step-2 Using TFN fuzzify the collected data in order to consider the vagueness of the raw data.

Step-3 Develop mathematical modelling of the system as per the series—parallel arrangement with AND/OR gate. The basic expression for developing the equations for the AND/OR gate are given represented by Eqs. 5–7.

AND gate transition expression [20]

$$\lambda^{\alpha} = \left[\prod_{i=1}^{n} \{(\lambda_{i2} - \lambda_{i1})\alpha + \lambda_{i1}\} \cdot \sum_{j=1}^{n} \left[\prod_{\substack{i=1 \\ i \neq j}}^{n} \{(\tau_{i2} - \tau_{i1})\alpha\} + \tau_{i1} \right], \right.$$

$$\left. \prod_{i=1}^{n} \{-(\lambda_{i3} - \lambda_{i2})\alpha + \lambda_{i3}\} \cdot \sum_{j=1}^{n} \left[\prod_{\substack{i=1 \\ i \neq j}}^{n} \{(\tau_{i3} - \tau_{i2})\alpha\} + \tau_{i3} \right] \right] \tag{5}$$

$$\tau^{\alpha} = \left[\frac{\prod_{i=1}^{n} \{(\tau_{i2} - \tau_{i1})\alpha + \tau_{i1}\}}{\sum_{j=1}^{n} \left[\prod_{\substack{i=1 \\ i \neq j}}^{n} \{-(\tau_{i3} - \tau_{i2})\alpha\} + \tau_{i3} \right]}, \frac{\prod_{i=1}^{n} \{(\lambda_{i3} - \lambda_{i2})\alpha + \lambda_{i3}\}}{\sum_{j=i}^{n} \left[\prod_{\substack{i=1 \\ i \neq j}}^{n} \{(\tau_{i2} - \tau_{i1})\alpha\} + \tau_{i1} \right]} \right] \tag{6}$$

OR gate transition expression [20]

$$\lambda^{\alpha} = \left[\sum_{i=1}^{n} \{(\lambda_{i2} - \lambda_{i1})\alpha + \lambda_{i1}\}, \sum_{i=1}^{n} \{-(\lambda_{i3} - \lambda_{i2})\alpha + \lambda_{i3}\} \right]$$

$$\tau^{\alpha} = \frac{\sum_{i=1}^{n} [\{(\lambda_{i2} - \lambda_{i1})\alpha + \lambda_{i1}\}.\{(\tau_{i2} - \tau_{i1})\alpha + \tau_{i1}\}]}{\sum_{i=1}^{n} \{-(\lambda_{i3} - \lambda_{i2})\alpha + \lambda_{i3}\}},$$

$$\frac{\sum_{i=1}^{n} [\{-(\lambda_{i3} - \lambda_{i2})\alpha + \lambda_{i3}\}.\{-(\tau_{i3} - \tau_{i2})\alpha + \tau_{i3}\}]}{\sum_{i=1}^{n} \{(\lambda_{i2} - \lambda_{i1})\alpha + \lambda_{i1}\}} \tag{7}$$

Step-4 Compute various reliability parameters at different spreads by using the relation as shown in Table 1.

Step-5 Defuzzify the reliability parameters values and analyze the failure dynamics of the study.

Table 1 Various reliability parameters

Reliability parameters	Formulas
Mean time to failure	$MTTF_s = \frac{1}{\lambda_s}$
Mean time to repair	$MTTR_s = \frac{1}{\mu_s}$
Mean time between failure	$MTBF_s = MTTF_s + MTTR_s$
Availability	$A_s = \frac{\mu_s}{\mu_s+\lambda_s} + \frac{\lambda_s}{\mu_s+\lambda_s} e^{-(\mu_s+\lambda_s)t}$
Reliability	$R_s = e^{-\lambda_s t}$
Expected number of failures	$ENOF = \frac{\lambda_s \mu_s t}{\mu_s+\lambda_s} + \frac{\lambda_s^2}{(\mu_s+\lambda_s)^2}\left[1 - e^{-(\mu_s+\lambda_s)t}\right]$

5 Case Study

In the present study cooling tower system of a medium size coal fired thermal power plant located in northern part of India has been considered. Cooling tower system of the thermal power plant consists of following four main subsystems.

(i) **Cooling water pump**: It is used to deliver cooling tower water and make up water to cooling water intake tunnel. The system consists of three Cooling Water Pump (CWP) in parallel configuration, failure of one result in reducing the efficiency of the system.
(ii) **CW intake tunnel**: Used to collect the cooled water. It is connected in series configuration with the considered system and the failure of this sub-system will result in complete system failure.
(iii) **Condenser tube box**: It is used to increase the cooled water temperature and is arranged in series configuration.
(iv) **CW discharge tunnel**: It is used for collecting the water which is further passed to cooling tower and is arranged in series configuration.

5.1 *Reliability Analysis*

PN model has been developed for the real structure of cooling tower and is represented in Fig. 2.

Fig. 2 PN model

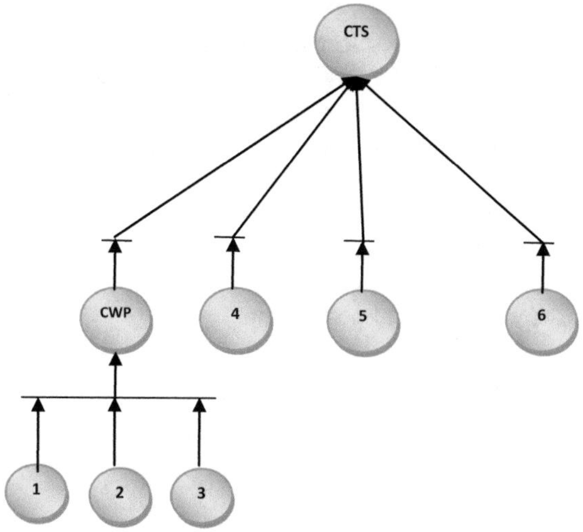

CTS→ Cooling Tower System; CWP→ Cooling Water Pump

Table 2 Failure and repair time data

Component	Failure rate (λ_i) (Failures/h)	Repair time (τ_i) (h)
Cooling water pump ($i = 1, 2, 3$)	3.85×10^{-5}	12
Cooling water intake tunnel ($i = 4$)	1.15×10^{-4}	10
Condenser tube box ($i = 5$)	1.15×10^{-4}	12
Cooling water discharge tunnel ($i = 6$)	1.15×10^{-4}	10

Failure rate and repair time data has been collected from the maintenance log book and discussed with experts for more accuracy. The collected data has been represented in Table 2.

The collected crisp data has been fuzzified using TMF and used in Eqs. 5–7 as per the PN modelling and using the relation as represented in Table 1 various reliability parameter has been tabulated at ±15, ±25% and ±60% spreads for 0–1 degree of freedom. The tabulated values for left side and right side at ±15% spread of a triangle are represented in Table 3. Due to space limitations ±25 and ±60% values are not shown here. However, graphs as shown in Fig. 3 shows the spread values behavior for different reliability parameters.

Reliability parameters at different spreads for Cooling Tower System.

The computed fuzzy values at different degree of freedom was defuzzify and the crisp values so obtained are represented in Table 4.

5.2 *Result Discussion*

From Table 4 it is clear that the repair time for the considered system increase with increase in the spread. With the increase in the repair time, availability of the system also shows decreasing trend with the increase in the spread. On the other hand reliability parameters such as MTBF and reliability of the system shows increasing trend with the increase in the spread values. As availability of the considered system shows decreasing trend this shows that the chances of sudden failure in the plant operation are quite high and hence the current trend of the result is clearly very helpful for the maintenance engineer to take necessary steps for scheduling the optimum RCM for the considered system.

Table 3 Spread values for 15%

DOF	Failure rate	Repair time	Availab.	Unavailab.	Reliab.	Unreliab.	MTBF	ENOF
At 15% left hand side								
1	0.000270	11.142900	0.997001	0.002999	0.955657	0.044343	3715.1209	0.045230
0.9	0.000266	10.651350	0.996816	0.002825	0.955007	0.043693	3659.8908	0.044558
0.8	0.000262	10.178985	0.996622	0.002659	0.954357	0.043042	3606.2742	0.043886
0.7	0.000258	9.724979	0.996418	0.002501	0.953708	0.042391	3554.2016	0.043214
0.6	0.000254	9.288553	0.996203	0.002352	0.953059	0.041739	3503.6075	0.042541
0.5	0.000250	8.868971	0.995976	0.002210	0.952411	0.041087	3454.4299	0.041867
0.4	0.000246	8.465537	0.995737	0.002076	0.951763	0.040434	3406.6105	0.041193
0.3	0.000242	8.077594	0.995485	0.001948	0.951116	0.039781	3360.0939	0.040519
0.2	0.000238	7.704519	0.995220	0.001827	0.950469	0.039128	3314.8278	0.039844
0.1	0.000234	7.345723	0.994941	0.001713	0.949823	0.038474	3270.7625	0.039169
0	0.000229	7.000648	0.994646	0.001604	0.949177	0.037819	3227.8511	0.038494
At 15% right hand side								
1	0.000270	11.142900	0.997001	0.002999	0.955657	0.044343	3715.1209	0.045230
0.9	0.000274	11.654512	0.997175	0.003184	0.956307	0.044993	3772.0383	0.045901
0.8	0.000278	12.187116	0.997341	0.003378	0.956958	0.045643	3830.7212	0.046571
0.7	0.000282	12.741702	0.997499	0.003582	0.957609	0.046292	3891.2527	0.047241
0.6	0.000286	13.319322	0.997648	0.003797	0.958261	0.046941	3953.7215	0.047910
0.5	0.000290	13.921096	0.997790	0.004024	0.958913	0.047589	4018.2217	0.048578
0.4	0.000294	14.548219	0.997924	0.004263	0.959566	0.048237	4084.8538	0.049246
0.3	0.000298	15.201966	0.998052	0.004515	0.960219	0.048884	4153.7249	0.049913
0.2	0.000302	15.883698	0.998173	0.004780	0.960872	0.049531	4224.9497	0.050579
0.1	0.000306	16.594870	0.998287	0.005059	0.961526	0.050177	4298.6504	0.051245
0	0.000310	17.337042	0.998396	0.005354	0.962181	0.050823	4374.9583	0.051910

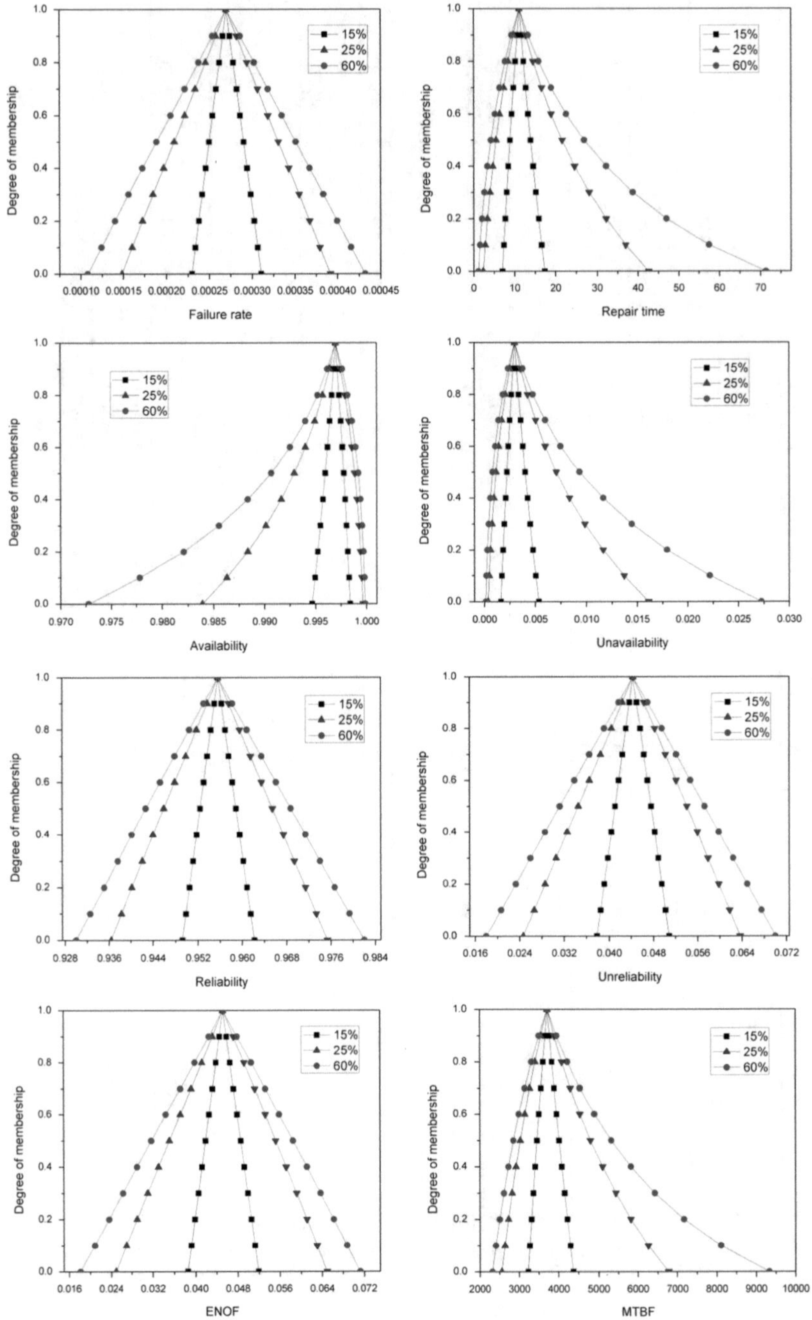

Fig. 3 Fuzzy values representation for cooling tower system parameters

Table 4 Crisp and defuzzified value cooling tower system

System parameters	Crisp value	Defuzzified value (±15% spread)	Defuzzified value (±25% spread)	Defuzzified value (±60% spread)
Failure rate	0.00026998	0.00026998	0.00027001	0.00027005
Repair time	11.1428998	11.8268631	18.6879404	27.8572693
Availability	0.99700066	0.99668103	0.99351713	0.98987691
Unavailability	0.00299934	0.00331897	0.00648288	0.01012309
Reliability	0.95565660	0.95567134	0.95578931	0.95589253
Unreliability	0.04434340	0.04432866	0.04421069	0.04410747
MTBF	3715.12094	3772.64345	4349.67167	5120.82708
ENOF	0.04522960	0.04521101	0.04503994	0.04486210

6 Conclusion and Limitation of the Work

Performance of cooling tower of a coal fired thermal power was analyzed under fuzzy environment. TFM was used to consider the vagueness of the failure and repair time data. Various reliability parameters were tabulated and their trend of failure was studied in order to minimize or eliminate the sudden failure of the plant. Plant availability shows decreasing trend for different spreads. The results are useful in framing the optimum maintenance interval for the considered system for improving plant availability. As the analysis results are totally based on the collected failure and repair time data therefore, the correctness of the results is totally depends upon the availability of correct information. However, the author has used TMF for considering the vagueness of the collected data for high accuracy of the results.

References

1. Panchal D, Kumar D (2017) Risk analysis of compressor house unit of thermal power plant. Int J Ind Syst Eng 25(2):228–250
2. Sharma RK, Sharma P (2012) Integrated framework to optimize RAM and cost decision in process plant. J Loss Prevent Process Ind 25:883–904
3. Panchal D, Singh AK, Chatterjee P, Zavadskas EK, Ghorabaee MK (2019) A new fuzzy methodology-based structured framework for RAM and risk analysis. Appl Soft Comput 74:242–254
4. Panchal D, Srivastva P (2018) Qualitative analysis of CNG dispensing system using fuzzy FMEA–GRA integrated approach. Int J Syst Assur Eng Manag. https://doi.org/10.1007/s13198-018-0750-9
5. Arora N, Kumar D (1997) Availability analysis of steam and power generation systems in thermal power plant. Int J Microelectr Reliab 37:795–799
6. Arora N, Kumar D (2000) Stochastic analysis and maintenance planning of ash handling system in thermal power plant. Int J Microelectr Reliab 37(5):819–834
7. Gupta P, Lal A, Sharma R, Singh J (2005) Numerical analysis of reliability and availability of series processes in butter oil processing plant. Int J Qual Reliab Manag 22(3):303–306

8. Gupta P, Lal AK, Sharma RK, Singh J (2007) Analysis of reliability and availability of serial processes of plastic-pipe manufacturing plant: a case study. Int J Qual Reliab Manag 24(4):404–419
9. Sharma RK, Kumar D, Kumar P (2007) Modeling system behavior for risk and reliability analysis using KBARAM. Qual Reliab Eng Int 23:973–998
10. Panchal D, Kumar D (2014) Reliability analysis of CHU system of coal fired thermal power plant using fuzzy λ-τ approach. Proc Eng 97:2323–2332
11. Garg H (2014) Performance and behavior analysis of repairable industrial systems using vague Lambda-Tau methodology. Appl Soft Comput 22:323–328
12. Panchal D, Kumar D (2016) Integrated framework for behavior analysis in a process plant. J Loss Prevent Process Ind 40:147–161
13. Kokso B (1999) Fuzzy engineering. Prentice Hall, Englewood Cliffs
14. Tanaka K (2001) An introduction to fuzzy logic for practical applications. Springer, New York
15. Zadeh LA (1996) Fuzzy sets, fuzzy logic, fuzzy systems: selected papers. World Scientific, Singapore
16. Zimmermann H (1996) Fuzzy set theory and its applications, 3rd edn. Kluwer Academic Publishers
17. Ross TJ (2000) Fuzzy logic with engineering applications. McGraw-Hill, New York, NY
18. Panchal D, Kumar D (2016) Stochastic behaviour analysis of power generating unit in thermal power plant using fuzzy methodology. OPSEARCH 53(1):16–40
19. Panchal D, Mangala S, Tyagi M, Mange R (2018) Risk analysis for clean and sustainable production in a urea fertilizer industry. Int J Qual Reliab Manag 35(7):1459–1476
20. Knezevic J, Odoom ER (2001) Reliability modeling of repairable systems using petri nets and fuzzy lambda-tau methodology. Reliab Eng Syst Saf 73(1):1–17

Performance Analysis of Suspension Bridge: A Reliability Approach

Amit Kumar, Mangey Ram, Monika Negi and Nikhil Varma

Abstract This paper addresses the overflowing issue of a suspension bridge in context of system's reliability measures. Suspension Bridges have proved to be the most stable structure in the never-ending list of bridge constructions. Unfortunately, various effects of operational and environmental variability (factors) have posed several challenges to the reliability of the whole structure. History has encountered various failures of the suspension bridges with no early prediction and aftermath analysis. This paper investigates the ability to use Markov process for degradation modelling of suspension bridges by taking some of its important section namely Tower Foundation, Tower, Anchor, Cable, Deck along with human error. Here we identify various factors responsible for deterioration of the major components of bridge, which further affects the working of the mainframe structure.

Keywords Suspension bridge system (SBS) · Markov process · Multi-state analysis · Degraded state · Sensitivity analysis

1 Introduction

In this paper, the structural analysis of suspension bridge is conducted using mathematical modelling/reliability theory. The analysis is based upon various failures and

A. Kumar (✉)
Department of Mathematics, Lovely Professional University, Phagwara, Punjab, India
e-mail: amit.20445@lpu.co.in

M. Ram
Department of Mathematics, Graphic Era Deemed to be University, Dehradun, Uttrakhand, India
e-mail: mangeyram@gmail.com

M. Negi
Department of Civil Engineering, Graphic Era Deemed to be University, Dehradun, Uttrakhand, India

N. Varma
Department of Mechanical Engineering, Graphic Era Deemed to be University, Dehradun, Uttrakhand, India

M. Ram and H. Pham (eds.), *Advances in Reliability Analysis and its Applications*, Springer Series in Reliability Engineering,
https://doi.org/10.1007/978-3-030-31375-3_13

their maintenance which may occur during its life spam. Practically there exist many things/structures which are complex in nature and whose states are defined in terms of Boolean functions of an individual unit event such as the existences of system failure modes and the failures of constituent subsystems [1]. A sound decision making on structural layout and repairs strategies are crucial to stipulate such structure actions in an effective way. However, calculating the possibility/reliability of such a system event is often a tedious task lieu to complexity of the system and/or insufficient knowledge regarding the system. A sample cable-stayed bridge scheme is used by Bruneau [2] for assessing the workability of the same, for this he take ductile and brittle cables in distinct configurations. From these finding many valuable insights regarding potential failures exposed also it provides how one can enhance overall reliability by using additional strength in the best way.

Much elite work done in past regarding structural reliability but still there is an enormous need for the well-organized use for structural monitoring. Bridge reliability assessment on the basis of monitoring were obtained in past by Frangopol et al., Kwon and Frangopol, Catbas et al., Chae et al., [3–6]. The expert inclusion of monitoring data in the structural reliability assessment was investigated by Frangopol et al. [3] for the demonstration of the use of monitored data in the development of predicted models. The methodology is explained on a highway bridge. Fatigue is one of the defects faced by many bridges during its life. Fatigue cracks experienced by many of existing bridge structure. Due to this bridge structural integrity may not be preserved safely up to its expected service spam. For this reason, it is essential to calculate and forecast bridge fatigue reliability. Kwon and Frangopol [4] used field monitoring data and probability density function of equivalent stress ran for fatigue reliability assessment for steel bridge systems. Robelin and Madanat [7] investigated the problem of improving bridge maintenance and restoration results for a heterogeneous system. The objective of their research is to find optimum repair and replacement strategies based on the information of the current conditions and on the forecast of future circumstances.

Bridge inspection is a serious accountability. If it is not done in a proper manner then it resulted in unexpected bridge collapses. Thus it is gradually essential to obtain a more efficient and reasonable method to maintain the bridges through precise analysis. An inspection regarding bridge safety analysis through robotic system was proposed by Oh et al. [8].

Suspension bridge is a type of bridge in which the load bearing component known as deck is suspended over by the vertical suspenders, which are connected by the main suspension cables. These suspension cables go over the tower and are secured at each end of the bridge by an anchor. The modern design of suspension bridge is a repercussion of various improvements and innovation throughout the history of bridges.

The first idea of suspension bridge comes back in 15th century, when a Tibetan saint, Thengtong Gyalpo used various iron chain to build suspension bridges in Tibet and Bhutan [9]. These bridges mostly used chains as the main cables while his early work used ropes. Following the course of history, during 1551–1617, few iron bridges also built but they were more of something in between suspension bridge and cable stayed bridge.

In 1830 another innovation by a French engineer who used cables consisting of various strands of wires instead of chains, brought the design closer to that of modern era.

A German born American engineer and inventor, John Roebling, gave his contribution to the then design by inventing the rigid deck platform stiffened with trusses in a web formation. His invention during mid-18th century made bridges more stable and capable enough to withstand high velocity winds. His another contribution was to introduce a technique of spinning the cable in space, rather than making a pre-fabricated cable that needed to be lifted to place of installation.

In following years, another development of technique pneumatic caisson, which permitted Pier foundation to much greater depth, augmented the design many folds. In following years after successive testing, modern steel alloys were preferred over iron as they proved to be much more capable of withstanding enormous loads and allowed construction of bridges with greater spans.

History of suspension bridges saw various failures as well, whose study made engineers vary of the flaws and the factors that tend to hinder the working of the whole structure. With continuous improvements and innovations, the modern design was fabricated which made suspension bridge as one of the most reliable and stable bridge capable enough to withstand greater loads with the lengths reaching much longer spans. However, does a stable design means that it is not susceptible to damage? Again, history answered this question that no matter how perfect a design may be, it can still affected and can fails.

In this paper, we have analyzed different factors that can affect suspension bridge (directly, indirectly). Various reliability measures e.g. reliability, availability, mean time to failure, sensitivity analysis of the suspension bridge are calculated by using Markov process. For this suspension bridge is critically analyzed to figure out different factors e.g. environmental factors, Human error, component degradation, Intrinsic factors etc. that can affect the functioning of the same. Along these affects the five main components of suspension bridge as per modern design are Tower Foundation, Tower, Anchor, Cable and Deck. These components are susceptible to damage caused by the above-mentioned factors.

2 Suppositions

The following assumptions are taken throughout the functioning of SBS.

Assumption 1: The working stints of each unit of the SBS are supposed to be independent. Maintenance team come in action as any of working component breaks down.

Assumption 2: Various process (Stochastic/Markov) involved in this paper are taken to be independent from each other's.

Assumption 3: The SBS goes in failed/degraded state as soon as the unit's failure/failures occur. In this research shut-off rule is the suspended animation.
Assumption 4: Sufficient repair always available.
Assumption 5: Average failures rate are taken as constant.

3 Nomenclature

t	Time scale
s	Laplace transformation variable
$P_i(t); i = 1, 2, 3, 4, 5$	Probability of the system being in state S_i at instant t
$\overline{P}_i(s)$	Laplace transform of $P_i(t)$
$P_j(x, t)$; $j = 6, 7, 8, 9, 10, 11, 12, 13, 14, 15, 16$	Probability density function of system being in completely failed state at instant t with elapsed repair time x
$\overline{P}_j(x, s)$;	Laplace transform of $P_i(x, t)$
$\alpha_1/\beta_1/\eta_1/\delta_1/\lambda_1$	Partial failure rate of Tower foundation/Tower/Anchor/Cable/Deck respectively
$\alpha_2/\beta_2/\eta_2/\delta_2/\lambda_2$	Complete failure rate of Tower foundation/Tower/Anchor/Cable/Deck respectively
λ_h	Failure due to human error
$\phi(x)/\psi(x)/\gamma(x)/\omega(x)/\varepsilon(x)$	Repair rate of Tower foundation/Tower/Anchor/Cable/Deck respectively
$\mu(x)$	Simultaneous repair rate of any two components of structure
$\theta(x)$	Human error recovery rate.

4 States Narrative

S_0	Good state: All the components of SBS are in good condition
S_1	Degraded state: State in which Tower foundation is degraded
S_2	Degraded state: State in which Tower is degraded
S_3	Degraded state: State in which Anchor is degraded
S_4	Degraded state: State in which Cable is degraded
S_5	Degraded state: State in which Deck is degraded

S_6 Failed state: State in which Tower foundation and Tower are completely failed
S_7 Failed state: State in which Tower and Anchor are completely failed
S_8 Failed state: State in which Anchor and Cable are completely failed
S_9 Failed state: State in which Cable and Deck are completely failed
S_{10} Failed state: State in which Tower foundation and Deck are completely failed
S_{11} Failed state: State in which Tower and Deck are completely failed
S_{12} Failed state: State in which Tower foundation and Anchor are completely failed
S_{13} Failed state: State in which Tower and cable are completely failed
S_{14} Failed state: State in which Anchor and Deck are completely failed
S_{15} Failed state: State in which Tower foundation and Cable are completely failed
S_{16} Failed state: Completely failed due to Human error.

5 State Transition Diagram

By carefully analyzing the effects of these factors on suspension bridge a transition state diagram formulated which successfully depicts the various stages of it throughout the operation. (Fig. 1).

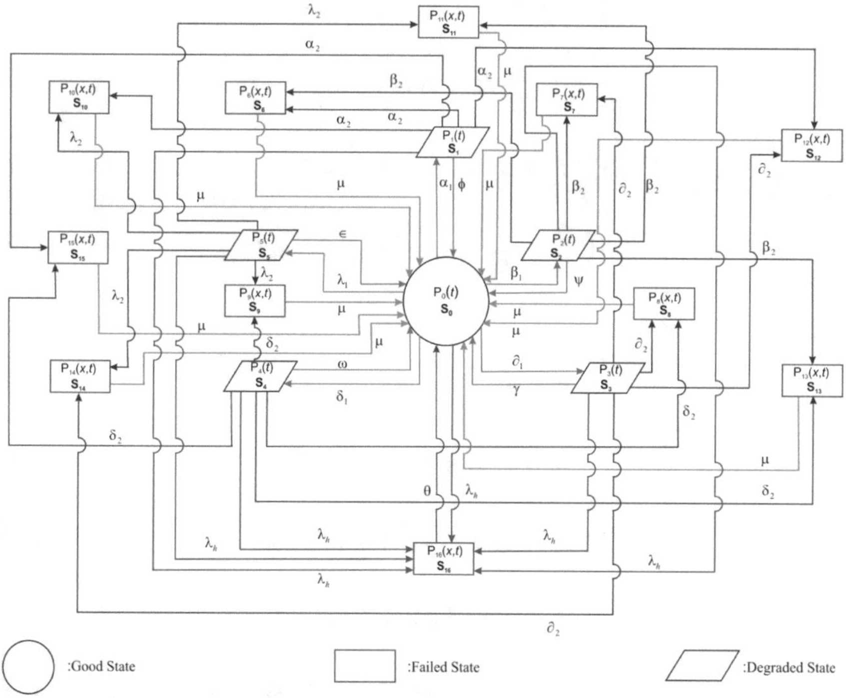

Fig. 1 State transition diagram

6 Analysis and Methodology

In this section we have develop a set of equations corresponding to the considered system, with the aid supplementary variable technique and Markov process.

$$\left(\frac{\partial}{\partial t}+\beta_1+\eta_1+\delta_1+\alpha_1+\lambda_1+\lambda_h\right)P_0(t)=\phi(x)P_1(t)+\psi(x)P_2(t)+\gamma(x)P_3(t)$$
$$+\omega(x)P_4(t)+\varepsilon(x)P_5(t)+\int_0^\infty\theta(x)P_{16}(x,t)dx$$
$$+\sum_i\int_0^\infty\mu(x)P_i(x,t)dx;\quad i=6,7,8,9,10,11,12,13,1,4,15; \tag{1}$$

$$\left(\frac{\partial}{\partial t}+4\alpha_2+\lambda_h+\phi(x)\right)P_1(t)=\alpha_1 P_0(t) \tag{2}$$

$$\left(\frac{\partial}{\partial t}+4\beta_2+\lambda_h+\psi(x)\right)P_2(t)=\beta_1 P_0(t) \tag{3}$$

$$\left(\frac{\partial}{\partial t}+4\eta_2+\lambda_h+\gamma(x)\right)P_3(t)=\eta_1 P_0(t) \tag{4}$$

$$\left(\frac{\partial}{\partial t}+4\delta_2+\lambda_h+\omega(x)\right)P_4(t)=\delta_1 P_0(t) \tag{5}$$

$$\left(\frac{\partial}{\partial t}+4\lambda_2+\lambda_h+\varepsilon(x)\right)P_5(t)=\lambda_1 P_0(t) \tag{6}$$

$$\left(\frac{\partial}{\partial x}+\frac{\partial}{\partial t}+\mu(x)\right)P_i(x,t)=0;\ i=6,7,8,9,10,11,12,13,14,15 \tag{7}$$

$$\left(\frac{\partial}{\partial x}+\frac{\partial}{\partial t}+\theta(x)\right)P_{16}(x,t)=0 \tag{8}$$

Boundary conditions

$$P_6(0,t)=\alpha_2 P_1(t)+\beta_2 P_2(t) \tag{9}$$

$$P_7(0,t)=\beta_2 P_2(t)+\eta_2 P_3(t) \tag{10}$$

$$P_8(0,t)=\eta_2 P_3(t)+\delta_2 P_4(t) \tag{11}$$

$$P_9(0,t)=\delta_2 P_4(t)+\lambda_2 P_5(t) \tag{12}$$

$$P_{10}(0,t)=\alpha_2 P_1(t)+\lambda_2 P_5(t) \tag{13}$$

$$P_{11}(0,t) = \beta_2 P_2(t) + \lambda_2 P_5(t) \tag{14}$$

$$P_{12}(0,t) = \alpha_2 P_1(t) + \eta_2 P_3(t) \tag{15}$$

$$P_{13}(0,t) = \beta_2 P_2(t) + \delta_2 P_4(t) \tag{16}$$

$$P_{14}(0,t) = \eta_2 P_3(t) + \lambda_2 P_5(t) \tag{17}$$

$$P_{15}(0,t) = \alpha_2 P_1(t) + \delta_2 P_4(t) \tag{18}$$

$$P_{16}(0,t) = \lambda_h [P_1(t) + P_2(t) + P_3(t) + P_4(t) + P_5(t)] \tag{19}$$

Initial condition

$$P_0(t) = \begin{cases} 1; \ t = 0 \\ 0; \ \text{otherwise} \end{cases} \tag{20}$$

By the aid of Laplace transform, the above set of equations can be rewritten as

$$\begin{aligned}(s + \beta_1 + \eta_1 + \delta_1 + \alpha_1 + \lambda_1 + \lambda_h)\overline{P}_0(s) &= \phi(x)\overline{P}_1(s) + \psi(x)\overline{P}_2(s) + \gamma(x)\overline{P}_3(s) \\ &+ \omega(x)\overline{P}_4(s) + \varepsilon(x)\overline{P}_5(s) + \int_0^\infty \theta(x)\overline{P}_{16}(x,s)dx \\ &+ \sum_i \int_0^\infty \mu(x)\overline{P}_i(x,s)dx; \ i = 6, 7, 8, 9, 10, 11, 12, 13, 1, 4, 15;\end{aligned} \tag{21}$$

$$(s + 4\alpha_2 + \lambda_h + \phi(x))\overline{P}_1(s) = \alpha_1 \overline{P}_0(s) \tag{22}$$

$$(s + 4\beta_2 + \lambda_h + \psi(x))\overline{P}_5(s) = \beta_1 \overline{P}_0(s) \tag{23}$$

$$(s + 4\eta_2 + \lambda_h + \gamma(x))\overline{P}_3(s) = \eta_1 \overline{P}_0(s) \tag{24}$$

$$(s + 4\delta_2 + \lambda_h + \omega(x))\overline{P}_4(s) = \delta_1 \overline{P}_0(s) \tag{25}$$

$$(s + 4\lambda_2 + \lambda_h + \varepsilon(x))\overline{P}_5(s) = \lambda_1 \overline{P}_0(s) \tag{26}$$

$$\left(\frac{\partial}{\partial x} + s + \mu(x)\right)\overline{P}_i(x,s) = 0; \ i = 6, 7, 8, 9, 10, 11, 12, 13, 14, 15 \tag{27}$$

$$\left(\frac{\partial}{\partial x} + s + \theta(x)\right)\overline{P}_{16}(x,s) = 0 \tag{28}$$

Boundary conditions

$$\overline{P}_6(0,s) = \alpha_2 \overline{P}_1(s) + \beta_2 \overline{P}_2(s) \tag{29}$$

$$\overline{P}_7(0,s) = \beta_2 \overline{P}_2(s) + \eta_2 \overline{P}_3(s) \tag{30}$$

$$\overline{P}_8(0,s) = \eta_2 \overline{P}_3(s) + \delta_2 \overline{P}_4(s) \tag{31}$$

$$\overline{P}_9(0,s) = \delta_2 \overline{P}_4(s) + \lambda_2 \overline{P}_5(s) \tag{32}$$

$$\overline{P}_{10}(0,s) = \alpha_2 \overline{P}_1(s) + \lambda_2 \overline{P}_5(s) \tag{33}$$

$$\overline{P}_{11}(0,s) = \beta_2 \overline{P}_2(s) + \lambda_2 \overline{P}_5(s) \tag{34}$$

$$\overline{P}_{12}(0,s) = \alpha_2 \overline{P}_1(s) + \eta_2 \overline{P}_3(s) \tag{35}$$

$$\overline{P}_{13}(0,s) = \beta_2 \overline{P}_2(s) + \delta_2 \overline{P}_4(s) \tag{36}$$

$$\overline{P}_{14}(0,s) = \eta_2 \overline{P}_3(s) + \lambda_2 \overline{P}_5(s) \tag{37}$$

$$\overline{P}_{15}(0,s) = \alpha_2 \overline{P}_1(s) + \delta_2 \overline{P}_4(s) \tag{38}$$

$$\overline{P}_{16}(0,s) = \lambda_h \left[\overline{P}_1(s) + \overline{P}_2(s) + \overline{P}_3(s) + \overline{P}_4(s) + \overline{P}_5(s) \right] \tag{39}$$

The transition state probabilities of the considered system were obtained by analytically solving Eqs. 21–39 with the help of initial condition as follow

$$\overline{P}_0(s) = \frac{1}{\left[\begin{array}{l} s + \beta_1 + \eta_1 + \lambda_h + \delta_1 + \lambda_1 + \alpha_1 - \dfrac{\alpha_1\{\phi(x) + 4\alpha_2 \overline{T}_\mu(s) + \lambda_h \overline{T}_\mu(s)\}}{(s + \phi(x) + 4\alpha_2 + \lambda_h)} \\ - \dfrac{\beta_1\{\psi(x) + 4\beta_2 \overline{T}_\mu(s) + \lambda_h \overline{T}_\mu(s)\}}{(s + \psi(x) + 4\beta_2 + \lambda_h)} - \dfrac{\eta_1\{\gamma(x) + 4\eta_2 \overline{T}_\mu(s) + \lambda_h \overline{T}_\mu(s)\}}{(s + \gamma(x) + 4\eta_2 + \lambda_h)} \\ - \dfrac{\delta_1\{\omega(x) + 4\delta_2 \overline{T}_\mu(s) + \lambda_h \overline{T}_\mu(s)\}}{(s + \omega(x) + 4\delta_2 + \lambda_h)} - \dfrac{\lambda_1\{\varepsilon(x) + 4\lambda_2 \overline{T}_\mu(s) + \lambda_h \overline{T}_\mu(s)\}}{(s + \varepsilon(x) + 4\lambda_2 + \lambda_h)} \end{array}\right]}$$

$$\overline{P}_1(s) = \frac{\alpha_1 \overline{P}_0(s)}{(s + \phi(x) + 4\alpha_2 + \lambda_h)}$$

$$\overline{P}_2(s) = \frac{\beta_1 \overline{P}_0(s)}{(s + \psi(x) + 4\beta_2 + \lambda_h)}$$

$$\overline{P}_3(s) = \frac{\eta_1 \overline{P}_0(s)}{(s + \gamma(x) + 4\eta_2 + \lambda_h)}$$

$$\overline{P}_1(s) = \frac{\delta_1 \overline{P}_0(s)}{(s + \omega(x) + 4\delta_2 + \lambda_h)}$$

$$\overline{P}_5(s) = \frac{\lambda_1 \overline{P}_0(s)}{(s + \varepsilon(x) + 4\lambda_2 + \lambda_h)}$$

$$\overline{T}_i(s) = \frac{i}{(s + i)}$$

As in transition state diagram it has been seen that the up state $P_{up}(t)$ (working/partially failed) and down state $P_{down}(t)$ (failed state) probability of the SBS is given as:

$$P_{up}(t) = P_0(t) + P_1(t) + P_2(t) + P_3(t) + P_4(t) + P_5(t) \quad (40)$$

$$P_{down}(t) = P_6(x,t) + P_7(x,t) + P_8(x,t) + P_9(x,t) + P_{10}(x,t) + P_{11}(x,t) + P_{12}(x,t) + P_{13}(x,t) + P_{14}(x,t) + P_{15}(x,t) + P_{16}(x,t) \quad (41)$$

Also

$$P_{up}(t) + P_{down}(t) = 1 \quad (42)$$

7 Mathematical Computation

7.1 Availability

The availability is a significant performance indicator of any system. It shows that how much a system is available for functioning with respect to time by considering different failure and repair which may occur during functioning. Availability, $P_{up}(t)$, of a system is the probability that the system is operating satisfactorily up to a specified time 't'. Using the obtained state probabilities in Eq. (40) and putting the different parameters as $\lambda_1 = 0.30$, $\lambda_2 = 0.19$, $\lambda_h = 0.001$, $\delta_1 = 0.31$, ,$\eta_1 = 0.005$, $\eta_2 = 0.004$, $\delta_2 = 0.16$, $\beta_1 = 0.09$, $\beta_2 = 0.05$, $\alpha_1 = 0.01$, $\alpha_2 = 0.008$, $\varphi(x) = \psi(x) = \gamma(x) = \omega(x) = \varepsilon(x) = \mu(x) = \theta(x) = 1$. Now using the Laplace transforms, the availability of SBS is obtained as follows

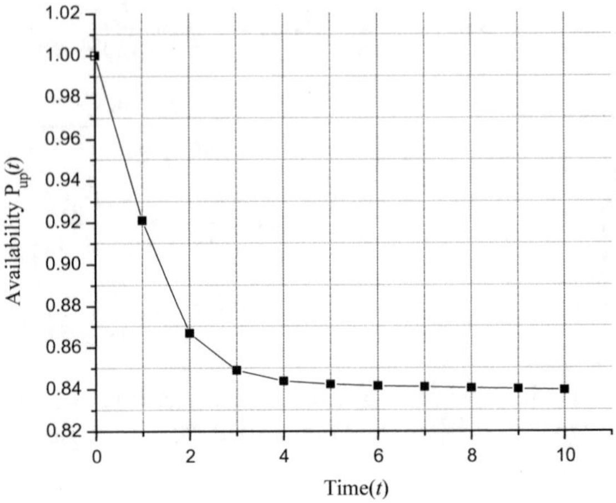

Fig. 2 Availability versus variation in time unit (t)

$$P_{up}(t) = \left\{ \begin{array}{l} 1.342918\, e^{(-1.715417t)} + 0.844469\, e^{(-0.000582t)} + 0.000119\, e^{(-1.017t)} \\ + 1.627769\, e^{(-1.641t)} + 0.029294\, e^{(-1.201t)} + 0.000468\, e^{(-1.033t)} \\ - 2.845037\, e^{(-1.761t)} \end{array} \right\} \tag{43}$$

The behaviour of availability of SBS is obtained by varying time unit t in (43) and it is shown in Fig. 2.

7.2 *Reliability Analysis*

Reliability of a system is the probability for performing its function for a given period of time under some specific condition and operating conditions. It depends on the collective performance of the all components of a system. It means that each and every component of the system contributes in its reliability. Here it is calculated by putting the value of various failures as $\lambda_1 = 0.30, \lambda_2 = 0.19, \lambda_h = 0.001,\ \delta_1 = 0.31,$ $\eta_1 = 0.005,\ \eta_2 = 0.004, \delta_2 = 0.16, \beta_1 = 0.09,\ \beta_2 = 0.05, \alpha_1 = 0.01,\ \alpha_2 = 0.008$ and repair as zero. The reliability of the suspension bridge system is obtained as follow

$$R(t) = \left\{ \begin{array}{l} 0.029282\, e^{(-0.3745t)} \sinh(0.3415\, t) + e^{(-0.716t)} + 0.371134\, e^{(-0.9585t)} \sinh(0.2425\, t) \\ +0.014061\, e^{(-0.3665t)} \sinh(0.3495\, t) + 8.266667\, e^{(-0.6785t)} \sinh(0.0375\, t) \\ +13.3334 e^{(-0.7385t)} \sinh(0.0225\,00\, t) \end{array} \right\} \tag{44}$$

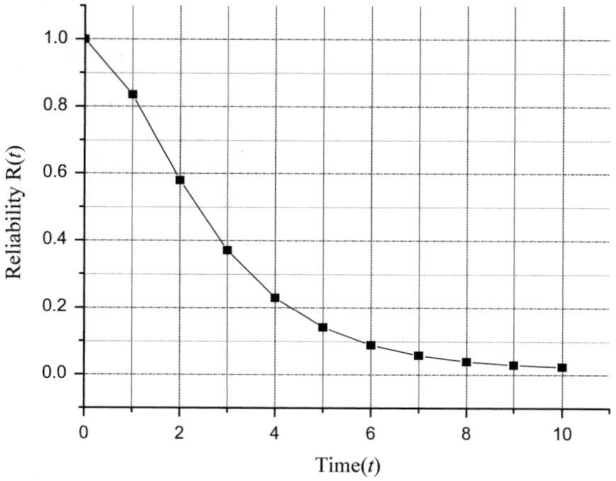

Fig. 3 Reliability versus variation in time unit (t)

The behaviour of reliability for the SBS is obtained by varying time unit t in (44) and it is shown in Fig. 3.

7.3 Mean Time to Failure (MTTF)

Basically MTTF is the average failure time for an individual component. Mathematically it is calculated as

$$MTTF = \int_0^{\infty} tf(t)dt = \int_0^{\infty} R(t)dt = \lim_{s \to 0} R(s); \tag{45}$$

By using Eqs. (40) and (45), the MTTF of SBS is obtained as follow

$$MTTF = \frac{1}{(\alpha_1 + \beta_1 + \eta_1 + \delta_1 + \lambda_1 + \lambda_h)} \left[\begin{array}{l} 1 + \frac{\alpha_1}{(4\alpha_2 + \lambda_h)} + \frac{\beta_1}{(4\beta_2 + \lambda_h)} + \frac{\eta_1}{(4\eta_2 + \lambda_h)} \\ + \frac{\delta_1}{(4\delta_2 + \lambda_h)} + \frac{\lambda_1}{(4\lambda_2 + \lambda_h)} \end{array} \right] \tag{46}$$

The nature of MTTF of the system with respect to different failure rates is obtained by varying failures in Eq. (46) as shown in Fig. 4.

7.4 Sensitivity Analysis

It is a procedure which is used to decide how the different values of an independent failure influence a specific dependent parameter under some limitations [10–12]

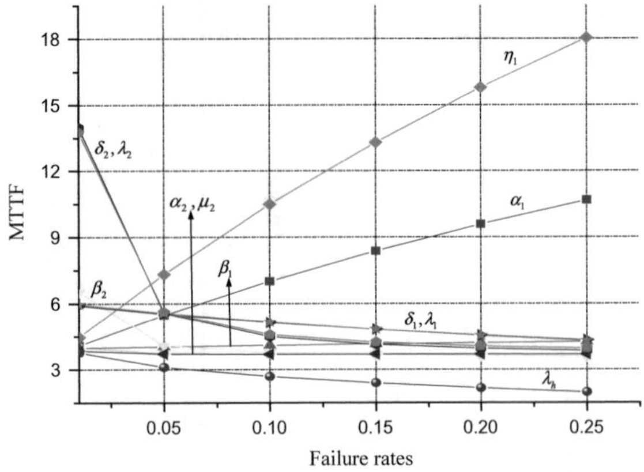

Fig. 4 MTTF versus variation in Failure rates

or in other way it can be used as an indicator by which one can inspect that which failure effects the systems performance most. Here author's performed the sensitivity analysis for system reliability and MTTF for finding that which failure affects the system reliability and MTTF most.

7.4.1 Sensitivity of Mean Time to Failure

Sensitivity analysis of the SBS with respect to MTTF is performed by differentiating the MTTF expression obtained in Eq. (46) with respect to various failure rates and then placed the values of various failure rates as $\lambda_1 = 0.30$, $\lambda_2 = 0.19$, $\lambda_h = 0.001$, $\delta_1 = 0.31$, $\eta_1 = 0.005$, $\eta_2 = 0.004$, $\delta_2 = 0.16$, $\beta_1 = 0.09$, $\beta_2 = 0.05$, $\alpha_1 = 0.01$, $\alpha_2 = 0.008$ in these partial derivatives. Now varying the failure rates one by one respectively, one can obtain Fig. 5 for sensitivity of MTTF for suspension bridge system.

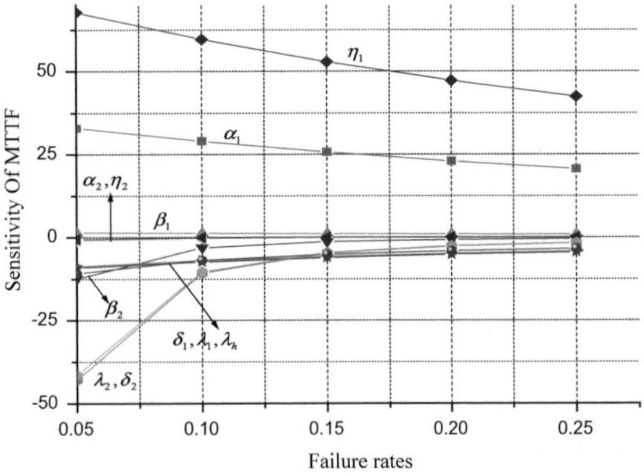

Fig. 5 Sensitivity of MTTF versus Failure rates

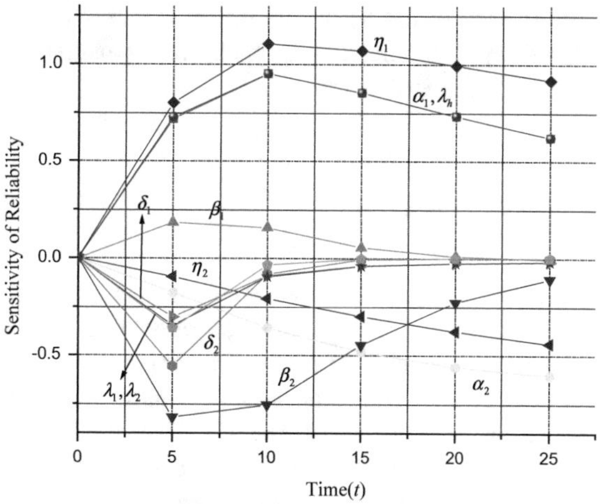

Fig. 6 Sensitivity of reliability versus time (t)

7.4.2 Sensitivity of Reliability

Sensitivity of reliability is performed in the same manner as sensitivity of MTTF and then placed the values of various failure rates as $\lambda_1 = 0.30$, $\lambda_2 = 0.19$, $\lambda_h = 0.001$, $\delta_1 = 0.31$, $\eta_1 = 0.005$, $\eta_2 = 0.004$, $\delta_2 = 0.16$, $\beta_1 = 0.09$, $\beta_2 = 0.05$, $\alpha_1 = 0.01$, $\alpha_2 = 0.008$ in these derivatives. Now varying the time unit t in these partial derivatives, one can obtain Fig. 6 for sensitivity of reliability of SBS.

8 Results Discussion

One can say that the performance of a system directly depends on the performance of each of its subsystems. So to increase the performance of the system, more attention should be given to their corresponding subunits for the effectiveness of the maintenance program. An investigation is carried out for suspension bridge system, in order to find its most critical component, which affects the performance of the system. The effect of various failure rates on system availability, reliability, mean time to failure (MTTF) are analyzed and depicted with graphs. Availability and reliability of the SBS with respect to time is shown in Figs. 2 and 3 respectively. These graphs show that the availability and reliability of the SBS decreasing with time passes. The difference between these graphs reflects importance of maintenance policy. MTTF of SBS with respect to different failures shown in Fig. 4. It reflects that MTTF of the partial failure of Tower foundation and partial failure of Anchor is increased. It is also seen that MTTF for human error is decrease. Sensitivity analysis has been done for SBS reliability and MTTF. Sensitivity of SBS MTTF is shown in Fig. 5. It reflects that the systems MTTF is most sensitive with respect to partial failure of Anchor, Tower foundation, complete failure of Deck, complete failure of cable. Sensitivity of reliability of the SBS is shown in Fig. 6. By critically analyzing Fig. 6 we can say that system reliability is much more sensitive with respect to human error and complete failure of Tower.

9 Conclusion

The present paper discussed about a suspension bridge system with the help of Markov process and mathematical modeling for analyzing its various reliability indices. The behavior of different performance measures of the system is presented with the help of graphs to critically analyzing them. On the basis of above result discussion we can say that the MTTF of the SBS for the partial failure of Tower foundation and partial failure of Anchor is increasing as compare to others. SBS reliability is much more sensitive with respect to Tower foundation, human error and complete failure of Tower in comparison with other failures. Also systems MTTF is most sensitive with respect to partial failure of Anchor, Tower foundation, complete failure of Deck, complete failure of cable. So we identify that the failure of Anchor, Tower foundation, human error and Deck affect the performance suspension bridge system most. Hence in order to increase the reliability of SBS one has to control/restrict these failures. It asserts that these finding are quit helpful for the engineers for planning and designing of suspension bridge system.

References

1. Kang WH, Song J, Gardoni P (2008) Matrix-based system reliability method and applications to bridge networks. Reliab Eng Syst Saf 93(11):1584–1593
2. Bruneau M (1992) Evaluation of system-reliability methods for cable-stayed bridge design. J Struct Eng 118(4):1106–1120
3. Frangopol DM, Strauss A, Kim S (2008) Bridge reliability assessment based on monitoring. J Bridge Eng 13(3):258–270
4. Kwon K, Frangopol DM (2010) Bridge fatigue reliability assessment using probability density functions of equivalent stress range based on field monitoring data. Int J Fatigue 32(8):1221–1232
5. Catbas FN, Susoy M, Frangopol DM (2008) Structural health monitoring and reliability estimation: Long span truss bridge application with environmental monitoring data. Eng Struct 30(9):2347–2359
6. Chae MJ, Yoo HS, Kim JY, Cho MY (2012) Development of a wireless sensor network system for suspension bridge health monitoring. Autom Constr 21:237–252
7. Robelin CA, Madanat SM (2008) Reliability-based system-level optimization of bridge maintenance and replacement decisions. Transp Sci 42(4):508–513
8. Oh JK, Jang G, Oh S, Lee JH, Yi BJ, Moon YS, Choi Y (2009) Bridge inspection robot system with machine vision. Autom Constr 18(7):929–941
9. Abdullah AA (2012) Analysis and Design of Suspension Bridge. Doctoral dissertation, University of Baghdad

10. Kumar A, Ram M, Pant S, Kumar A (2018) Industrial system performance under multistate failures with standby mode. In: Modeling and simulation in industrial engineering. Springer, Cham, pp 85–100
11. Ram M, Kumar A (2014) Performance of a structure consisting a 2-out-of-3: F substructure under human failure. Arab J Sci Eng 39(11):8383–8394
12. Kumar A, Ram M (2016) System reliability measures in the presence of common cause failures. Int J Ind Syst Eng 24(1):44–61

Printed in the United States
by Baker & Taylor Publisher Services